LEED–NEW CONSTRUCTION
PROJECT MANAGEMENT

McGRAW-HILL'S GREENSOURCE SERIES

Attmann
Green Architecture: Advanced Technologies and Materials

Gevorkian
Alternative Energy Systems in Building Design
Solar Power in Building Design: The Engineer's Complete Design Resource

GreenSource: The Magazine of Sustainable Design
Emerald Architecture: Case Studies in Green Building

Haselbach
The Engineering Guide to LEED–New Construction: Sustainable Construction for Engineers, 2d ed.

Luckett
Green Roof Construction and Maintenance

Melaver and Mueller (eds.)
The Green Building Bottom Line: The Real Cost of Sustainable Building

Nichols and Laros
Inside the Civano Project: A Case Study of Large-Scale Sustainable Neighborhood Development

Winkler
Recycling Construction & Demolition Waste: A LEED-Based Toolkit

Yellamraju
LEED–New Construction Project Management

Yudelson
Green Building Through Integrated Design
Greening Existing Buildings

About *GreenSource*

A mainstay in the green building market since 2006, *GreenSource* magazine and GreenSourceMag.com are produced by the editors of McGraw-Hill Construction, in partnership with editors at BuildingGreen, Inc., with support from the U.S. Green Building Council. *GreenSource* has received numerous awards, including American Business Media's 2008 Neal Award for "Best Website" and 2007 Neal Award for "Best Start-up Publication," and FOLIO magazine's 2007 Ozzie Awards for "Best Design, New Magazine" and "Best Overall Design." Recognized for responding to the needs and demands of the profession, *GreenSource* is a leader in covering noteworthy trends in sustainable design and best practice case studies. Its award-winning content will continue to benefit key specifiers and buyers in the green design and construction industry through the books in the *GreenSource* Series.

About McGraw-Hill Construction

McGraw-Hill Construction, part of The McGraw-Hill Companies (NYSE: MHP), connects people, projects, and products across the design and construction industry. Backed by the power of Dodge, Sweets, *Engineering News-Record* (*ENR*), *Architectural Record*, *GreenSource*, *Constructor*, and regional publications, the company provides information, intelligence, tools, applications, and resources to help customers grow their businesses. McGraw-Hill Construction serves more than 1,000,000 customers within the $4.6 trillion global construction community. For more information, visit www.construction.com.

LEED–NEW CONSTRUCTION PROJECT MANAGEMENT

VIJAYA YELLAMRAJU, LEED AP

New York   Chicago   San Francisco   Lisbon   London   Madrid
Mexico City   Milan   New Delhi   San Juan   Seoul
Singapore   Sydney   Toronto

The **McGraw·Hill** Companies

Cataloging-in-Publication Data is on file with the Library of Congress

ISBN 978-0-07-174445-4
MHID 0-07-174445-2

Sponsoring Editor: Joy Bramble
Editing Supervisor: Stephen M. Smith
Production Supervisor: Richard C. Ruzycka
Acquisitions Coordinator: Alexis Richard
Project Managers: Nancy Dimitry and Joanna V. Pomeranz, D&P Editorial Services
Copy Editor: Nancy Dimitry
Proofreaders: Donald Dimitry and Donald Paul Pomeranz
Indexer: Seth Maislin
Art Director, Cover: Jeff Weeks
Cover Image Design: Laura Carter
Composition: D&P Editorial Services

Printed and bound by RR Donnelley.

McGraw-Hill books are available at special quantity discounts to use as premiums and sales promotions, or for use in corporate training programs. To contact a representative, please e-mail us at bulksales@mcgraw-hill.com.

The pages within this book were printed on acid-free paper containing 100% postconsumer fiber.

Information contained in this work has been obtained by The McGraw-Hill Companies, Inc. ("McGraw-Hill") from sources believed to be reliable. However, neither McGraw-Hill nor its authors guarantee the accuracy or completeness of any information published herein, and neither McGraw-Hill nor its authors shall be responsible for any errors, omissions, or damages arising out of use of this information. This work is published with the understanding that McGraw-Hill and its authors are supplying information but are not attempting to render engineering or other professional services. If such services are required, the assistance of an appropriate professional should be sought.

To Sai, my parents, and my family.

About the Author

Vijaya Yellamraju, LEED AP, is Principal at Green Potential, a multinational *green building* and LEED consulting firm based in Austin, Texas. She is responsible for leading the technical and operational aspects of the business, providing direction to developers, architects, contractors, and other project team members for successful integration of green strategies into the design process. Ms. Yellamraju has worked on numerous green and LEED projects, with a focus on multifamily and commercial buildings. She is a trained architect and earned an M.S. in architecture, specializing in *Energy* and *Sustainable Design*. She has several years of experience in the architecture and construction industries and has managed many multimillion-dollar projects in the United States and India.

The **U.S. Green Building Council®** is committed to a prosperous and sustainable future for our nation through cost-efficient and energy-saving green buildings. With a community comprising 78 local affiliates, more than 20,000 member companies and organizations, and more than 140,000 LEED® Professionals™, USGBC® is the driving force of an industry that is projected to contribute $554 billion to the U.S. gross domestic product from 2009 to 2013. USGBC leads an unlikely diverse constituency of builders and environmentalists, corporations and nonprofit organizations, elected officials and concerned citizens, and teachers and students. USGBC is the developer of the LEED green building certification program and the convenor of the Greenbuild® International Conference & Expo.

LEED®—an acronym for the phrase "Leadership in Energy and Environmental Design"—is a registered trademark of the U.S. Green Building Council. The LEED green building certification program is the nationally accepted benchmark for the design, construction, and operation of green buildings.

CONTENTS

FOREWORD

JERRY YUDELSON, PE, MS, MBA, LEED AP BD+C/O+M

In the second half of the 1990s, the nascent U.S. Green Building Council (USGBC) began to formulate a new rating system that would define a *green building* so that building teams could actually design and construct one. That system, piloted in 1998 and 1999, was introduced in March 2000 as LEED for New Construction (LEED-NC), version 2.0. In the annals of sustainable design, this date will be remembered for a long time. It was a revolution in green building thinking and a major incentive for progressive architects, engineers, and contractors to push forward with new initiatives in creating sustainable buildings.

Within the span of a decade, the product of that group effort, the LEED®—Leadership in Energy and Environmental Design—green building rating system, had become by 2010 the de facto national sustainable design standard in the United States and had been formally adopted as the national certification standard by green building councils in Canada, India, Italy, and several other countries. By insisting on a rigorous, third-party-verified minimum standard, LEED effectively took the guesswork out of designing and certifying a green building, one that would use fewer scarce natural resources and generate less waste than a conventional building and be more *user friendly* in the process.

LEED defined a major challenge to the conventional practice of architecture, engineering, and construction. It said, effectively, that a *standard* building, one built merely to meet building code minimums, was *not good enough* to prevent unnecessary resource depletion and environmental contamination. As uncomfortable as this was initially for the construction industry professions, by 2005 many large architecture, engineering, and building firms had begun to compete on the basis of how many LEED-registered projects they had begun and how many LEED Accredited Professionals they employed.[1] By 2010, it was impossible for most design and construction firms to compete for sustainable design projects without having completed a substantial number of

LEED-certified projects. Effectively, within 10 years, LEED certification had over-whelmed most initial opposition from building owners, developers, architects, engineers, and builders to become the widely expected outcome of a high-performance construction project.

I first became aware of the USGBC during 1997, helped organize the first USGBC chapter in the Pacific Northwest in 1998, and attended my first USGBC national meeting in 1999. At that time, all attendees could introduce themselves to each other during the course of a few hours, and they did. In 2000, I visited my first LEED-certified project, the LEED Platinum Chesapeake Bay Foundation headquarters in Annapolis, Maryland. By 2001, it had become impossible for LEED members to self-introduce themselves at national meetings, as the USGBC enjoyed the beginnings of a decade of rapid growth that reached 20,000 corporate members by 2008. Contrast those early meetings of a small cabal of green building zealots to the USGBC's 2009 annual *Greenbuild* conference and expo, attended by more than 25,000 people, and you will see that what was a fringe movement 15 years ago has become the *new normal* for the work of designing, constructing, owning, and operating commercial buildings and homes.

In 2000, I helped organize a green building consulting group, one of the first in the country, within a large electric utility in the Pacific Northwest. Our group managed the certification application for the first LEED Gold project in the western United States (and only the second in the country) and the first LEED-certified project in Oregon. In 2001, I became one of the original ten national LEED trainers for the USGBC. At the time, we had a rating system that included requirements for achieving various credits, but no detailed manual on how to proceed and very little in the way of organized procedures. All LEED project documentation was submitted in hard-copy form, in very large three-ring binders, to the USGBC and was then sent to outside reviewers, who were not identified to project participants or allowed to communicate with anyone on the design/construction team. By the end of 2004, only 118 projects had been officially certified, and many professionals complained about the cost and complexity of LEED certification.[2]

Starting in 2001, the USGBC began certifying LEED Accredited Professionals (LEED APs), who had to demonstrate knowledge of one of the four major LEED rating systems, particularly LEED-NC, laying the groundwork for the rapid expansion of the LEED certification process. Mighty oaks from little acorns grow—by 2010, more than 150,000 LEED APs!

By early 2010, LEED certification had grown tremendously! More than 5,000 projects had already been certified, and more than 25,000 had registered to receive eventual certification.[3] LEED Platinum (the highest rating) projects now span the globe, from the United States and Canada, to Dubai (UAE), India, Australia, Brazil, China, Spain, Korea, Saudi Arabia, Sweden, and Germany.[4] Leading U.S. multinationals routinely request LEED-certified office space in other countries and often build their own U.S. headquarters now to LEED Gold and Platinum standards.[5]

In 2009, my firm helped to certify the first LEED retail project in Arizona, using the LEED for Core & Shell (LEED-CS) rating system, one that I had helped develop during my tenure on the national *core committee* for LEED-CS from 2004 to 2006. By 2009, certifying a project under a LEED system had become a well-organized process, and the USGBC had spun off the Green Building Certification Institute to manage the system for project certification and LEED AP accreditation.

LEED version 3, introduced in 2009, contained many upgrades and adjustments that have made LEED a workable method for certifying green building performance claims. Nevertheless, based on my own experience, LEED certification is still an intricate dance, one with its own melodies, lyrics, scales, steps, and rules of engagement. Even as more than 150,000 LEED APs have joined the roster since the first 100 (of which I was one) were accredited in 2001, the process of communicating LEED requirements to building teams remains one of hands-on engagement.

So, the question remains: how do you actually make the LEED certification process work? My firm is currently managing the LEED certification for a new 250,000-ft^2 corporate headquarters building in Arizona, and I can attest to the fact that the system is still cumbersome in actual application, requiring continual monitoring and management through the entire design, construction, and start-up process.

That's why a book such as this one is badly needed.

Each project team struggles with managing the design, construction, and certification process, as LEED requirements and submittals are laid on top of an already intricate building process. Even when teams are using BIM (building information modeling) tools such as Revit®, the LEED process is not well integrated with the balance of the building team's work. For this reason, many building teams find it useful to hire specialists to manage the LEED certification process, just as they might hire experts in geotechnical analysis, vertical transportation, code compliance, and cost estimating. Even when the work is done *in-house* at larger architectural firms, the LEED project management specialist has typically been hired just to perform that function.

In 2008, I wrote *Green Building Through Integrated Design* (McGraw-Hill) as an attempt to show how to manage an integrated design process through asking the right question at the right time, so as not to preclude good solutions from consideration. I've also written extensively on the use of project *eco-charrettes* as an early-stage design tool to bring sustainable design considerations to the forefront early enough in the process to have an influence on architectural and engineering choices. But nowhere in my research and writing have I tackled how to manage the LEED process in such detail and so carefully laid out a manner as in *LEED–New Construction Project Management*, a masterful effort by the Texas-based expert Vijaya Yellamraju.

Judging by the numbers, most of the 155,000-plus LEED Accredited Professionals have yet to take a project all the way to LEED certification (5,000 projects certified by April 2010), so they truly need a good guide to LEED project management.[6] This book should fill that requirement for the foreseeable future. It's worth a place on every LEED management professional's bookshelf.

Notes

[1] See, for example, my book, *Marketing Green Buildings: Guide for Engineering, Construction and Architecture*, 2006 (Taunton, GA: The Fairmont Press), 128.

[2] USGBC LEED Matrix, personal communication, USGBC staff, June 2010.

[3] Ibid.

[4] www.usgbc.org/LEED/Project/CertifiedProjectList.aspx (accessed June 3, 2010).

[5] www.environmentalleader.com/2009/04/22/mcdonalds-hq-achieves-leed-platinum-status/ describes how McDonald's built its own LEED Platinum headquarters building in Oak Brook, IL in 2009 (accessed June 3, 2010).

[6] http://www.gbci.org/org-nav/announcements/10-04-11/GBCI_Certifies_5_000th_LEED_Project .aspx (accessed June 3, 2010).

ACKNOWLEDGMENTS

This book would not have been possible without the guidance, contribution, advice, and encouragement of many friends, family members, peers, and colleagues.

I am grateful to the team involved with Harvard Divinity School's Rockefeller Hall—Ralph DeFlorio, Nancy Rogo Trainer, Cecilia Wan, and Nathan Gauthier —for sharing information and granting permission to feature the school as a case study. I am also thankful to the team involved with GRE headquarters, specifically Michael Finley and Douglas Pierce, for granting me access to their project. I would also like to acknowledge Barbara Erwine, Maggie Santello, and Tom Paladino at Paladino and Company for their case study contribution. I am also thankful to my colleagues—Dan Alexander, Sam Pate, Lindsey Rose, Anica Landreneau, and Don Brooks—for extending their support.

I would like to specially thank my friends—Shilpa Malik for inspiring me to write a book in the first place; Karen Poff for patiently reviewing my manuscript and providing valuable feedback that helped shape the book; Laura Carter for designing the cover photograph in an amazingly short time frame; Varun Chandra for offering timely advice regarding the content of the book; and Shilpi Burman and Rohit Dhamankar for their encouragement as I worked my way through the book.

I am also deeply indebted to my family—parents, parents-in-law, siblings, and siblings-in-law—for their support throughout the book-writing process. They were my *cheerleaders*. I am also thankful to my husband, Kaushik Valluri, for his unconditional support, love, advice, encouragement, and belief in me.

Finally, I am grateful to Joy Bramble at McGraw-Hill for responding to my proposal within hours of receiving it and agreeing to publish my book. Her encouragement and direction were invaluable throughout the process of writing this book. I would also like to thank Stephen Smith at McGraw-Hill and Nancy Dimitry at D&P Editorial Services for patiently responding to my queries and helping me through the editorial and production processes.

Last but not least, it is a privilege to have the book open with a Foreword by renowned green building authority Jerry Yudelson. I am very thankful to Jerry for his kind words.

Vijaya Yellamraju, LEED AP

LEED–NEW CONSTRUCTION
PROJECT MANAGEMENT

PROCESS

INTRODUCTION

From *Green* to *LEED*

Green building design or, at least, the principles of building green have been around for thousands of years. Examples in history and vernacular architecture all over the globe show that buildings were designed and built in direct response to natural climatic conditions. These structures were *green* even before the term was coined. Right from caves to tents to igloos, people adapted their shelter structures to the natural surroundings. Pueblo Indians, for instance, used a southern exposure plus overhanging cliffs to warm their adobes in winter, while sunlight struck less directly during the hot summer.[1] The ancient Romans similarly employed southern facing houses to trap heat inside during the cold winter months, using clear materials like mica or glass, which acted as solar heat traps.[2]

In fact, good building design should be inherently *green*—oriented to maximize the benefits of sunlight and wind, with an envelope that is designed in response to the local climate and with building systems and components that are energy, water, and resource efficient. These should be fundamental to the way buildings are designed. However, with the advent of heating and air conditioning systems in the twentieth century, there was a dramatic shift in the way buildings were designed. It became easy to alter the internal environment rather than designing in response to external climate conditions. As green building pioneer William McDonough puts it in the Foreword to *Big and Green*, "Most conventional practitioners of modern design and construction find it easier to make buildings as if nature and place did not exist. In Rangoon or Racine, their work is the same."[3] Modern buildings became isolated from the immediate environment and started consuming energy created by burning fossil fuels, like coal and oil. Buildings became one of the major sources of consumption of energy in the world, in the United States in particular.

During the 1970s, because of the energy crisis, there was a push to build energy-efficient and green buildings. Incentives were provided to use solar technologies, and there was a growing concern about environmental issues arising from depletion of resources. But the movement was considered more of a trend and it soon died out in the 1980s with the boom in the economy, falling oil prices, and changes in government regulation. Despite these early efforts, *green* never really became a part of modern architectural vocabulary. With no serious efforts made toward energy efficiency, the last thirty years have seen a tremendous increase in consumption of energy and resources by the building sector. This has escalated to the point that, now, buildings account for about 48% of the total energy consumption in the United States.[4]

Today, the large energy footprint of buildings and their role in contributing to climate change are finally being realized, and the building community is slowly reinventing itself to design and operate buildings in more energy-efficient ways. The next four decades present to us an amazing opportunity to redefine the path of building design and reverse the negative impact of buildings on the environment. There have been many reasons attributed to this resurgence of green building, such as:

1 Increased awareness of the role of buildings in contributing to global climate change
2 Emergence of green building rating systems
3 Realization of cost-effectiveness of building green
4 Several local and federal incentives
5 Availability of new resources, materials, and technologies

One of the most significant factors that has transformed the building industry is the emergence of several green building rating systems, such as Leadership in Energy and Environmental Design (LEED®), Building Research Establishment Environmental Assessment Method (BREEAM™), GreenGlobes™, Green Star™, etc.[5] Refer to Table 1.1 for examples of some green-building rating systems across the globe. Of these, the LEED rating system developed by the U.S. Green Building Council® (USGBC) has become one of the most recognized and internationally accepted benchmarks. The rating system provides a framework to design, build, and operate green buildings and presents metrics to

TABLE 1.1 EXAMPLES OF GREEN BUILDING RATING SYSTEMS

COUNTRY	RATING SYSTEM
United States	Leadership in Energy and Environmental Design (LEED)
Canada	LEED Canada
India	LEED India and Indian Green Building Council (IGBC) Rating System
United Kingdom	Building Research Establishment's Environmental Assessment Method (BREEAM)
Australia	Green Star
New Zealand	Green Star NZ
South Africa	Green Star SA
Germany	German Sustainable Building Certification
Canada	Green Globes
Japan	Comprehensive Assessment System for Building Environmental Efficiency (CASBEE)
Malaysia	Green Building Index
Singapore	GreenMarks

measure their performance. Since its inception in 1998, the LEED rating system has played a major role in catalyzing the green building movement and bringing it to the mainstream. The proof lies in the thousands of buildings that are registered to be certified, the growing number of members in the USGBC organization, the advancement in the products and technologies available to build green, the number of governments mandating LEED certification, the increasing number of LEED projects outside the United States, and the list goes on.[6] LEED has permeated the industry to such an extent that financiers for new development projects now have LEED as a prerequisite for funding a project. The environmental, economic, and social benefits of green buildings along with the brand value associated with LEED will only ensure that green buildings continue to grow. Critics argue that LEED takes a mechanical approach to green building design and that certification is not really necessary to design green buildings. However, it must be noted that LEED provides a broad framework that serves as a starting point for green building design, and the certification process provides a measure of success and ensures accountability.

In summary, the building community needs to revive some of the traditional principles of good design, which were lost or disregarded, and combine them with the many modern-day solutions, so that buildings can truly make an impact in reducing energy use, conserving resources, and creating healthy environments. Tools such as the LEED rating system augment the process of green building design. Today, LEED buildings have become synonymous with green buildings. Figure 1.1 illustrates the evolution of green buildings.

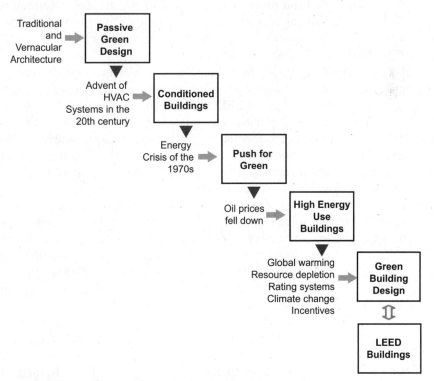

Figure 1.1   Evolution of green buildings.

The Fundamentals of the LEED Rating System

Since the first LEED pilot project program was released in 1998, the rating system has undergone many revisions—version 1.0, 2.0, 2.1, 2.2—to version 3 or LEED 2009, which is the current rating system. The system has evolved over the last decade and grown to include rating systems to cater to various building types, sectors, and project scopes. Examples include LEED for Existing Buildings: Operations & Maintenance (LEED-EBOM), LEED for Core & Shell (LEED-CS), LEED for New Construction (LEED-NC), LEED for Schools, LEED for Homes, and LEED for Commercial Interiors. There are several pilot programs that are under way and expected to be released to the public in the next year or so. These include LEED for Neighborhood Development, LEED for Retail, LEED for Healthcare, and LEED Portfolio program. The first rating system was the LEED-NC system, which was written with a fairly broad scope of requirements that allowed a number of building types to get certified. Until 2008, USGBC administered the certification process for projects. Now the registration and certification processes are handled by the Green Building Certification Institute (GBCI), which also administers credentialing of LEED Accredited Professionals.

The LEED rating systems are voluntary, consensus-based, and market-driven. Based on existing and proven technology, they evaluate environmental performance from a whole building perspective over a building's life cycle, providing a definitive standard for what constitutes a green building in design, construction, and operation.[7]

In 2009, the latest version of LEED—version 3 (LEED v3)—was released by the USGBC. The first key difference between this version and earlier versions is that the maximum number of points available and the number of points per credit have changed. Earlier versions of LEED had varying number of points across the various rating systems, and all credits were weighed equally at 1 point. LEED v3 has standardized the number of total points available across all rating systems to 100 base points. In addition, the allocation of points among credits is based on potential environmental impacts and human benefits of each credit with respect to a set of impact categories. This means that the credits are weighted according to the relative importance of the building-related impacts they address, thereby making some credits worth more than 1 point. Another significant difference between LEED v3 and the earlier versions is that the intent and requirements governing the LEED-NC, LEED for Schools, and LEED for Core & Shell rating systems have been combined into a single LEED reference book titled, *LEED Reference Guide for Green Building Design and Construction*. The guide provides various credit criteria, implementation strategies, and documentation requirements for each of these rating systems.

The focus of this book is on the LEED-NC rating system. The project management processes, implementation tools, and documentation samples included in this book are written with the LEED-NC rating system in mind. Projects using other rating systems may also leverage information as applicable to their projects. Figure 1.2 presents a timeline of the different versions and types of LEED rating systems based on their year of release.

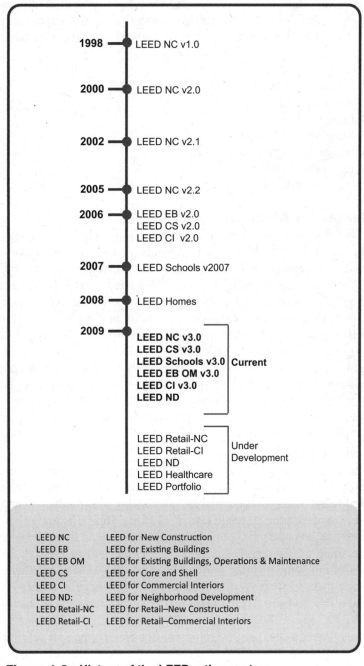

Figure 1.2 History of the LEED rating system.

LEED-NC CREDIT CATEGORIES

The LEED-NC rating system provides a framework for design, construction, and operation of buildings from a whole building perspective and recognizes performance in seven key areas. Points are awarded for achievement of credit requirements in each of these areas. In addition, there are certain prerequisites under each category that the project must meet in order to qualify for consideration of other credits. The seven categories are:

1 Sustainable Sites
2 Water Efficiency
3 Energy and Atmosphere
4 Materials and Resources
5 Indoor Environmental Quality
6 Innovation in Design
7 Regional Priority

Sustainable Sites (SS)

This category of credits provides guidance on how to select, design, and manage project sites. It specifically encourages building on previously developed land, designing regionally appropriate landscape, and utilizing best management practices for stormwater quality and quantity control. The category also rewards smart transportation choices and reduction in erosion, light pollution, heat island effect, and construction activity pollution. The credits, in essence, try to mitigate some of the negative effects buildings have on the local and regional environment. Table 1.2 lists the credits that fall under this category of the LEED-NC system. A total of up to 26 points may be achieved in this category. The credits in this category are typically referenced in this book as "SS Cx" or "SS Px" where "SS" represents "Sustainable Sites," "C" indicates "Credit," "P" indicates "Prerequisite," and "x" represents the credit or prerequisite number. For instance, Credit 1 in the "Sustainable Sites" category would be referred to as SS C1.

Water Efficiency (WE)

The increasing use of potable water by buildings has caused a severe drop in the water level in underground aquifers and led to a strain on the country's water resources.[8] The credits and prerequisites in this category outline water reduction targets and encourage the use of various strategies, such as using high efficiency fixtures, reusing rainwater, etc. to reduce the amount of potable water used in buildings, thereby minimizing the strain on municipal water supply sources and systems. It also encourages the use of efficient irrigation and landscape design strategies to reduce the amount of water used for irrigation purposes. Table 1.3 lists the LEED-NC credits that fall under this category. These credits are typically referenced in this book as "WE Cx" or "WE Px" where "WE" represents "Water Efficiency," "C" indicates "Credit," "P" indicates "Prerequisite," and "x" represents the credit or prerequisite number. For instance, Credit 1 in the "Water Efficiency" category would be represented as WE C1. A maximum of 10 points may be achieved in this category.

TABLE 1.2   LEED-NC SUSTAINABLE SITES CREDITS

CREDIT	TITLE	POINTS
Prerequisite 1	Construction Activity Pollution Prevention	
Credit 1	Site Selection	1
Credit 2	Development Density and Community Connectivity	5
Credit 3	Brownfield Redevelopment	1
Credit 4.1	Alternative Transportation—Public Transportation Access	6
Credit 4.2	Alternative Transportation—Bicycle Storage and Changing Rooms	1
Credit 4.3	Alternative Transportation—Low-Emitting and Fuel-Efficient Vehicles	3
Credit 4.4	Alternative Transportation—Parking Capacity	2
Credit 5.1	Site Development—Protect or Restore Habitat	1
Credit 5.2	Site Development—Maximize Open Space	1
Credit 6.1	Stormwater Design—Quantity Control	1
Credit 6.2	Stormwater Design—Quality Control	1
Credit 7.1	Heat Island Effect—Non-roof	1
Credit 7.2	Heat Island Effect—Roof	1
Credit 8	Light Pollution Reduction	1

Energy and Atmosphere (EA)

Buildings are known to use 39% of the total energy and 74% of the total electricity produced annually in the United States.[9] This category of credits and prerequisites sets standards for minimum energy efficiency of buildings and encourages the use of various strategies to reduce their overall energy consumption. It also rewards monitoring the energy used and supplementing it with on-site renewable energy systems. Table 1.4 identifies the credits in this category. Up to 35 points can be achieved by implementing various credit requirements in this category. The credits are typically referenced in this book as "EA Cx" or "EA Px" where "EA" represents "Energy and

TABLE 1.3   LEED-NC WATER EFFICIENCY CREDITS

CREDIT	TITLE	POINTS
Prerequisite 1	Water Use Reduction—20% Reduction	
Credit 1	Water Efficient Landscaping	2 to 4
Credit 2	Innovative Wastewater Technologies	2
Credit 3	Water Use Reduction	2 to 4

TABLE 1.4   LEED-NC ENERGY AND ATMOSPHERE CREDITS

CREDIT	TITLE	POINTS
Prerequisite 1	Fundamental Commissioning of Building Energy Systems	
Prerequisite 2	Minimum Energy Performance	
Prerequisite 3	Fundamental Refrigerant Management	
Credit 1	Optimize Energy Performance	1 to 19
Credit 2	On-Site Renewable Energy	1 to 7
Credit 3	Enhanced Commissioning	2
Credit 4	Enhanced Refrigerant Management	2
Credit 5	Measurement and Verification	3
Credit 6	Green Power	2

Atmosphere," "C" indicates "Credit," "P" indicates "Prerequisite" and "x" represents the credit or prerequisite number. For instance, Credit 1 in the "Energy and Atmosphere" category is referred to as EA C1.

Materials and Resources (MR)

This category of credits focuses on reducing the amount of waste that buildings generate during both their construction and operation cycles. It encourages the selection of sustainable materials that incorporate recycled content, are made from rapidly renewable materials, and are sourced regionally. The category also rewards minimization of landfill and incinerator disposal for materials that leave the building. Up to 14 points can be achieved in this category (Table 1.5). The credits in this category are typically referenced in this book as "MR Cx" or "MR Px" where "MR" represents "Materials and Resources," "C" indicates "Credit," "P" indicates "Prerequisite," and "x" represents the credit or prerequisite number. For instance, Credit 1 in the "Materials and Resources" category is represented as MR C1.

Indoor Environmental Quality (EQ)

This category of credits is written with the intent of improving the overall indoor environmental quality of buildings. On an average, Americans spend about 90% of their day indoors, making it necessary to design indoor environments that are healthy and comfortable.[10] The EQ credits reward projects that provide improved ventilation, monitor outdoor air, are designed with occupant thermal comfort in mind, and provide controllability of lighting and thermal comfort systems by individual occupants. It also encourages the use of daylighting and improving access to outdoor views for occupants. A total of 15 points can be achieved in this category (Table 1.6). The

TABLE 1.5 LEED-NC MATERIALS AND RESOURCES CREDITS		
CREDIT	**TITLE**	**POINTS**
Prerequisite 1	Storage and Collection of Recyclables	
Credit 1.1	Building Reuse—Maintain Existing Walls, Floors, and Roof	1 to 3
Credit 1.2	Building Reuse—Maintain 50% of Interior Non-Structural Elements	1
Credit 2	Construction Waste Management	1 to 2
Credit 3	Materials Reuse	1 to 2
Credit 4	Recycled Content	1 to 2
Credit 5	Regional Materials	1 to 2
Credit 6	Rapidly Renewable Materials	1
Credit 7	Certified Wood	1

credits in this category are typically referenced in this book as "EQ Cx" or "EQ Px" where "EQ" represents "Environmental Quality," "C" indicates "Credit," "P" indicates "Prerequisite," and "x" represents the credit or prerequisite number. For instance, Credit 1 in the "Indoor Environmental Quality" category is referred to as EQ C1.

Innovation in Design (ID)

The Innovation in Design credits offer projects a chance to gain extra points in addition to the credit points offered in the previous categories. Up to 6 points can be achieved under this category. There are two ways by which these credits can be achieved: one is by demonstrating use of innovative strategies to improve the building's performance beyond what is required by the LEED credit (these are referred to as *exemplary performance* points) and the second way is to pursue innovative ideas that are not specifically addressed by any credit in the rating system. There is also one point available for having a LEED Accredited Professional (LEED AP) on the project team. Table 1.7 lists the number of ID credits available for a project. The credits in this category are typically referenced in this book as "ID Cx" where "ID" represents "Innovation in Design," "C" indicates "Credit," and "x" represents the credit number. For instance, Credit 1 in the "Innovation in Design" category is referred to as ID C1.

Regional Priority (RP)

This last category of credits was introduced in 2009 in the new version LEED v3. This category promotes credits that have been identified as important to a particular region and are a priority for that region. In order to achieve these credits, project teams are not required to do anything other than achieve the credits that are marked as *regional priority* for their region. Six credits for each zip code have been identified and made

TABLE 1.6	LEED-NC INDOOR ENVIRONMENTAL QUALITY CREDITS	
CREDIT	TITLE	POINTS
Prerequisite 1	Minimum Indoor Air Quality Performance	
Prerequisite 2	Environmental Tobacco Smoke (ETS) Control	
Credit 1	Outdoor Air Delivery Monitoring	1
Credit 2	Increased Ventilation	1
Credit 3.1	Construction IAQ Management Plan—During Construction	1
Credit 3.2	Construction IAQ Management Plan—Before Occupancy	1
Credit 4.1	Low-Emitting Materials—Adhesives and Sealants	1
Credit 4.2	Low-Emitting Materials—Paints and Coatings	1
Credit 4.3	Low-Emitting Materials—Flooring Systems	1
Credit 4.4	Low-Emitting Materials—Composite Wood and Agrifiber Products	1
Credit 5	Indoor Chemical and Pollutant Source Control	1
Credit 6.1	Controllability of Systems—Lighting	1
Credit 6.2	Controllability of Systems—Thermal Comfort	1
Credit 7.1	Thermal Comfort—Design	1
Credit 7.2	Thermal Comfort—Verification	1
Credit 8.1	Daylight and Views—Daylight	1
Credit 8.2	Daylight and Views—Views	1

available for project teams. Up to four points may be achieved under this category. These credits are typically referenced in this book as "RP Cx" where "RP" represents "Regional Priority," "C" indicates "Credit," and "x" represents the credit number. For instance, Credit 1 in the "Regional Priority" category is referred to as RP C1.

TABLE 1.7	LEED-NC INNOVATION IN DESIGN CREDITS	
CREDIT	TITLE	POINTS
Credit 1.1	Innovation in Design (specify title)	1
Credit 1.2	Innovation in Design (specify title)	1
Credit 1.3	Innovation in Design (specify title)	1
Credit 1.4	Innovation in Design (specify title)	1
Credit 1.5	Innovation in Design (specify title)	1
Credit 1.2	LEED Accredited Professional (LEED AP)	1

Impact of LEED Buildings

Since its inception in 1998, there have been several discussions among the building community regarding the real benefit of LEED buildings. There was no raw data available to verify the tangible benefits offered by LEED buildings, but in the past two years, a number of studies have provided significant evidence regarding the impact of LEED buildings, specifically in the areas of:

- Energy Performance
- Financial Returns
- Occupant Productivity
- CO_2 Emissions Reduction

ENERGY PERFORMANCE

A 2008 study[11] conducted by the New Building Institute (NBI) provides critical information regarding the energy performance of LEED buildings and whether they are performing as they were intended or designed to. Very rarely do building facility managers or design teams evaluate the performance of the building after construction is complete, and it is never really known if the systems are performing as they were originally meant to. The study by NBI analyzed the measured energy performance for 121 LEED-NC projects, and the results showed that LEED projects averaged *substantial energy performance* over non-LEED building stock. The metrics analyzed were (a) Energy Use Intensity (EUI), (b) ENERGY STAR ratings of LEED buildings, and (c) measured results compared to initial energy modeling.[12]

Energy Use Intensity (EUI)

The energy use intensity is a measure of total energy use normalized for floor area. This is used to compare the energy use of different buildings. Whole-building energy use is measured in kBtu (1000 British thermal units) per square foot, per year, to standardize units between fuels. The Commercial Building Energy Consumption Survey (CBECS) is a survey of building energy characteristics that is completed every four years by the U.S. Energy Information Administration (EIA). The CBECS provides information about the energy use intensities of the national building stock, and one way to know if a building is performing better, from an energy use perspective, is to compare it with this national building EUI data. The NBI study compared the EUI (in kBtu/sf/yr) to the CBECS national EUI data and found that "for all 121 LEED buildings the median measured EUI was 69 kBtu/sf, 24% below (better than) the CBECS national average for all commercial building stock." Comparisons by building activity type showed similar relationships. For offices, the single most common building type, "LEED EUIs averaged 33% below CBECS."[13]

ENERGY STAR Ratings of LEED Buildings

The ENERGY STAR program is a joint program run by the U.S. Environmental Protection Agency (EPA) and U.S. Department of Energy (DOE) that rates the energy performance of a building. To determine the performance of a facility, EPA compares its energy use with other, similar types of facilities on a scale of 1–100; buildings that achieve a score of 75 or higher may be eligible for the ENERGY STAR. The EPA rating system accounts for differences in operating conditions, regional weather data, and other important considerations. The NBI study found that "the average ENERGY STAR rating of LEED buildings was 60, compared with the median rating of 50 for the complete building stock." However, the study also showed that nearly half of the LEED buildings had an ENERGY STAR rating of 75, meeting the qualification level of an ENERGY STAR building. Although there was a significant gap between the number of LEED buildings qualifying for ENERGY STAR and those that didn't, it must be realized that at least 50% of the LEED buildings were performing better than non-LEED buildings. It obviously reflects the potential for improvements in the LEED rating system but signals that it is a step in the right direction.

Measured Results Compared to Energy Modeling

The LEED-NC rating system awards points in the EA category on the basis of the results of an energy simulation that compares the proposed building's energy use with a baseline building. The simulation is done using the performance rating method— Appendix G of ASHRAE 90.1-2007 standard. The calibration of measured performance with the energy model predictions is a healthy indicator of whether the building is performing as predicted. The NBI study found that "measured energy savings averaged 28% compared to code baselines, which was close to the average 25% savings predicted by energy modeling in LEED submittals."[14]

FINANCIAL RETURNS

Green buildings also afford substantial direct financial returns in terms of higher occupancy and rental rates, as was found by several recent studies. According to a 2008 study by the CoStar group, "LEED buildings command rent premiums of $11.33 per square foot over their non-LEED peers and have 4.1% higher occupancy. Rental rates in ENERGY STAR buildings represent a $2.40 per square foot premium over comparable non-ENERGY STAR buildings and have 3.6% higher occupancy." It further states that "LEED buildings command $171 more per square foot than non-LEED buildings." The study also found that "LEED buildings had 3.5% lower vacancy rates and 13% higher rental rates than the market."[15]

Another research study done by CB Richard Ellis and the University of San Diego in 2009 found that green buildings outperformed their non-green peers in key areas, such as occupancy, sale price, and rental rates, sometimes by wide margins. The results of the study indicated that there was a greater demand by property investors and tenants for buildings that had earned either LEED certification or the ENERGY STAR label.[16]

The present market downturn and recession in the economy have had an impact on the overall rental and sale values of properties, and the numbers, as reported by these studies. may have fallen. However, the business case for LEED buildings remains strong, especially because of the long-term benefits of reduced operating costs and the several government incentives for building green.

OCCUPANT PRODUCTIVITY

There have been many studies in the past that have linked improved indoor air quality and daylight to improved health and productivity of building occupants. In a study[17] done by William Fisk for Lawrence Berkeley National Laboratory (LBNL), a survey of 100 U.S. office buildings, found that 23 percent of office workers experienced frequent symptoms of sick building syndrome (SBS) such as respiratory ailments, allergies, and asthma. The study indicated that the impact was usually hidden in sick days, lower productivity, and medical costs; but the economic impact was enormous, with an estimated decrease in productivity around 2 percent nationwide, resulting in an annual cost to the United States of approximately $60 billion.

There are several other studies done by LBNL on the effect of daylighting on student performance. One case study, from a school located in San Juan Capistrano in southern California, indicated that students with the most daylighting in their classrooms progressed 20% faster on math comprehension and 26% faster on reading comprehension tests in one year than those with minimal exposure to natural light. Similarly, students in classrooms with the largest window surface areas were found to progress 15% faster in math and 23% faster in reading than those located in classrooms with a minimal window surface area. Students with well-designed skylights in the space improved 20% faster than students without skylights.[18]

While the above studies indirectly indicate that buildings that are designed with better indoor air quality and daylighting have a positive, direct impact on the productivity of occupants, the study by CB Richard Ellis (CBRE) and the University of San Diego provides information linking LEED buildings directly with occupant productivity. The CBRE report states that "tenants in green buildings experience increased productivity and fewer sick days." The report backs up the results with numbers generated as a result of surveying 154 buildings, totaling more than 51.6 million square feet and housing 3000 tenants in ten markets across the United States. The report also found that tenants "reported an average of 2.88 fewer sick days in their current green office versus their previous non-green office, and about 55% of respondents indicated that employee productivity had improved. Based on the average tenant salary, an office space of 250 square feet per worker and 250 workdays a year, the decrease in sick days translated into a net impact of nearly $5.00 per square foot occupied, and the increase in productivity translated into a net impact of about $20 per square foot occupied." [19] Although the study included both ENERGY STAR rated and LEED buildings, what it found was indeed noteworthy in terms of the impact of *green* buildings.

CO_2 EMISSIONS REDUCTION

One of the primary needs for designing *green* buildings is to reduce their impact on the environment. According to the U.S. Green Building Council, buildings account for more than 72% of all electricity consumption, 39% of energy use, 38% of CO_2 emissions, 40% of raw materials use, 30% of waste output and 14% of potable water consumption in the United States.[20] The increasing use of energy by buildings has led to a depletion of fossil-fuel based resources as well as generation of greenhouse gas

(GHG) emissions. These GHG emissions have resulted in global warming that is causing the earth's temperatures to rise at a dramatic pace. It is believed that the current global warming is at 0.7°C above preindustrial levels. Scientists predict that in order to avoid dangerous climate change, global warming must remain under 2°C above preindustrial levels. Higher temperatures could cause catastrophic climate change.[21]

LEED buildings can help achieve significant reductions in GHG emissions, thereby controlling global warming and mitigating the effects of serious climate change. They are designed and operated to reduce electricity comsumption, energy, water, and materials use. The Green Building Impact Report published in 2009 states that "annual CO_2 savings from LEED buildings was approximately 2.9 million tons from energy efficiency and renewables. This figure was expected to grow to 130 million tons per year by 2020 and almost 320 million tons annually by 2030."[22]

In summary, it can be seen that LEED buildings have had a significant impact in the areas of energy savings, financial returns, and occupant productivity. Figure 1.3 summarizes the findings of all the studies described in this section.

Energy Performance of LEED Buildings*	
Energy Use Intensity (EUI)	24% better than national average
Energy Star Ratings	50% of buildings have rating > 75
Comparison with Simulation	Measured energy reduction: 28% Simulated energy reduction: 25%

Financial Returns on LEED Buildings**	
Rent Premium	$11.33 more per square foot (sf)
Resale Value	$171 more per sf
Occupancy	3.6% higher occupancy

Occupant Productivity in LEED Buildings ***	
Sick Days	2.88 fewer sick days; $5.00 per sf
Productivity	55% increase; $20.00 per sf

* Cathy Turner, Mark Frankel, *Energy Performance of LEED for New Construction Buildings* (NBI,USGBC,2008)
** Andrew C. Burr, CoStar Study Finds Energy Star, LEED Bldgs. Outperform Peers, CoStar Group News (2008)
***Norm G. Miller, Dave Pogue, *Do Green Buildings Make Dollars and Sense?* (Burnham-Moores Center for Real Estate University of San Diego, CB Richard Ellis, 2009)

Figure 1.3 Impact of LEED buildings: energy performance, financial returns and occupant productivity.

What This Book Is About

This book is a beginner's guide for people who are new to LEED project management. It can also be used as a companion handbook by more seasoned professionals. The intent of this book is to become a go-to guide for architects, engineers, consultants, contractors, and LEED Accredited Professionals (APs) who are managing a LEED project. It is also an indirect resource for building owners, developers, and managers to understand what it takes to get a building certified.

Realizing the significance of the LEED rating system in shaping our modern-day building vocabulary, where the terms *green* and *LEED* are used interchangeably, we (members of the building community) must become better equipped to integrate the requirements of this rating system with our traditional building design processes. The LEED process requires high level planning, attention to detail, coordination, and management. In addition, the success of the certification is dependent on the integrity of the documentation provided to support the project's compliance with the rating system's requirements.

There has been an exponential increase in the number of LEED APs and an increasing number of projects that are registered with the USGBC to become LEED certified. A member update by USGBC in March 2010 stated that there were 27,358 registered projects of which a total of 4,890 had been certified. It was also noted that there were almost 140,218 LEED APs globally. These numbers indicate the tremendous potential of the LEED rating system in transforming and influencing the building community across the world. The fact that only a small fraction of the projects that were registered attained certification tells us the rigor that is needed by teams to achieve certification. This book gives readers a framework to take their projects to successful certification.

One of the many reasons that can be attributed to why projects fail to get certification is that they have not been managed any differently from traditional projects. The *LEED* element adds another dimension to the traditional project management model and requires careful integration into the overall design process. The growth in the number of LEED Accredited Professionals indicates that there will be a lot of freshly minted LEED APs trying to manage a LEED project without being equipped with tools for implementing a LEED project. Many design teams struggle with understanding the specifics of the LEED process. There is no defined methodology or roadmap available for actual implementation of a LEED project. While the USGBC provides a number of resources, including reference guides and workshops, to understand the LEED rating system, it does not necessarily outline steps or implementation processes. Most other available resources on LEED are books and online resources that are great references on general concepts of green, strategies to implement green, and also how to become LEED accredited. But many of these books fall short—there is no one "go-to" book that demystifies the LEED process, lays out a roadmap defining the various steps involved in a simplistic manner, and provides the necessary tools to implement a LEED project.

This book will provide step-by-step guidance to manage a LEED project to ensure that green requirements are integrated into the design and construction process. It will include steps, workflow diagrams, schedules, checklists, and sample documents—all the necessary tools that will provide a framework for the LEED project manager to start, manage, and implement a LEED project. This book will tie the building design process with the GBCI's LEED certification process. This will, hopefully, generate widespread implementation of the LEED rating system, thereby bridging the gap between registered and certified projects.

What This Book Is Not

The intent of this book is to serve as a complimentary resource to the LEED Reference Guide provided by USGBC. Information regarding a credit's official intent and requirements should be sourced from the Reference Guide. This book is written assuming that the readers or users of the book have some level of familiarity with a credit's intent and requirements. Care has been taken not to repeat information that is already available in the Reference Guide, except for sections where further explanation of a particular process is required.

The book primarily focuses on implementation procedures for new construction projects that are specifically pursuing LEED-NC certification. The book is not directly applicable to LEED for Schools, LEED for Core & Shell (LEED-CS), LEED for Existing Buildings; Operations & Management (LEED EBOM), or any other green-building rating system. However, projects that are pursuing other types of certification may certainly be able to leverage information related to the processes, procedures, and general project management. The framework presented in this book is one of many that can be used to implement a LEED project. The author recognizes that there is no "one size fits all" solution on how to manage a LEED project. Every project is unique; and designs, schedules, teams, and delivery mechanisms vary with each project. The intent of this book is to present *a* defined set of implementation strategies, and there could be many variations of the method presented here.

The green building movement is rapidly evolving, and so are the different rating systems. This book is based on the LEED rating system that is applicable at the time of publication—LEED-NC v3. The intent is to present a set of management tools that can be adapted to changes in the rating system. Often the changes are made to the rating system requirements and not necessarily to the processes that define it; so this book should be scalable for rating system revisions in the near future.

Why Manage LEED Projects Differently?

The book is based on the premise that LEED projects should be managed differently from non-LEED projects. Before proceeding with the descriptions on how to manage LEED projects, it is important to understand why we need to manage them differently. For many decades, the process of building design and construction has been honed to a science where each team member is cognizant of their roles in the overall process. However, the LEED process demands a commitment to questioning techniques and seeking new answers. Green building takes a whole system approach to design and requires early involvement of the design and construction disciplines to be successful. LEED is intended to improve building practices through transformation and innovation, and therefore requires that traditional methods of design and construction be revised and made more efficient. A LEED project requires additional considerations during the design, construction, and operation of the building. There are unique responsibilities and documentation requirements that are expected from the design team. For instance, a LEED project requires the use of energy modeling throughout the project to make informed design decisions. This can be considered as a *non-traditional*, yet useful practice that needs to be incorporated into the design process. By integrating these non-traditional practices into the design procedures, it is hoped that, eventually, these added responsibilities and requirements will become part of the regular design process. That will mark the true success of a rating system such as LEED in transforming the way buildings are designed. Below are some areas where a LEED project process differs from that of a traditional or a non-LEED building.

Design Decisions

A LEED project warrants a different approach to making design decisions, using tools that consider the interactions of the whole building and its impact on the environment. Decisions related to envelope, choice of HVAC systems, choice of materials, and choice of lighting are influenced by the LEED credit framework. Simulating the building's energy performance by means of a computer-generated model at every stage of design to come up with a design solution that optimizes the energy performance of the building is a requirement that is not usually part of the design process. Simulating the daylight levels inside a building to design the best possible fenestration strategy and coordinating it with interior lighting systems are other such examples. A LEED project requires preparing a comprehensive water budget that takes into account indoor water consumption along with outdoor irrigation needs as well as stormwater availability. For a building to achieve the true benefits of a green building, LEED requirements need to be integrated within the design from the beginning instead of being forced as an afterthought.

Construction Practices

The LEED rating system has given rise to a number of new construction practices and created a whole new set of responsibilities for general contractors and sub-contractors. A LEED project has significant impact on project costs, schedules, durations, and even project administration and contracts. In addition, there are procedures to be implemented on-site that are not typical construction procedures, such as recycling construction waste or implementing measures to mange indoor air quality.

Building Operations

In a LEED project, there are many aspects from an operational perspective that need to be considered early-on. The maximum benefit of LEED buildings is derived from the operational savings associated with energy-efficient systems. Decisions during the design process need to be made with long-term operational benefits in mind. The rating system itself gives credit to projects that institute post-occupancy evaluations and measurement as well as verification procedures after construction is complete. Many management policies related to recycling, smoking, and alternative transportation may need to be developed and implemented by the building management. Traditionally, the building management's role in the building design process is fairly limited. However, while designing a LEED building, the management's input during design is important to ensure that the building is able to derive all the benefits associated with operational savings.

Team Member Responsibilities

The various team members working on a LEED project have additional layers of responsibility compared to those of a traditional project. The civil engineer and landscape architect may need to perform additional stormwater or irrigation calculations, the architect may need to write *green* specifications, or the mechanical engineer may need to create an energy model for the building. While some team members may lead the implementation of some credits, there are other team members who might need to support the implementation of that credit. Special consultants, such as LEED consultants, energy analysts, lighting consultants, commissioning authorities, IAQ specialists, and others, may be involved with the project. This requires increased level of coordination between the various consultants and team members to ensure the project's success.

Documentation

Finally, a LEED project requires diligent documentation of all activities to show that the project has met all the LEED credit requirements. Lack of clear direction as to what is required, from whom, and within what time frame can easily derail the process and delay certification. It has been observed that the necessity of providing documentation is a reason why project teams are wary of the LEED certification process. In reality, setting up documentation deliverables from the team right from the onset of the project and periodically reviewing them at various design and construction stages will simplify the process and make it more easily manageable.

How to Use This Book

This book is divided into two parts and each part is self-explanatory. The first part, Process, comprises the various steps and activities involved in managing a LEED project from pre-schematic design to construction completion and submission of the project to GBCI. The second part, Implementation, identifies activities that need to be completed by design team members to implement and document a specific credit. It also includes some credit sample documents that were submitted to USGBC by the teams of the case study projects described in Chapter 8.

PART 1: PROCESS

Chapter 1   Introduction

Chapter 1 provides a brief history of the green building movement, introduces the LEED Rating System developed by the U.S. Green Building Council (USGBC), and describes the impact it has had on the green building market in the United States and the world over. It also discusses the goals and aspirations of this book and how the readers can use and benefit from it.

Chapter 2   Project Management Process

Chapter 2 describes in detail the essential components of project management—process, people, and planning—the three pillars for a LEED project. It discusses the traditional design process, the GBCI certification process, and how the LEED project management process described in this book aligns with both. Three distinct stages are identified to manage the LEED process, and each of these stages is explained in detail in Chapters 3, 4, and 5 respectively. The chapter further identifies key roles and responsibilities of various design team members and outlines how LEED milestones may be integrated with the design schedule.

Chapter 3   Stage I: Project Definition and Goal Setting

Chapter 3 elaborates on the first stage of the process, which is to define the project and set goals. It identifies all key activities, tasks, and objectives of this first stage of the process.

Chapter 4   Stage II: Design Phase Integration

Chapter 4 describes the second stage of the process, which is to integrate green building and LEED requirements into the design process. The implementation steps and activities are aligned with the schematic design, design development, and construction documents phases of the design process.

Chapter 5   Stage III: Construction Phase Implementation

Chapter 5 describes the final stage of the process, which is to implement LEED credit requirements during construction. It identifies specific activities that need to be completed to document the project, and submit to GBCI.

Chapter 6   LEED Process: Adaptability, Applicability, and Best Practices

Chapter 6 identifies the benefits of implementing this project management process and provides best management practices for LEED projects. It also offers advice on how the processes described in this book may be adapted to different rating systems and building types.

Chapter 7   Future of Green: 2010 and Beyond

Chapter 7 identifies the importance of the LEED rating system in catapulting *green* to the spotlight over the last decade and provides a glimpse into the potential strategies, technologies, and processes that could shape the green building movement in the coming decades.

Chapter 8   Success Stories

Chapter 8 provides an insight into how various, well-known projects have successfully achieved LEED certification. The chapter highlights the design process and green building strategies that were followed and offers key lessons learned from each project.

PART 2: IMPLEMENTATION

Section 1   Credit Implementation—Sustainable Sites

This section presents activities that the project team members may follow during the various phases of design to implement the requirements of credits in the *Sustainable Sites* category of the LEED-NC rating system. It also identifies team member responsibilities, documentation required, as well as tips and strategies to achieve a particular credit. The section also includes some sample documents from the case study projects described in Chapter 8.

Section 2   Credit Implementation—Water Efficiency

This section presents activities that the project team members may follow during the various phases of design to implement the requirements of credits in the *Water Efficiency* category of the LEED-NC rating system. It also identifies team member responsibilities, documentation required, as well as tips and strategies to achieve a particular credit.

Section 3   Credit Implementation—Energy and Atmosphere

This section presents activities that the project team members may follow during the various phases of design to implement the requirements of credits in the *Energy and Atmosphere* category of the LEED-NC rating system. It also identifies team member responsibilities, documentation required, as well as tips and strategies to achieve a particular credit. The section further includes some sample documents from the case study projects described in Chapter 8.

Section 4   Credit Implementation—Materials and Resources

This section presents activities that the project team members may follow during the various phases of design to implement the requirements of credits in the *Materials and Resources* category of the LEED-NC rating system. It also identifies team member responsibilities, documentation required, as well as tips and strategies to achieve a particular credit. The section further includes with sample documents from the case study projects described in Chapter 8.

Section 5   Credit Implementation—Indoor Environmental Quality

This section presents activities that the project team members may follow during the various phases of design to implement the requirements of credits in the *Indoor Environmental Quality* category of the LEED-NC rating system. It also identifies team member responsibilities, documentation required, as well as tips and strategies to achieve a particular credit. The section further includes sample documents from the case study projects described in Chapter 8.

Section 6   Credit Implementation—Innovation in Design
This last section provides some examples of innovation credits achieved by the case study projects described in Chapter 8.

Notes

[1] Ethan Goffman, *Green Buildings: Conserving the Human Habitat* (CSA—Discovery Guides, 2006). http://www.csa.com/discoveryguides/green/review2.php (accessed February 8, 2010).

[2] Ibid.

[3] David Gissen, *Big and Green: Toward Sustainable Architecture in the 21st Century* (New York: Princeton Architectural Press, 2002), 9.

[4] Architecture 2030, "The Building Sector: A Hidden Culprit," http://www.architecture2030 .org/current_situation/ building_sector.html (accessed February 10, 2010).

[5] LEED is a registered trademark of the U.S. Green Building Council (USGBC); BREEAM is a registered trademark of Building Research Establishment (BRE); GreenGlobes is a registered trademark of ECD Energy and Environment Canada; GreenStar is a registered trademark of Green Building Council of Australia.

[6] *USGBC Member Update* of March 2010 reported 18,000 Members; 140,218 LEED APs and Green Associates; and 27,358 Registered and 4,890 Certified LEED projects.

[7] USGBC, "LEED Rating Systems," http://www.usgbc.org.

[8] USGBC, *LEED Reference Guide for Green Building Design and Construction* (USGBC, 2009), 161.

[9] Ibid., 213.

[10] Ibid., 401.

[11] Cathy Turner, Mark Frankel, *Energy Performance of LEED for New Construction Buildings* (NBI, USGBC, 2008).

[12] ENERGY STAR is a joint program of the U.S. Environmental Protection Agency and the U.S. Department of Energy.

[13] Cathy Turner, Mark Frankel, *Energy Performance of LEED for New Construction Buildings* (NBI, USGBC, 2008), 2.

[14] Ibid.

[15] Andrew C. Burr, *CoStar Study Finds Energy Star, LEED Bldgs. Outperform Peers* (CoStar Group News, March 26, 2008). http://www.costar.com/News/Article.aspx?id=D968F1E0DCF737 12B03A099E0E99C679. (accessed March 15, 2010).

[16] Norm G. Miller, Dave Pogue, *Do Green Buildings Make Dollars and Sense?* (Burnham-Moores Center for Real Estate University of San Diego, CB Richard Ellis, 2009).

[17] William J. Fisk, "Health and Productivity Gains from Better Indoor Environments," in *The Role of Emerging Energy-Efficient Technology in Promoting Workplace Productivity and Health*, (Lawrence Berkeley National Laboratory, 2002).

[18] Heschong Mahone Group, Lisa Heschong, "Daylighting and Human Performance," in *The Role of EmergingEnergy-Efficient Technology in Promoting Workplace Productivity and Health* (Lawrence Berkeley National Laboratory, 2002).

[19] Norm G. Miller, Dave Pogue, *Do Green Buildings Make Dollars and Sense?* (Burnham-Moores Center for Real Estate University of San Diego, CB Richard Ellis, 2009).

[20] USGBC, "Green Building Research," http://www.usgbc.org/DisplayPage.aspx?CMSPage ID=1718.

[21] Architecture 2030, "The Science: Converging Events," http://www.architecture2030.org/cur rent_situation/science.html.

[22] Rob Watson, *Green Building Market and Impact Report 2009* (Greener World Media, Inc., 2009).

LEED PROJECT MANAGEMENT PROCESS

Introduction

The three major components that are critical to managing any project are process, people, and planning. Understanding the process, coming up with a strategic plan, and effectively utilizing the skills of people to implement a project are important for any project, be it a LEED project or not. However, as explained in Chapter 1, a LEED project requires additional levels of input at all stages of design, and therefore, these three factors assume an even more important role than in traditional projects. This chapter will focus on these three *Ps*—process, people, and planning—as they are applicable to a LEED-NC project. Figure 2.1 illustrates these three aspects.

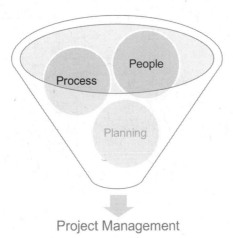

Figure 2.1 The three *Ps* of project management.

Process

In a LEED project, there is a need to overlay the traditional architectural design process with the Green Building Certification Institute's (GBCI) certification process. GBCI administers certification of buildings and has outlined a streamlined series of steps to obtain LEED certification. The traditional design process and the GBCI processes need to be married together in a manner that is unified and integral to the way buildings are designed. This can be achieved by introducing a set of definitive steps and milestones into the traditional architectural design process that also align with the GBCI certification process, thus creating a project management process specific to LEED projects. Figure 2.2 describes the overlaying of the architectural design process with the GBCI certification process.

THE TRADITIONAL ARCHITECTURAL DESIGN PROCESS

Traditional architectural design process is a linear progression of activities that takes the project from an initial concept to actual realization. While every project has its own set of variations and subphases, in essence, building design projects typically follow the broad phases outlined below. Figure 2.3 shows the various phases of a typical architectural design process.

Pre-Schematic Design Phase

The pre-schematic design phase of the project involves gathering information, establishing project goals, budgeting, programming, master planning, and creating a conceptual design for the project. Some project teams may divide these activities into separate phases of programming and conceptual design. The programming phase involves arriving at a set of criteria, preferences, and requirements for the project. This phase helps establish a basis of design to create a conceptual design idea for the project.

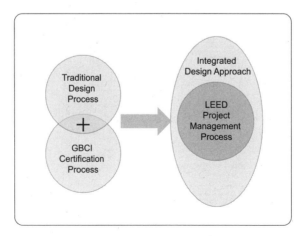

Figure 2.2   Overlaying the architectural design process with GBCI certification process.

Architectural Design Process

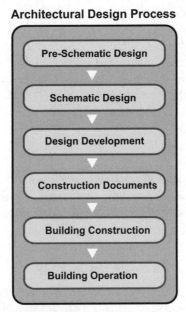

Figure 2.3 Typical architectural design process.

Schematic Design Phase

The schematic design phase of the project involves the exploration and testing of various design schemes according to the programming needs identified in the pre-schematic design phase. By the end of this phase, a preliminary design and layout is established, and drawings are prepared to reflect this schematic design. Preliminary cost estimates and outline specifications are prepared as well in this phase. Projects with large scopes of work may have two subphases within the schematic design phase—one at 50% completion and one at 100% completion of the schematic design phase.

Design Development Phase

The design development phase is the process of refining the design and working out all the details, including the selection of materials and the engineering systems. The goal is to finalize all design decisions before proceeding with construction documents. A more detailed cost estimate may also be prepared at this phase. This phase also involves coordination between the engineering consultants and the architects to ensure that there are no design conflicts.

Construction Documents Phase

The construction documents are detailed working drawings of the project, which present the design with a level of detail for all spaces, systems, and materials so that the contractors can build the project based on these drawings. Detailed specifications are prepared, and the construction may also be put out for bids by contractors and subcontractors. Usually at the end of this phase, the construction documents are sent to the local authorities for obtaining permission to build.

Construction Phase

The construction phase begins when the contractor starts building the project according to the construction documents. Site visits, construction administration, material procurement, subcontractor management, and budget control—all become pivotal activities during construction. Usually, projects have a milestone of *substantial completion,* which implies the project is finished to a point where a client could occupy the nearly completed space. After final completion of the project, a *certificate of occupancy* is obtained, which is the official recognition by a local building department that the building has been built according to applicable codes and standards.

Building Operation and Maintenance Phase

Once construction is complete and the certificate of occupancy is obtained, the building is officially ready for occupancy and can be operated and maintained by the facilities personnel.

THE INTEGRATED DESIGN PROCESS

Integrated Design Process (IDP) is a design approach that, in recent years, has gained importance among project teams that want to achieve high performance *green* design. What differentiates integrated design from the traditional design is that instead of being purely linear with increasing layers of information being added at each design stage, it is an iterative process where the whole building design is considered right from the beginning of the project. The *Roadmap for the Integrated Design Process,* prepared for the BC Green Building Roundtable by Busby Perkins + Will, defines the integrated design process as "an approach to building design that seeks to achieve high performance on a wide variety of well-defined environmental and social goals, while staying within budgetary and scheduling constraints. It relies upon a multidisciplinary and collaborative team whose members make decisions together based on a shared vision and a holistic understanding of the project. It follows the design through the entire project life, from pre-design through occupancy, and into operation. IDP is a term that is not exclusively associated with high-performance building design; in principle it is a flexible approach that can be applied to almost any type of design or decision-making process."[1]

In order to execute the project, the framework that the integrated design approach utilizes is similar to the phases of the traditional design process, but the difference lies in the nature of activities in those phases and the roles of people involved, as is explained below.

Pre-Schematic Design Phase

During the pre-schematic design phase, project goals, objectives, and the direction of the project are established through a *visioning* session. It involves exploration of the project's surroundings, site, climate, etc. to identify optimum design choices and involves many experts from the design team at the outset.

Schematic Design Phase

The schematic design phase involves refinement of the vision from the previous phase along with exploration of various innovative ideas, technologies, and methods in achieving the broad vision of the project. It also presents the opportunity for different disciplines to collectively explore synergies and trade-offs to come up with a preferred schematic design.

Design Development Phase

During the design development phase, the schematic design idea is further firmed up and optimized for best performance from a whole building design perspective. All aspects—architectural, mechanical, electrical, site, and landscape—are taken into consideration holistically and tested to validate the choices made. Final approval is sought from the owner/developer to ensure that the proposed design meets the overall objectives and vision of the project.

Construction Documents Phase

The construction documents phase involves the preparation of drawings based on the refined and approved design. As the project progresses with interactions between various groups, coordination and finalization of all specifications and drawings are completed.

Construction Phase

During construction, qualified contractors are chosen and communication procedures are established to implement construction phase activities. Partial commissioning of the building systems during construction and final commissioning, testing, and validation also become integral to the construction phase. The goal of the construction phase is to realize the original vision of the project and deliver a functional and well-commissioned building.

Building Operation and Maintenance Phase

This final stage involves activities to ensure that the transition from the design and construction team to the building owners, occupants, and operating staff occurs smoothly. The phase may involve preparation of a training manual for the operating staff in line with the commissioning activities. Post-occupancy evaluations may be completed to optimize the building's performance and give feedback to the design team for future projects.

The *Roadmap for the Integrated Design Process* summarizes the key principles that differentiate the integrated design approach from traditional design as illustrated in Figure 2.4 and described below:

1 Integrated design involves a broad collaboration of team members right from the beginning of the project in comparison to traditional design where team members are involved on an as-needed basis.
2 Life-cycle costing is emphasized in an integrated design approach in comparison to the emphasis placed on first-costs in the traditional design approach. Life-cycle costing is a way of assessing total building cost over time, taking into account the initial costs, operating costs, and maintenance costs along with social and environmental benefits.
3 The whole building and all the systems are taken into consideration in an integrated design, as opposed to conventional design where systems are considered in isolation.
4 There is more opportunity to seek synergies between various disciplines and optimize the building design, and decisions are influenced by a broader team in an integrated design approach. On the other hand, the conventional design process does not seek out synergies or optimization, and decisions are made by fewer people in isolation.

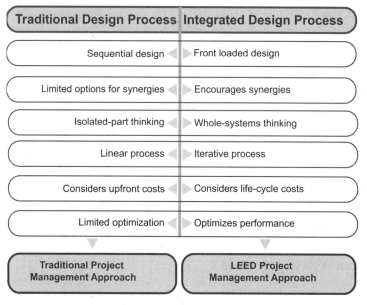

Figure 2.4   Differences between traditional design and the integrated design approach.

5 The integrated design approach is iterative in nature, where there is constant refinement of the design based on feedback obtained from various decision-making tools.

6 The integrated approach is front-loaded. A lot of time and energy is invested upfront in the early stages of design, where there is maximum flexibility to impact the design. On the other hand, the conventional design process is linear and sequential, with limited investment during the early stages of design.

THE GBCI LEED CERTIFICATION PROCESS

The entire process of certification is managed by GBCI through a web interface portal known as *LEED Online* (www.leedonline.org). All communication with GBCI occurs through this online portal. There are no in-person meetings/reviews/inspections scheduled for the LEED-NC rating system. Project teams are expected to use LEED Online to compile all documentation information, which can then be submitted to GBCI for certification reviews. Until March 2009, the certification process was administered by USGBC, but now GBCI manages the review and certification process. The GBCI utilizes the services of a network of 10 *certification bodies* that are accredited to International Standards Organization (ISO) standard 17021. These organizations are known for their role in certifying organizations, processes, and products to ISO and other standards.[2] The certification bodies perform the review and verification of the project documentation. However, all communication and approvals occur via LEED Online with GBCI directly. Figure 2.5 gives an overview of the certification process as defined by GBCI. The various steps of this process are explained in detail below.

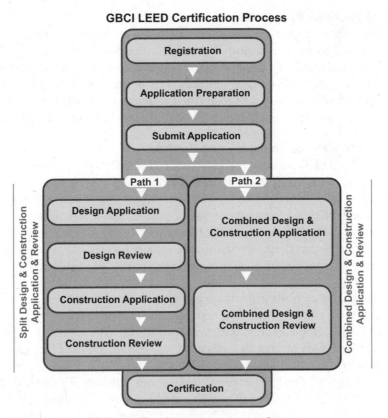

Figure 2.5   GBCI certification process overview.

Registration

This is the first step that teams interested in achieving LEED certification for their projects must complete. It establishes contact with GBCI and declares the intent of the project team to pursue LEED certification. The process involves completing an online form with basic project information and payment of the appropriate registration fees. Table 2.1 describes the fees for registration listed on the GBCI website as of April 2010.[3] Teams are advised to refer to the GBCI website for most current fee information. Once the project is registered, the project team is given access to LEED Online to complete the certification application.

TABLE 2.1 LEED PROJECT REGISTRATION FEES	
PROJECT REGISTRATION FEES	
USGBC Members	$ 900
Non-Members	$1200

Application Preparation

The next step after registration is to prepare application materials for submission to GBCI. This involves identifying the credits that will be pursued, performing calculations, and compiling documentation to demonstrate that the project meets the requirements of the credits or prerequisites. This documentation is then submitted to GBCI for review along with the credit forms available on LEED Online. Note that previous versions of the rating system referred to these credit *forms* as credit *templates.*

Application Submission and Review

For LEED-NC projects, two options are available to project teams for submission and review of the application material. The first option is *Split Design and Construction Review*; the second option is *Combined Design and Construction Review*. The fundamental difference between the two options is that if a project team chooses the first review path, then they have the option of splitting the submission of the application materials—part during design and the remaining later after construction. On the other hand, if a project team chooses the second path, all application materials have to be submitted together at the end of construction. In a split review, GBCI reviews the design phase credits and lets the project teams know of the credits that are *anticipated* and those that are *denied.* In both options, the final award of certification is granted only after all credit documentation is submitted. The advantage of the split review is that it enables project teams to assess the likelihood of achieving credits and to get valuable technical advice and feedback from GBCI. It also gives project teams an opportunity to make changes, if needed, to the design. This would not be possible if teams wait until the end of construction to submit all documentation.

Certification fees are split between the design and construction phases for teams that elect the *Split Review* path. On the other hand, all certification fees are paid at one time by teams that elect to pursue the *Combined Review* path. Table 2.2 describes the fees associated with certification for both options, as stated on the GBCI website.[4] Note that these fees are current as of April 2010. Teams must refer to the GBCI website for the most current fee information.

The submission and review options are explained in detail below.

Option 1: Split Design and Construction Review

Using the Split Design and Construction Review application materials may be submitted in up to 4 steps as described below:

1 Design application.

This is the first time the design credit documentation is submitted by the project team for review by GBCI. GBCI will then review the materials submitted and respond with preliminary Design Review comments and a description of credit points that are *anticipated* or *denied* or those that are *pending* for clarifications.

2 Response to preliminary Design Review

This step is optional to project teams. The team can accept the results of the preliminary Design Review as final, or they can respond to the Review and provide clarifications on credits that were marked as *pending* or *denied.* The response from the design team should occur within 25 business days of receiving the preliminary Design

TABLE 2.2 LEED PROJECT CERTIFICATION FEES			
OPTION 1: SPLIT DESIGN AND CONSTRUCTION REVIEW			
	LESS THAN 50,000 SQUARE FEET	**50,000–500,000 SQUARE FEET**	**MORE THAN 500,000 SQUARE FEET**
Design Review: USGBC Members	$2,000	$0.04/sf	$20,000
Design Review: Non-Members	$2,250	$0.045/sf	$22,500
Construction Review: USGBC Members	$500	$0.010/sf	$5,000
Construction Review: Non-Members	$750	$0.015/sf	$7,500
OPTION 2: COMBINED DESIGN AND CONSTRUCTION REVIEW			
USGBC Members	$2,250	$0.045/sf	$22,500
Non-Members	$2,750	$0.055/sf	$27,500

Review comments. GBCI will review the responses and issue *final Design Review comments*. Note that in a split design and construction review, one must either accept or appeal the result for a design credit comment that has been submitted for Design Review and on which comments have been received. The credit may not be deferred to the Construction Review. For instance, if there is a design credit for which comments were received and it is realized by the project team that more information will be available for that credit after construction is complete, they cannot defer the response until the construction phase. The design credit attempted and reviewed at Design Review must be either accepted or appealed, as deemed appropriate by the design team.

3 Construction application

This is the final application after construction is complete and all credits have been documented by the project team using LEED Online. The team must provide confirmation that none of the previously reviewed design credits have changed. If the construction process has impacted any of the design credits, then those credits must be resubmitted with the appropriate changes. GBCI will review all the information submitted and designate credits as *anticipated* or *pending* or *denied*, accompanied with technical advice regarding the credits.

4 Response to preliminary Construction Review

The results of the preliminary Construction Review may be accepted as final, or the team has the option to respond to the Construction Review comments and results within 25 business days of receipt of the comments. GBCI will review all submitted documentation and give a final review with credits designated as either *awarded* or *denied*.

Option 2: Combined Design and Construction Review

The Combined Design and Construction Review is the second application and review path that project teams may choose. This option involves submiting all documentation—

for both design and construction phases—at one time after construction is complete. Information may be submitted in two steps:

1 Preliminary Review

If the project team elects to pursue this path, then documentation for all credits and prerequisites (both design and construction) is submitted to GBCI at the same time. This usually occurs after construction is complete. GBCI will review all documentation submitted and designate credits as *anticipated, denied,* or *pending* and provide technical advice and feedback.

2 Response to Preliminary Review

The results of the Preliminary Review may be accepted as final by the project team or they have the option to respond to the Preliminary Review comments for the combined submittal within 25 business days of receipt of the comments. GBCI will review all submitted documentation and give a final review with credits designated as either *awarded* or *denied.*

Appeals

If, for some reason, the final review decisions are not found satisfactory during any of the reviews, then the team has an option to appeal the GBCI decision. Each appeal has a $500 payment fee associated with it and must be made within 25 business days of receiving a final review decision.

Credit Interpretation Requests and Rulings

A credit interpretation ruling process has been established by GBCI to address specific issues or challenges that a project team may encounter when interpreting the requirements of a prerequisite or credit for their project. Project teams have the option to submit throughout the application process what are known as Credit Interpretation Requests. GBCI reviews the submitted requests and proposed interpretations and gives feedback to the project teams in the form of Credit Interpretation Rulings (CIRs). The ruling alone is not sufficient to achieve a credit; project teams must still demonstrate and document the credit requirement during the regular reviews.

Prior to the release of LEED v3, the CIRs were considered precedent setting, and teams could use rulings from other projects to demonstrate achievement of credits. All CIRs were made available to teams through an online CIR database. However, now all CIRs are project specific and no longer regarded as precedent setting.

Certification

The culmination of the application submittals and review is the certification awarded by GBCI. Once the final application review is obtained from GBCI, either through the split or combined review path, the project team can either accept or appeal the final decision. If the final decision is accepted, then the LEED certified project receives a formal certificate of recognition and information on how to order plaques, certificates, photo submissions, and other marketing materials. If the owner/developer desires, the project may be included in the online directory of registered and certified projects and/or in the U.S. Department of Energy's High Performance Buildings Database. Table 2.3 illustrates the various levels of certification that may be achieved by projects, depending on the total points they achieve.

TABLE 2.3   LEVELS OF LEED CERTIFICATION

LEED CERTIFICATION LEVEL	POINTS
Certified	40–49
Silver	50–59
Gold	60–79
Platinum	80–110

THE LEED PROJECT MANAGEMENT PROCESS

In order to effectively manage a LEED project, all the LEED-related activities (from the GBCI certification process) need to be aligned with the different phases of the architectural design process. This alignment of activities leads to a process that follows the phases of the traditional design procedure but, at the same time, incorporates activities that follow the integrated design approach. Figure 2.6 describes how the LEED project management process aligns with the architectural design phases and the GBCI process. For the sake of simplicity, the split design and construction review path (option 1) is shown to represent the GBCI process in the figure.

By aligning various LEED activities according to the design phases, three distinct stages can be identified in the LEED project management process:

Stage I: Project Definition and Goal Setting

Stage II: Design Phase Integration

Stage III: Construction Phase Implementation

Stage I: Project Definition and Goal Setting

The first stage of the LEED project management process involves *defining the project and setting goals*. This comprises collecting information, conducting a preliminary assessment, establishing a LEED target, and creating a roadmap for LEED certification. These activites align with the pre-schematic design phase of the architectural process and the registration phase of the GBCI process. These activities should be completed during the pre-schematic design phase. However, if for some reason, the decision to pursue LEED certification is made at a later time, it is recommended that these activities be completed anyway in order to save time (that may need to be spent later). This stage comprises five primary steps shown below. Each of these steps along with its recommended activities is presented in much more detail in Chapter 3.

Step 1: Collect and Compile Preliminary Data

Step 2: Prepare a Preliminary LEED Assessment

Step 3: Register the Project on LEED Online

Step 4: Organize a LEED Vision Workshop

Step 5: Prepare a LEED Project Workbook

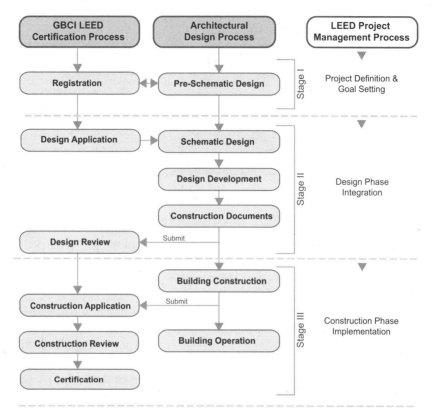

Figure 2.6   Alignment of the architectural design phases with GBCI certification process.   *Credit: Green Potential LLC (www.green-potential.com).*

Stage II: Design Phase Integration

The second stage of the LEED project management process involves the integration of *green* building principles, strategies, and products into the design using the LEED-NC credits as a guiding framework. The activities in this stage are aligned with the schematic-design, design-development, and construction-documents phases of the architectural design process. As the design progresses from one phase to the next, there is an iterative process of refining the design to ensure that the building is optimized and on path to achieve certification. The various steps and activities of this stage have been described in detail in Chapter 4 of this book. There are three phases and 10 steps within this stage:

1 Schematic Design (SD) LEED Phase
2 Design Development (DD) LEED Phase
3 Construction Documents (CD) LEED Phase

Schematic Design (SD) LEED Phase

The activities in this phase align with the schematic design phase of the architectural process and include reviewing LEED credit requirements, completing preliminary calculations, and determining the *green* design decisions that need to be integrated into the schematic design drawings. Broadly speaking, this phase consists of four steps:

Step 1: Perform Credit Analyses

Step 2: Research and Identify *Green* Products

Step 3: Integrate *Green* Requirements into Design

Step 4: Review Schematic Design Drawings

Design Development (DD) LEED Phase

The activities in this phase are aligned with the design development phase of the architectural process and include refining the design, updating calculations, as well as optimizing building systems for maximum energy performance. Key decisions regarding materials and construction implementation procedures are made during this phase, followed by a review of the drawings to ensure that the project is on path to achieve the desired credits. This phase comprises four steps:

Step 5: Update Credit Analyses and Drawings

Step 6: Prepare for Construction Phase

Step 7: Develop Building Management Policies

Step 8: Review Design Development Drawings

Construction Documents (CD) LEED Phase

This phase aligns with the construction documents phase of the architectural design process and involves the compilation, final review, and submittal of all credit documentation to GBCI for Design Review. The activities in this phase are described based on the assumpion that the Split Design and Construction review path will be chosen by the project team. However, even if a project team chooses the Combined Design and Construction Review path, it is recommended that all activities be completed up until the final submission to GBCI. This will ensure that all documentation is completed and uploaded to LEED Online while the design team members are actively involved with the project. It has been observed that tracking and compiling design-related information after construction is complete becomes difficult, as the design team may no longer be available or as deeply involved due to the significant lapse of time. The two steps that constitute this stage are:

Step 9: Prepare Documentation for *Design Credits*

Step 10: Submit to GBCI for Design Review

Stage III: Construction Phase Implementation

This is the final stage of the LEED project management process and includes activities that are integral to the successful implementation of *green* construction practices on-site, such as construction waste recycling and air quality management. The activities in this stage will help the construction team with implementation and documentation of LEED construction credits for GBCI's final Construction Review. This stage has five primary steps and each step has been further described in detail in Chapter 5 of this book.

Step 1: Organize a Pre-Construction LEED Kick-Off Meeting

Step 2: Review Progress of Credit Implementation

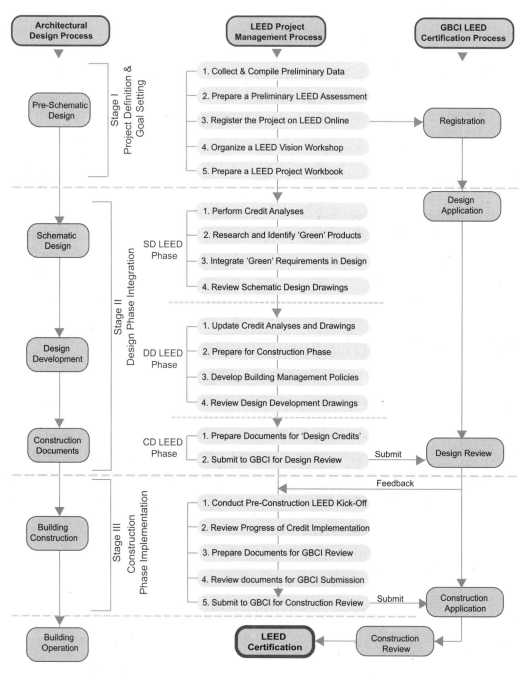

Figure 2.7   LEED project management process overview.
Credit: Green Potential LLC (www.green-potential.com).

Step 3: Prepare Documents for GBCI Construction Review

Step 4: Review Final Documents for GBCI Submission

Step 5: Submit to GBCI for Construction Review

The different stages—Stage I, Stage II, Stage III—of the LEED project management process are described in detail in Chapters 3, 4, and 5, respectively. Figure 2.7 presents a big picture overview of the LEED project management process.

People

The second component of project management, *people,* play a very important role in a LEED project. It is only through the successful effort and contribution of a diverse group of people with varying expertise that a LEED project comes together. Before delving into the details of the specific roles and responsibilities of various team members, it is important to understand the project team model for a LEED project. It must be noted that every project is unique and will have different team models based on the project type, scope, and delivery mechanism. Due to the sheer number of cross-functional interactions, it is critical for all team members to be open to dialogue and explore synergies between various disciplines to achieve the best results for the project.

PROJECT TEAM

Project Team—Traditional Architectural Design Process

In a traditional project, the involvement of the various team members is on an as-needed basis, with very little opportunity for interaction and discussion of synergistic solutions. Communication occurs in a linear progression from the architect to the engineers and finally to the contractors. There is usually a hierarchy associated with the communications and rarely are inputs sought from all team members in the beginning of the project. The result is often a building product that has not explored the synergies between various building systems, has not been optimized for its performance, and many times, is not fully coordinated between the various disciplines. This can often lead to costly modifications in the later stages of the design or increased costs in the operation of the building because of lack of optimization of the various building systems.

Project Team—Integrated Design Process/LEED Process

The integrated design approach, on the other hand, encourages engagement of all team members right from the beginning of the project, so there is scope for ongoing learning and the capacity to address a range of strategies. This also allows for synergistic interactions and coordinated efforts as the design progresses, where each team member feels a sense of ownership of the project. Although it might appear that the involvement of too many people might have a negative impact on the process, it has been found that clearly defined roles and communication protocols can help avert any conflicts. A LEED project is a classic example of a project where the team model of an integrated design

Project Team - Traditional Design

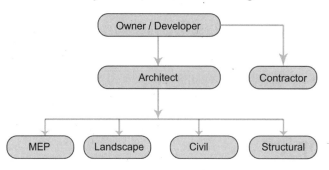

Project Team - Integrated Design

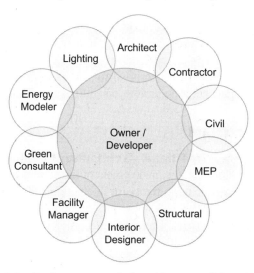

Figure 2.8 Project team relationships—traditional and integrated design.

approach is most appropriate. This is due to the fact that the project involves the integration of *green* building principles into the design from the very beginning and requires input from various team members to generate a building solution that is both optimized for performance and also cost-effective. Figure 2.8 shows a comparison between project team relationships in traditional design and integrated design.

LEED PROJECT TEAM ROLES AND RESPONSIBILITIES

The best results for any LEED project may be achieved by an integrated design approach where there is quality interaction between various team members. In order for this to occur smoothly, it is important to understand who the key members of a LEED project team are and to define their roles and responsibilities. It is hard to draw a line on where one team member's responsibility ends and another's begins, as it would appear to defeat the purpose of the integrated design approach. However, it is still essential to do so. This way, tasks do not get left out because of lack of clear direction. Team members should be aware of their respective responsibilities and, at the same time, collaborate to find synergistic solutions. As explained in Chapter 1, a LEED project warrants the involvement of many specialist consultants in addition to the core project design team, and team members have responsibilities that are not typical to project design. This makes the definition of roles and responsibilities even more important.

The LEED Project Team Composition
Core Project Team

In a LEED project the client or owner plays an active role throughout the design process. The owner should engage a core group of team members who are familiar with the integrated design process. A LEED team leader, champion, or project manager should be identified to manage and facilitate the LEED project management process.

The core LEED project team comprises:

- Client—Owner, Developer, Owner's Representative
- Architect
- Civil engineer
- Structural engineer
- Mechanical engineer
- Electrical engineer
- Plumbing engineer
- Landscape architect
- Facilities manager/building management representative
- General contractor or construction manager
- Interior designers
- Cost estimator/consultant
- Green/LEED consultant
- Energy simulation expert
- Lighting consultant
- Daylighting simulation expert
- Commissioning Authority
- LEED project manager

The composition of the core team will change from project to project based on the expertise of the team members and the project type. For instance, if the mechanical engineer has expertise in energy modeling, then an energy simulation expert need not be engaged. Similarly, if the green consultant has expertise in daylighting simulations, then a daylight simulation expert need not be engaged. The purpose of the core team composition presented above is to reflect the need for expertise in more areas for a LEED project than for a non-LEED project. The core team can be assembled by the owner in consultation with the architect. It will also be beneficial for the team members to be effective communicators with a willingness to collaborate, in addition to being

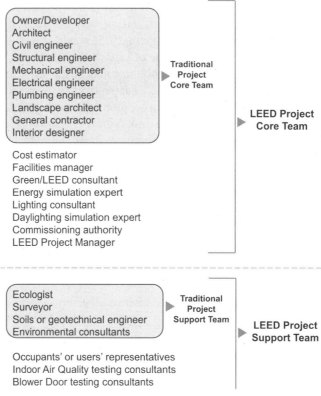

Figure 2.9   Team members of LEED projects and non-LEED projects.

technically competent. Figure 2.9 identifies the differences between the core team for a LEED project and a non-LEED project.

Other team members/specialists

Additional specialist consultants may need to be brought in for the project during specific times or for the entire duration of the project. The core team members may advise the client in identifying the additional team members required, depending on the project type. Some of the additional members who may need to be included in the project are:

- Ecologist
- Surveyor
- Soils or geotechnical engineer
- Environmental assessment consultants
- Valuation/appraisal professional
- Occupants' or users' representatives
- Planning/regulatory/code approvals agencies representatives

- Marketing expert
- Indoor air quality testing consultants
- Other experts as required (e.g., natural ventilation, thermal storage, acoustic)

The LEED Project Manager

It is imperative to designate a specific individual as the LEED project manager (PM) for the project. The LEED PM has a unique role of spearheading, guiding, managing, and taking the project to successful certification. A member of the design team may be designated as the LEED PM; or if the project has a green/LEED consultant, then that individual could be designated as the PM. The goal is to designate someone (who is conversant with the LEED rating system) to champion the project, motivate the team, and manage the process proactively.

Who Can Be the LEED Project Manager?

It is easy to assume that any LEED Accredited Professional (LEED AP) could be designated as the LEED PM. However, having the LEED AP designation alone is not sufficient reason to designate someone as the LEED PM. The ideal candidate should be someone who is a LEED AP at a minimum, but also has the skills of an effective manager who can motivate people, explore innovative solutions, and facilitate processes to achieve certification. Since the LEED professional accreditation program was introduced, thousands of professionals have taken the exam to become LEED Accredited Professionals. While the exam helps individuals understand the requirements of the rating system, it does not necessarily prepare them to manage a LEED project. Although a LEED AP may have the knowledge of the rating system, he or she may not be aware of the desired practical implementation methods. A real-world situation to illustrate this is a project where a novice LEED AP was engaged to facilitate and manage the LEED certification process. Even though the LEED AP was familiar with the rating system, he faced challenges when it came to implementing the process and struggled to keep up with the documentation. The owner realized his inadequacies and replaced him with another qualified LEED consultant to facilitate the rest of the process.

Different architectural firms have varying level of expertise and resources to manage LEED projects. Larger architectural firms usually separate the roles of project architect from the project manager and designate one individual as a project architect and another individual as a project manager. In this case, the individual designated as the project manager can be the LEED PM, as he or she is disengaged from the responsibilities of the project architect. On the other hand, smaller architectural firms, due to smaller capital budgets, usually have a single individual serving as both the project architect and project manager. Over time this individual may be inundated with many responsibilities, and managing the LEED process becomes challenging. In such situations, the firm may choose to hire an independent consultant to facilitate the LEED process. In many cases, the owner may even choose to directly hire a LEED consultant.

In summary, a LEED PM could be either a member of the design team or a third party LEED consultant. The crucial factor to consider is that this individual has sufficient experience in managing LEED projects, is familiar with the LEED rating system, is an effective communicator, and has the capacity to manage diverse team members.

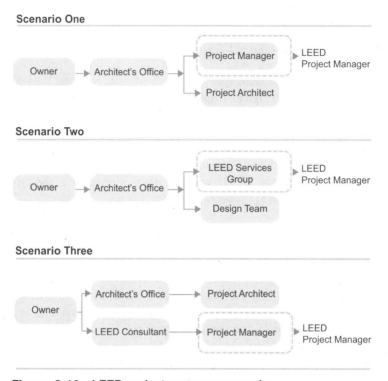

Figure 2.10   LEED project manager scenarios.

Figure 2.10 illustrates some examples of who can be the LEED PM, depending on the project structure.

Skills/Characteristics Desired of a LEED Project Manager

The LEED PM should ideally have the following characteristics:

- Expertise in green design principles and the LEED rating system
- Knowledge of the architectural design process, the integrated design process, and the GBCI certification process
- An understanding of how projects of the type under consideration work

The LEED PM also has the following functions:

- Is the steward of the LEED goals and targets, which are set during the charrettes and/or workshops and updated throughout the process.
- Helps align the team to a common vision while challenging members to push themselves to the highest level of performance possible for the project.
- Ensures a proper flow of information for all green design matters.
- Is responsible for keeping the team on time and on target for LEED milestones throughout the project.

Team Member Roles and Responsibilities

Even though the process of integrated design explores synergies and encourages inter-action between the various team members, it is important to understand each individual's role and their responsibility from a LEED perspective. This will assist in smooth communication between all team members. The roles and responsibilities of each of the core team members in this section are presented in the following format:

General Responsibilities

This section describes the general role and responsibilities of the team member in the LEED project.

LEED Credit Responsibilities

This section identifies the specific LEED credits that the team member might be responsible for. The credit responsibilities are further categorized based on whether the team member is to lead the implementation of the credit or support other team members. Note that these are typical responsibilities and will vary from project to project. They should not be regarded as a definitive list. The intent in providing these is to present information on *possible* responsibilities that the team members should be aware of.

Client—Owner, Developer, Owner's Representative
General Responsibilities

In a LEED project, the client has a significant role to play and needs to be actively involved throughout the design process. In general, there should be one person desig-nated from the ownership—whether it is a developer, public agency, or organization—who should be the point of contact for the design team and also the decision maker for the client. The general responsibilities of the owner are, but not limited to, the following:

- Hire a team of motivated and experienced professionals, preferably with prior LEED project experience.
- Communicate the vision for the project and the targeted LEED goals.
- Work with the design team in the decision-making process and provide clear direc-tion to the team on the viability of LEED credit-related strategies.
- Make decisions that reflect life-cycle thinking.
- Coordinate with building management to develop and institute operational policies, programs, and initiatives.
- Make decisions related to building operations monitoring and post occupancy evaluations.
- Provide direction to the team on building design and construction schedules, and budget.

LEED Credit Responsibilities

The owner's role is to assist the team in making decisions regarding strategies to use for implementing LEED credits. The owner must also provide relevant information to the team and LEED PM so that they are able to document the credit requirements. Table 2.4 describes the specific credits where the owner's input might be required.

TABLE 2.4   TYPICAL LEED CREDIT RESPONSIBILITIES OF OWNER [L=LEAD, S=SUPPORT]

L/S	CREDIT	RESPONSIBILITIES OF CLIENT/OWNER/DEVELOPER
S	SS C1	If the site has not been selected, then select a site according to the credit requirements. If site has been selected, then provide site-related information—surveys, drawings, soil reports, etc.—to the team member documenting the credit.
S	SS C2	If site has not been selected, then evaluate options to locate site based on development density and proximity to existing infrastructure.
S	SS C3	If site has not been selected, then evaluate options to choose a brownfield or contaminated site. Consult with environmental consultants for site assessment and remediation efforts.
S	SS C4.1–4.4	Select site based on proximity to existing or planned public transport network. Provide anticipated number of occupants to team for LEED occupancy calculations.
S	WE P1/C3	Assist in making decisions on the choice of plumbing fixtures.
S	WE C1/C2	Assist in making decisions on the use of non-potable sources of water—treating greywater, reusing rainwater, etc.
S	EA P1/C3	Designate Commissioning Authority (CxA) and prepare owner's project requirements (OPR).
S	EA C2	Decide on applicability, use, and type of on-site renewable energy system.
S	EA C5	Determine the feasibility of implementing a measurement and verification plan and engage third-party consultants, if required.
S	EA C6	Make decision on purchasing *green* power.
S	MR credits	Assist team in making decisions on materials and products.
S	EQ credits	Support team members with decision making.
S	ID C1	Decide on ID credits to be pursued.

Architect

General Responsibilities

The primary role of the architect in a LEED project is to design a building that takes into consideration the climate of the location, has an efficient envelope, and incorporates green materials that make it an inherently efficient building. The architect must collaborate with other team members to make decisions that are beneficial from a whole building perspective. The general responsibilities of the architect are, but not limited to, the following:

- Work with the LEED PM to schedule green charrettes and goal setting workshops early in the design process.
- Ensure all *green* design features required for compliance with LEED credits are incorporated in drawings and specifications.

TABLE 2.5 TYPICAL LEED CREDIT RESPONSIBILITIES OF THE ARCHITECT [L=LEAD, S=SUPPORT]

L/S	CREDIT	RESPONSIBILITIES OF ARCHITECT
L	SS C4.2	Provide spaces for bicycle storage and showers.
L	SS C4.3	Designate preferred parking spaces for low-emitting and fuel-efficient vehicles and provide signage information; identify locations for alternate-fuel charging stations.
L	SS C4.4	Design to minimize number of parking spaces; designate preferred parking spaces and provide signage details.
L	SS C7.1	Design to minimize hardscape surfaces on-site; specify materials with low emissivity or high solar reflectance index (SRI) for all non-roof surfaces.
L	SS C7.2	Specify materials with high SRI for the roof. Evaluate vegetated roofs as an option and coordinate with landscape architect for vegetation type and structural engineer for roof loads.
S	SS C8	Coordinate with lighting consultant to ensure lighting layout is in compliance with credit requirements.
S	WE credits	Coordinate with plumbing engineer to select and specify efficient fixtures.
S	EA P1/C3	Prepare basis of design document.
S	EA P2/C1	Consult with energy modeler/mechanical engineer during early design stages to ensure that the building is optimized for energy performance with the appropriate orientation, envelope composition, insulation, window U-Value, etc.
S	EA C2	Design to incorporate any on-site renewable energy system, such as solar panels on roof or other building integrated wind or solar technologies.
L	MR P1	Designate areas for collection and storage of recyclables in the building.
L	MR C1	Evaluate options of reusing existing building structure and interior elements.
S	MR C3-C7	Specify *green* building materials—salvaged materials, materials with recycled content, regional materials, products made with rapidly renewable materials, or FSC certified wood products.
L	EQ P1/C2	If building is naturally ventilated, then design natural ventilation system according to ASHRAE 62.1-2007, with operable window areas according to the prerequisite's requirements.
L	EQ P2	Provide designated smoking rooms with exhaust to outdoors, weather strip doors, seal penetrations in walls, etc., as needed.
S	EQ C4.1–4.4	Specify materials according to the volatile organic compounds (VOC) requirements and standards identified in the credits.

(continue)

TABLE 2.5   TYPICAL LEED CREDIT RESPONSIBILITIES OF THE ARCHITECT [L=LEAD, S=SUPPORT] *(continued)*

L/S	CREDIT	RESPONSIBILITIES OF ARCHITECT
L	EQ C5	Design permanent entryway systems for dirt capture; isolate spaces containing hazardous chemicals with deck-to-deck partitions, self-closing doors, and exhaust to the outside. Coordinate with mechanical engineer for exhaust incorporation.
S	EQ C6.1	Design lighting layout and controls in consultation with lighting consultant.
S	EQ C6.2	Design to maximize thermal comfort of occupants—operable windows, glare control, etc.
S	EQ C8.1–8.2	Design to maximize access to daylight and views for occupants.

- Coordinate *with* consultants, especially the energy simulation expert, to optimize the building orientation and envelope.
- Specify *green* materials and products, which also meet LEED credit requirements. Consult with *green*/LEED consultant in preparation of specifications.
- Perform or participate in post-occupancy evaluation of the project.

LEED Credit Responsibilities

The credits that the architect is responsible for will vary from project to project depending on the responsibilities of the LEED PM. While the final documentation may be done by another team member, the architect's input is required on a number of credits. Table 2.5 describes some typical credit-specific responsibilities of the architect.

Civil Engineer

General Responsibilities

The role of the civil engineer is to design the site's stormwater systems and hardscape in a manner that maximizes the pervious areas on-site and controls the quality and quantity of stormwater runoff from the site. The general responsibilities of the civil engineer are, but not limited to, the following:

- Provide input into site-specific opportunities regarding water conservation, reuse, and treatment.
- Incorporate stormwater quality and quantity control best management practices into the design.
- Perform stormwater calculations according to LEED credit requirements.
- Coordinate with landscape architect in the development of the site hardscape, any stormwater detention areas, bioswales, etc., as needed.

LEED Credit Responsibilities

The specific credits that will need input from the civil engineer are SS P1, SS C6.1, and SS C6.2. Additional support may be required from the civil engineer in the imple-

TABLE 2.6   TYPICAL LEED CREDIT RESPONSIBILITIES OF CIVIL ENGINEER [L=LEAD, S=SUPPORT]

L/S	CREDIT	RESPONSIBILITIES OF CIVIL ENGINEER
L	SS P1	Identify erosion-prone areas and outline soil stabilization measures. Create erosion and sedimentation control plans.
S	SS C1, SS C3	Provide site surveys, flood plain information, soil reports, and environmental site assessment reports.
S	SS C2, SS C4	Provide any site-vicinity related information, maps, zoning information, etc.
L	SS C5.1–Case 1	Create a construction disturbance limits drawing according to credit requirements.
S	SS C5.2	Provide information on open space requirements per local zoning.
L	SS C6.1	Perform calculations and design a stormwater management system to control stormwater quantity according to credit requirements; prepare detailed drawings, specifications and narrative describing approach.
L	SS C6.2	Implement best management practices to control stormwater quality; prepare detailed drawings, specifications, and narrative describing approach.
S	SS C7.1	Coordinate with architect and landscape architect to minimize hardscape surfaces; specify pervious and/or high SRI paving.
S	WE credits	Coordinate with landscape architect and plumbing engineer in the design of any rainwater harvesting systems, if needed.

mentation of credits by the landscape consultant and the architect. Table 2.6 describes the credit-specific responsibilities of the civil engineer.

Structural Engineer

General Responsibilities

The role of the structural engineer is to make appropriate choices for the structural systems, reduce material use, select green materials, and provide input on durability of materials. The general responsibilities of the structural engineer are, but not limited to, the following:

- Evaluate the impact of various structural choices on the overall form and massing of the project, which may indirectly impact the energy performance of the building.
- Provide input to architect regarding the durability of materials chosen and their life-cycle impacts.
- Provide innovative structural solutions that may help in conservation of resources.
- Evaluate the impact of various structural choices on daylighting potential and material selection.
- Ensure all durability and other *green* requirements are incorporated in drawings and specifications.

TABLE 2.7   TYPICAL LEED CREDIT RESPONSIBILITIES OF STRUCTURAL ENGINEER [L=LEAD, S=SUPPORT]

L/S	CREDIT	RESPONSIBILITIES OF STRUCTURAL ENGINEER
S	SS C5.1, C5.2, C7.2	Design roof to structurally support a vegetated roof system and coordinate with architect and landscape architect as needed.
S	MR credits	Specify materials according to the various credit requirements, such as materials with recycled content (concrete with fly-ash) or FSC certified wood.

LEED Credit Responsibilities

The structural engineer's role, from a LEED credit perspective is more of a support nature to the team. The decisions made by the structural engineer will have an impact on various credit requirements. Table 2.7 describes the specific credits that might need input of the structural engineer.

Mechanical, Electrical, and Plumbing (MEP) Engineers
General Responsibilities

The MEP engineering team has a significant role to play because the three prime areas—water, energy, and air quality—that a LEED building addresses require their input. Depending on the level of expertise, the mechanical engineer may also perform energy simulations to optimize the building performance. This section is written with the assumption that an energy simulation expert will be engaged for that purpose. The responsibilities of the energy simulation expert have been described separately in this chapter. The general responsibilities of the MEP engineering team (without energy simulation) are, but not limited to, the following:

■ Work with the energy simulation expert to design building HVAC and lighting systems that are optimized for the whole building's energy performance.
■ Work with the architect to ensure that opportunities related to climate, orientation, and massing are evaluated and implemented to ensure that the building's mechanical systems are sized appropriately.
■ Provide innovative cost-effective solutions that will make the building's systems and operations more efficient.
■ Design building systems in accordance with the microclimate of the building.
■ Provide input regarding the system choices based on life-cycle impact and energy use.
■ Coordinate with lighting consultant to develop appropriate lighting layouts and specifications.
■ Design building ventilation systems in accordance with ASHRAE 62.1-2007 and thermal comfort parameters in accordance with ASHRAE 55-2004.
■ Ensure that equipment selection, adhesive choices, material selection, and construction methods reflect LEED goals.

- Work with the CxA and the contractor to implement commissioning related activities.
- Engage in measurement and verification studies and building performance monitoring after occupancy, if required.
- Design plumbing systems and select fixtures to meet the targeted water reduction goals.

LEED Credit Responsibilities

The MEP engineering team is directly responsible for the implementation and documentation of a number of LEED credits. The design of the mechanical, electrical, and plumbing systems as well as the choice of equipment can contribute up to 40 points. Table 2.8 describes the critical credits that need input from the MEP team.

Landscape Architect

General Responsibilities

The role of the landscape architect is to explore site-specific opportunities that can help preserve or restore habitat, promote native vegetation, and reduce potable water consumption for irrigation. He or she also will need to coordinate with the civil engineer in developing best management practices, such as bioswales and detention ponds, etc., to manage stormwater. The general responsibilities of the landscape architect are, but not limited to, the following:

- Incorporate native and adapted vegetation and water-efficient irrigation systems into landscape design.
- Work with the civil engineer and architect to reduce the overall development footprint and maximize the area of the vegetated open space provided in the project.
- Evaluate the option of incorporating vegetated roofs for the building in consultation with the architect and the owners.
- Perform water irrigation calculations and incorporate LEED credit requirements into drawings and specifications.
- Explore the potential for inclusion of rainwater harvesting strategies to supplement irrigation water requirements.
- Coordinate with lighting consultant and/or electrical engineer in the design of site landscape lighting.

LEED Credit Responsibilities

The LEED credits that the landscape architect might be responsible for are SS C5.1, SS C5.2, and WE C1. Table 2.9 describes the specific responsibilities.

Building Management

General Responsibilities

It is very important for representatives from the building management to be involved in a LEED project from the onset. Decisions that are made during the design phase will have an impact on the building operations and management; therefore, having the input of the management team will ensure that they are comfortable with the operations of the buildings. The general responsibilities of the building management are, but not limited to, the following:

**TABLE 2.8   TYPICAL LEED CREDIT RESPONSIBILITIES OF MEP TEAM
[L=LEAD, S=SUPPORT]**

L/S	CREDIT	RESPONSIBILITIES OF MEP ENGINEERING TEAM
S	SS C8	Coordinate with lighting consultant, architect, and landscape architect to verify lighting layout and compliance of fixtures with credit requirements.
L	WE P1, WE C3	Perform water use reduction calculations; specify high efficiency fixtures to achieve targeted water reduction goals.
S	WE C1	Collaborate with landscape architect in design of any greywater reuse systems for irrigation.
L	WE C2	Perform water use calculations; specify high efficiency fixtures and/or design sewage treatment and greywater reuse systems.
S	EA P1/EA C3	Prepare basis of design (BOD) document and work with CxA on any commissioning-related activities.
S	EA P2/C1	Coordinate with energy simulation expert in the preparation of energy model and design efficient building HVAC and lighting systems.
L	EA P3, EA C4	Design building refrigeration systems according to credit requirements.
L	EA C2	Coordinate with renewable energy consultant or installer in the design and implementation of on-site renewable system; perform calculations.
L	EA C5	Develop Measurement and Verification (M&V) procedures; prepare plan for corrective action. (This is applicable only if the mechanical engineer is providing the M&V services; if a separate M&V consultant is hired, then the MEP team's role will be of a supporting nature.)
L	EQ P1, EQ C2	Design building mechanical ventilation systems in accordance with ASHRAE 62.1-2007 and according to the credit requirements.
L	EQ C1	Incorporate CO_2 monitors and outdoor air monitoring systems in design.
S	EQ C5	Provide exhausts, as needed, in rooms with hazardous chemicals or substances; coordinate with architect in the implementation of the credit.
S	EQ C6.1	Coordinate with lighting designer/consultant in ensuring the requirements of the credit are met.
L	EQ C6.2	Incorporate thermal comfort controls in the design according to the credit requirement.
L	EQ C7.1	Design building HVAC system and building envelope to meet ASHRAE 55-2004; coordinate with architect on the building envelope design.
S	EQ C7.2	Participate in preparing thermal comfort survey and implementing plan for corrective action.

TABLE 2.9 TYPICAL LEED CREDIT RESPONSIBILITIES OF LANDSCAPE ARCHITECT [L=LEAD, S=SUPPORT]

L/S	CREDIT	RESPONSIBILITIES OF LANDSCAPE ARCHITECT
L	SS C5.1–Case 2	Design landscape to protect or restore site with native and adapted vegetation.
L	SS C5.2	Design landscape to maximize vegetated open space; perform calculations.
S	SS C6.1, SS C6.2	Coordinate with civil engineer in the design of any vegetated filter strips, bioswales, or other stormwater management systems that are part of the landscape.
S	SS C7.1	Coordinate with civil engineer and architect in the design of the site hardscape; provide vegetation to shade non-roof parking spaces.
S	SS C7.2	Coordinate with architect in the design of vegetated roof system, if applicable.
S	SS C8	Coordinate with lighting consultant in the design of site landscape lighting.
L	WE C1	Design site landscape and irrigation systems to reduce the potable water use for irrigation purposes; coordinate with plumbing engineer if any treated greywater will be used for irrigation.

■ Provide input to design team regarding specific building management issues and concerns over system operations from previous projects or experiences.

■ Participate in design charrettes, workshops, and meetings to understand the design direction and express opinion on any concerns.

■ Provide input regarding possible management-related policies and programs that can be developed with respect to obtaining Innovation in Design (ID) credits.

■ Work with LEED consultant to develop policies and programs to reflect various LEED credit requirements.

■ Participate in training and workshops that may be scheduled by the CxA after construction is complete.

■ Confirm participation in any building performance monitoring efforts that may be required post-occupancy.

LEED Credit Responsibilities

Although there are no credits that the building management is responsible for directly implementing, their input is required in order to complete the documentation of many credits. Table 2.10 describes the typical credit responsibilities.

TABLE 2.10   TYPICAL LEED CREDIT RESPONSIBILITIES OF BUILDING MANAGEMENT [L=LEAD, S=SUPPORT]

L/S	CREDIT	RESPONSIBILITIES OF BUILDING MANAGEMENT
S	SS C4.1–SS C4.4	Provide occupancy data on estimated number of full-time occupants, part-time occupants, residents, visitors, etc.; help develop and create occupant transport-related programs, such as preferred parking policy or comprehensive transport management plans.
S	EA P1/EA C3	Work with CxA and contractors to implement any post-occupancy commissioning activities.
S	EA C5	Coordinate with mechanical engineer to implement M&V measures post-occupancy or work with third-party M&V providers, as applicable.
S	MR P1	Develop and implement occupant recycling program.
S	EQ P2	Develop and implement smoking policies, as applicable.
S	EQ C7.2	Conduct thermal comfort survey within 6–18 months of occupancy and implement corrective action measures, as necessary.
S	WE Credits	Provide occupancy data on estimated number of full-time occupants, part-time occupants, residents, visitors, etc., as requested, to help in water-use calculations.
S	ID Credits	Work with LEED consultant to develop innovative post-occupancy, operational and management programs.

General Contractor or Construction Manager
General Responsibilities

The general contractor (GC) or construction manger needs to be involved as early as possible in a LEED project. Decisions related to the building material choices and their procurement require the GC's input. Further, the contractor must provide feedback to the design team regarding constructability issues up front so that costly modifications need not be made at a later stage of design. The general responsibilities of the GC are, but not limited to, the following:

- Provide input to the project team on the procurement process and how things will be done during construction.
- Provide team feedback on constructability issues related to site or building program.
- Provide preliminary cost estimates of the project and work with design team in costing differences between various *green* strategies.
- Review the specifications provided by the team to understand LEED specific requirements and express concerns or issues in implementation of any credits.
- Ensure proper execution and documentation of *green* strategies and LEED credit requirements by subcontractors.

TABLE 2.11   TYPICAL LEED CREDIT RESPONSIBILITIES OF GENERAL CONTRACTOR [L=LEAD, S=SUPPORT]

L/S	CREDIT	RESPONSIBILITIES OF GENERAL CONTRACTOR
S	SS P1	Implement the erosion and sedimentation control measures detailed by the civil engineer; document the measures implemented via photographs and logs.
S	SS C5.1	Conform to the site disturbance limits set according to this credit's requirements by the civil engineer; ensure construction activities are not encroaching on the protected area; document via photographs and inspection logs.
S	EA P1/EA C3	Perform all responsibilities and testing of systems as outlined in the commissioning plan; coordinate with CxA on scheduling commissioning-related activities.
L	MR C2	Implement a construction waste management plan (draft may be provided by design team) to divert construction waste from landfills; document via photographs, waste diversion calculations, and receipts from waste hauler; coordinate with LEED consultant to ensure progress is according to targeted goals.
L	MR C3–C7	Collect, maintain, and compile information requested from subcontractors with respect to materials/products installed on-site.
L	EQ C3.1	Implement an indoor air quality management plan (draft may be provided by design team) during construction; document via photographs, inspection logs, and checklists.
L	EQ C3.2	Provide building flush-out before occupancy or coordinate with third party indoor air quality (IAQ) testing agency to meet the requirements of this credit; document via photographs, inspection logs, and checklists; report any conflicts with the schedule for occupancy.
L	EQ C4.1–EQ C4.4	Ensure all on-site adhesives, sealants, paints, coatings, and other materials used in the interior of the building are within the allowable VOC limits and standards defined by these credits; collect and compile documentation to demonstrate compliance.

- Work with CxA to ensure all systems are functioning according to design.
- Implement construction procedures on-site in accordance with LEED credit requirements, such as waste management and indoor air quality management.
- Work with design team to ensure smooth handover to the facilities staff.
- Provide appropriate documentation to demonstrate compliance with LEED credit requirements.

LEED Credit Responsibilities

Besides executing the LEED credits that have been integrated into the construction documents and specifications, the GC is also responsible for implementing and documenting specific credits that are exclusive to the construction process. Table 2.11 describes the contractor's LEED credit-specific responsibilities.

Interior Designer
General Responsibilities
The role of the interior designer in a LEED project is to incorporate green material choices and plan interior spaces to enhance the indoor environmental experience for the occupants. The general responsibilities of the interior designer are, but not limited to, the following:

- Work with architect to design interior spaces that will maximize daylighting and access to views for occupants.
- Provide input on material choices and selections based on *green* requirements.
- Ensure *green* requirements and specifications are incorporated into drawings.

LEED Credit Responsibilities
The interior designer typically has a support role in a LEED project and may not have direct credit documentation responsibilities. However, he/she may have to provide information to implement certain credit requirements. Table 2.12 describes the typical credits that need input from the interior designer.

Cost Estimator
General Responsibilities
The role of the cost estimator is to assist the design team and the owner in making decisions based on life-cycle costs of items. The general responsibilities of the cost estimator are, but not limited to, the following:

TABLE 2.12   TYPICAL LEED CREDIT RESPONSIBILITIES OF INTERIOR DESIGNER [L=LEAD, S=SUPPORT]

L/S	CREDIT	RESPONSIBILITIES OF INTERIOR DESIGNER
S	MR P1	Incorporate spaces for storage and collection of recyclables in building program and interior space design.
S	MR C3–C7	Specify materials and products that have recycled content, are available regionally, are made with rapidly renewable materials, or are FSC certified, depending on the requirements of the credits.
S	EQ C4.1–C4.4	Specify materials with VOC limits according to the requirements of the credits.
S	EQ C6.1	Coordinate interior space planning with lighting consultants to provide individual lighting control.
S	EQ C8.1, EQ C8.2	Design interior layouts/rooms to maximize access to daylight and views; choose interior finishes that will enhance daylighting within spaces.

- Provide a realistic budget to the team during the design phase for preliminary MR credit estimates and life-cycle cost analysis.
- Help the design team with life-cycle cost analysis.
- Evaluate both costs and benefits of the *green* features.
- Update cost estimates as design progresses and provide input to the team on issues and choices to keep costs under control.

Green/LEED Consultant or LEED Project Manager
General Responsibilities
This section is written with the assumption that a LEED consultant will be engaged to facilitate the LEED certification process and he/she will be the designated LEED PM. (Refer to an earlier section in this chapter for the skills desired of a LEED PM.) A design team member may be designated as the LEED PM as well. In either case, the general responsibilities of the LEED PM are, but not limited to, the following:

- Provide research support to the team and help them identify potential *green* strategies.
- Provide direction and guidance to the design team regarding *green* design resources, *green* material choices, and selection criteria.
- Assist design team in preparing *green* specifications and other documents including, but not limited to, contractor's toolkits, etc.
- Organize *green* charrettes and workshops at the beginning and during the design process to ensure LEED goals are met.
- Provide training to contractors and subcontractors regarding their LEED responsibilities.
- Review drawings and specifications periodically to ensure project is on path to achieve its goals.
- Ensure open communication exists between team members; assign responsibilities and coordinate the efforts of the team members.
- Help the team stay on schedule and budget.
- Manage the entire LEED project process, including reviewing, coordinating, compiling, and submitting documentation to GBCI.

LEED Credit Responsibilities
The specific credits that the LEED PM is responsible for may vary from project to project, depending on the involvement of other team members. Table 2.13 presents the credits typically documented by the LEED PM with support from other team members.

Energy Simulation Expert
General Responsibilities
The energy simulation expert may be a part of the MEP team, the *green* consultant's team, or may be an independent consultant engaged by the owner. In any scenario, the general responsibilities of the energy simulation expert are, but not limited to, the following:

TABLE 2.13   TYPICAL LEED CREDIT RESPONSIBILITIES OF LEED PM [L=LEAD, S=SUPPORT]

L/S	CREDIT	RESPONSIBILITIES OF LEED CONSULTANT/LEED PM
L	SS C1	Collect information to confirm the project meets the site characteristics according to credit requirements.
L	SS C2	Research site neighborhood; perform calculations; prepare site vicinity plan.
L	SS C4.1–C4.4	Perform occupancy count calculations; compile and highlight architectural/site drawings; prepare site vicinity drawings; coordinate with building management for developing transport policies.
S	SS C7.1	Collect information from the architect, the landscape architect, and the civil engineer; compile and prepare exhibit drawings.
S	SS C7.2	Collect information from the architect, the landscape architect, and the civil engineer; compile and prepare exhibit drawings.
S	WE credits	Support the landscape architect and the plumbing engineers in calculations, research of new products, and completion of credit documentation.
S	MR credits	Support the architect, the contractor, and the interior designers with product selection, developing *green* material specifications, construction waste management plans, instructions to subcontractors, and other such items related to the MR credits; assist with completing documentation.
S	EQ C4.1–C4.4	Support the architect, the contractor, and the interior designers with product selection and developing *green* material specifications.
L	EQ C8.2	Create exhibit plan drawings showing access to views; coordinate with the architect and the lighting consultant to obtain information.
S	ID credits	Work with the design team, owners, and building management in identifying ID credit opportunities and develop documentation.
S	All credits	Provide to the consultants constant research support and design direction for implementation of all credits; assist in completing online documents and narratives; review all information to demonstrate compliance.

- Analyze the impact of massing and orientation of the building on the mechanical systems and performance.
- Perform energy simulation of the project at all stages of the design to provide feedback to design team and ensure project is optimized for best performance.
- Recommend energy efficiency measures with respect to the building's envelope, HVAC systems, and lighting.
- Work with design team to propose innovative solutions and refine system choices to stay within the established energy targets and meet LEED credit requirements.
- Ensure the energy model assumptions are coordinated with the design drawings.

TABLE 2.14   TYPICAL LEED CREDIT RESPONSIBILITIES OF SIMULATION EXPERT [L=LEAD, S=SUPPORT]

L/S	CREDIT	RESPONSIBILITIES OF ENERGY SIMULATION EXPERT
L	EA P2, EA C1	Develop building energy simulation model to verify that the proposed building's energy performance exceeds ASHRAE 90.1-2007 baseline by at least 10% and identify energy efficiency measures that will help increase that percentage performance; coordinate with the MEP engineers and the architects to develop the model and propose design options; prepare documentation to demonstrate compliance with credit requirements.
S	EA C2	Provide simulated building annual energy cost to the mechanical engineer or the renewable energy consultant.
S	EA C5, EA C6	Provide data on proposed building's energy performance.

LEED Credit Responsibilities

The energy simulation expert is responsible for the documentation of the credits in the *Energy and Atmosphere* category—specifically EA P2, EA C1, and EA C2. Table 2.14 describes in detail the specific responsibilities with respect to these credits and prerequisites.

Lighting/Daylighting Consultant

General Responsibilities

The lighting consultant may be engaged by the owner to provide both lighting and daylighting simulation services. In some cases, the *green*/LEED consultant may be able to provide daylighting simulation, and the electrical engineer may be able to provide lighting simulation services. The owner may choose between a third party consultant and someone from the existing team, depending on their level of expertise. This section is written with the assumption that a lighting consultant will be engaged to provide the lighting and daylighting simulation services. The general responsibilities of the lighting consultant are, but not limited to, the following:

- Assist the design team in making decisions related to orientation, massing, and envelope based on their impact on daylight and lighting design.
- Prepare daylight simulation models from early design process to guide design team in making appropriate decisions with respect to window sizes, types, glazing choices, etc.
- Prepare site lighting photometric studies and design internal lighting systems and controls to meet LEED credit requirements.
- Coordinate with the energy simulation expert to understand the impact of lighting choices on the overall building energy performance.

TABLE 2.15   TYPICAL LEED CREDIT RESPONSIBILITIES OF LIGHTING CONSULTANT [L=LEAD, S=SUPPORT]

L/S	CREDIT	RESPONSIBILITIES OF LIGHTING/DAYLIGHTING CONSULTANT
L	SS C8	Design site lighting in accordance with the credit's requirements; create site photometric model; coordinate with the landscape architect, the architect, and the electrical engineer on the design of landscape lighting, façade lighting, and interior lighting layouts and controls.
S	EA P2/EA C1	Coordinate with the energy simulation expert and/or the mechanical engineer to design lighting systems that are optimized for whole building energy performance.
L	EQ C6.1	Incorporate lighting system controls in the design.
L	EQ C8.1	Create daylight simulation models and inform the design team regarding design decisions related to windows, glazing, etc.; if needed, perform lighting measurements to demonstrate compliance.
S	EQ C8.2	Coordinate with the architect and the LEED consultant regarding information related to window types, sizes, glazing type, etc.

- Incorporate innovative technologies, such as daylight sensors, Solatubes, etc., to maximize benefits of daylighting and reduce dependence on artificial lighting.
- Coordinate with electrical engineer in developing lighting layout drawings and specifications.

LEED Credit Responsibilities

The specific LEED credits that the lighting consultant might be responsible for documenting are described in Table 2.15.

Commissioning Authority

General Responsibilities

The role of the Commissioning Authority (CxA) is to ensure that the building's systems are installed and commissioned according to the owner's project requirements and the design. The process of commissioning is required for a project to achieve certification because it is a prerequisite (EA P1). The owner should designate a CxA as early as possible in the design process.

The *LEED Reference Guide* provides details on who can be a CxA and states that the individual must have experience working on at least two other projects of similar managerial and technical complexity. For projects larger than 50,000 square feet, the CxA must be independent of the project's construction and design teams. He or she may be a qualified staff member of the owner, owner's consultants, or an employee of the firm providing design and/or construction services. However, the CxA may not

TABLE 2.16   TYPICAL LEED CREDIT RESPONSIBILITIES OF CXA [L=LEAD, S=SUPPORT]

L/S	CREDIT	RESPONSIBILITIES OF COMMISSIONING AUTHORITY (CXA)
L	EA P1, EA C3	Develop and implement commissioning plan; complete all fundamental and enhanced commissioning activities; review drawings; coordinate with contractor on testing of systems; review contractor submittals; provide summary commissioning report and develop systems manual; perform post occupancy review.
S	EA C5	Coordinate post-occupancy building operation review with M&V team, if required.

have responsibility for design of the project. For projects that are smaller than 50,000 sf, the CxA may be a qualified staff member of the owner or owner's consultant, or he/she may be an individual on the design or the construction team and may have additional project responsibilities. If the team is pursuing the credit for enhanced commissioning (EA C3), then the CxA may not be an employee of the design firm but may be contracted through that firm.[5]

The general responsibilities of the Commissioning Authority are, but not limited to, the following:

- Work with the design team and the owner to develop the owner's project requirements and the basis of design documents to ensure that the project goals are incorporated into design.
- Review drawings to ensure proper integration of owner's needs and requirements.
- Develop commissioning plan and coordinate/oversee the commissioning testing.
- Review contractor submittals, if needed, for enhanced commissioning.
- Coordinate with contractor to verify installation and performance testing of commissioned systems.
- Develop systems manual for commissioned systems, if needed, for enhanced commissioning.
- Verify that the requirements for training are completed and review building operation within 10 months after substantial completion, if needed, for enhanced commissioning.
- Complete a summary commissioning report.
- Keep communication open between the owner, the contractor, and the design team.

LEED Credit Responsibilities

The specific credits that the CxA is responsible for implementing and documenting are described in Table 2.16.

Planning

The third and final component of the project management process is *Planning*. Establishing a plan for the execution of the LEED process is crucial to ensure that the project is able to achieve its targeted goals in the stipulated time. The three measures that typically define the successful planning of a project are schedule, budget, and quality.[6] In the case of LEED projects, a fourth measure is successful certification. Early investment of time and efforts in proper planning of the LEED process can help deliver the desired results and have better control over the outcome.

SCHEDULE

The schedule for all LEED related items may be prepared by integrating the various LEED milestones with the overall building design and construction schedule. The LEED PM may prepare a LEED schedule specific to the project in consultation with the owners and the project design team. The steps and activities leading up to these milestones are described in detail in Chapters 3, 4, and 5 along with a schedule highlighting the duration of each activity. The key LEED milestones that need to be integrated into the design and construction schedule are

- Completion of Preliminary LEED Assessment
- Completion of LEED Vision Workshop
- Completion of Schematic Design LEED Review
- Completion of Design Development LEED Review
- Submission to GBCI for Design Review
- Completion of Pre-Construction LEED Kick-Off Meeting
- Submission to GBCI for Construction Review

Preliminary LEED Assessment

The completion of a preliminary assessment using the LEED credit framework is the first milestone that needs to be integrated into the typical design schedule. The activities that are recommended in order to reach this milestone and complete the assessment are described in steps 1 and 2 of Stage I—Project Definition and Goal Setting in Chapter 3. This item may be included during the pre-schematic design stage or early schematic design stage, depending on when the decision to pursue LEED certification is made.

LEED Vision Workshop

This next milestone is to organize and conduct a visioning session/workshop to discuss the preliminary assessment and establish LEED-related goals for the project. The activities that are recommended in order to complete this milestone are described in step 4 of Stage I—Project Definition and Goal Setting in Chapter 3.

Schematic Design LEED Review

This milestone occurs at the end of the schematic design phase, after the design drawings have been reviewed for integration of *green* building principles and strategies.

The activities that are recommended in order to reach this milestone are described in steps 1 through 4 of Stage II—Schematic Design (SD) LEED Phase in Chapter 4.

Design Development LEED Review

This milestone occurs at the end of the design development phase, after the design drawings have been developed in more detail and reviewed for integration of LEED-related items. The activities that are recommended in order to reach this milestone are described in steps 5 through 8 of Stage II—Design Development (DD) LEED Phase in Chapter 4.

Submission to GBCI for Design Review

This milestone occurs at the end of the construction documents phase, after the construction documents and LEED Online documents have been reviewed to verify that all credit/prerequisite requirements are appropriately reflected. The activities that are recommended in order to reach this milestone are described in steps 9 and 10 of Stage II—Construction Documents (CD) LEED Phase in Chapter 4.

Pre-Construction LEED Kick-Off Meeting

This milestone involves organizing and conducting a pre-construction LEED kick-off meeting to discuss implementation and documentation of construction-related LEED credits with the construction team. The activities that are recommended in order to complete this milestone are described in step 1 of Stage III— Construction Phase Implementation in Chapter 5.

Submission to GBCI for Construction Review

The last milestone marks the completion of all documentation and its submission to GBCI for Construction Review. This is typically toward the end of construction and may coincide with substantial completion. The activities that are recommended in order to reach this milestone are described in steps 1 through 5 of Stage III—Construction Phase Implementation in Chapter 5.

Figure 2.11 illustrates how these LEED milestones may be integrated with a typical design schedule. Part A of the figure shows the various stages and steps of the LEED Project Management Process described in this book and identifies the key milestones described above. Part B of the figure shows an example of a typical design and construction schedule for a project and at what points the LEED milestones should be integrated. Finally, Part C shows the complete design schedule with the LEED milestones integrated. The LEED PM may follow a similar process and integrate the LEED milestones with the design schedule.

BUDGET/COSTING LEED

The next aspect of project planning involves establishing a methodology to keep a tight lid on the project budget. It is important for the LEED PM to be aware of the monetary impact of various *green* strategies and products, certification costs, testing costs, etc. that may be incurred by the project. The LEED PM should collaborate with the cost estimator to provide the owner with a detailed estimate identifying individual costs of various components. Further, this cost estimate should be used to prepare a

A. LEED Project Management Process

Stage I	Project Definition and Goal Setting
Step 1	Collect and Compile Preliminary Data
Step 2	Prepare a preliminary LEED assessment
Step 3	Register the project on LEED Online
Step 4	Organize a LEED Vision Workshop
Step 5	Prepare a LEED Project Workbook

Stage II	Design Phase Integration
	Schematic Design LEED Phase
Step 1	Perform Credit Analyses
Step 2	Research and Identify 'Green' Products
Step 3	Integrate 'Green' Requirements into Design
Step 4	Review Schematic Design Drawings
	Design Development LEED Phase
Step 5	Update Credit Analyses and Drawings
Step 6	Prepare for Construction Phase
Step 7	Develop Building Management Policies
Step 8	Review Design Development Drawings
	Construction Documents LEED Phase
Step 9	Prepare Documentation for 'Design Credits'
Step 10	Submit to GBCI for Design Review

Stage III	Construction Phase Implementation
Step 1	Organize a Pre-Construction LEED Kick-Off Meeting
Step 2	Review Progress of Credit Implementation
Step 3	Prepare Documents for GBCI Construction Review
Step 4	Review Final Documents for GBCI Submission
Step 5	Submit to GBCI for Construction Review

★ LEED Milestone to be incorporated into Design Schedule

B. Typical Design & Construction Schedule

Milestone	Date
Programming and Concept Design Begins	1/15/2009
Schematic Design Phase Begins	4/15/2009
Schematic Design Phase Complete	7/31/2009
Design Development Phase Complete	12/10/2009
Construction Documents Complete	6/1/2010
Bid & Negotiate Construction Contract	10/1/2010
Construction Work Commencement	11/15/2010
Construction Work Complete	8/15/2010

C. Design Schedule with LEED Milestones

Milestone	Date
Programming and Concept Design Begins	1/15/2009
Preliminary LEED assessment completed	**3/15/2009**
LEED Vision Workshop	**3/30/2009**
Schematic Design Phase Begins	4/15/2009
Schematic Design Phase Complete	7/31/2009
Schematic Design LEED Review complete	**8/21/2009**
Design Development Phase Complete	12/10/2009
Design Development LEED Review complete	**1/4/2010**
Construction Documents Complete	6/1/2010
Submission to GBCI for Design Review	**7/1/2010**
Bid & Negotiate Construction Contract	10/1/2010
LEED Pre-Construction Kick-Off Meeting	**10/15/2010**
Construction Work Commencement	11/15/2010
Construction Work Complete	8/15/2010
Submission to GBCI for Construction Review	**9/15/2010**

Figure 2.11 Integrating LEED milestones with the design schedule.

Credit: Green Potential LLC (www.green-potential.com).

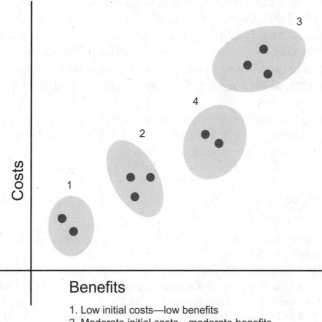

Benefits

1. Low initial costs—low benefits
2. Moderate initial costs—moderate benefits
3. High initial costs—high benefits
4. Moderate initial costs—high benefits

Figure 2.12   Example of a preliminary cost-benefit data plot.

cost-benefit analysis that will allow the owner to make prudent decisions based on the long-term benefits associated with implementing a particular strategy (refer to Chapter 4 for an example). Figure 2.12 shows a preliminary cost-benefit plot that was prepared for a project. The owner used this plot to make the final decisions on the choice of *green* strategies.

In addition, this estimate should be periodically reviewed and updated based on changes in the design. The role of the LEED PM should be to assist the team in finding synergies between various strategies so that the high costs associated with a particular strategy may be offset by reduced costs for another. For instance, the high costs of implementing an underfloor air distribution system (UFAD) may be compensated for by the lower cost of materials as a result of reduced floor-to-floor height.

QUALITY

The final aspect of the planning process involves ensuring proper control on the quality and integrity of the documentation presented to GBCI. Having fully coordinated, accurate, and complete documentation is critical to ensure a smooth review process with minimal clarifications and easy certification from GBCI. To make sure that the quality of the work submitted to GBCI is maintained, the LEED PM should conduct detailed reviews of all drawings, specifications, and online documentation and check them for clarity, consistency, accuracy, and completeness in demonstrating the project's intent to meet LEED credit requirements. The regular reviews will point out any issues or concerns before submission to GBCI. These will give team members the opportunity to make any changes and will also save the project team time that they would perhaps have to spend later, when correcting mistakes pointed out by the GBCI reviewers.

The various tools, such as process schedules, checklists, etc., that have been included throughout this book will help the LEED PM ensure that all LEED-related activities and tasks have been completed on time satisfactorily.

Notes

[1] Busby Perkins + Will, *Roadmap for the Integrated Design Process* (BC Green Building Roundtable, 2007).

[2] GBCI, "Certification Bodies," LEED Online, www.usgbc.org/Docs/News/CBs%20072908.pdf.

[3] GBCI, "LEED Project Registration and Certification Fees," LEED Online, http://www.gbci.org/main-nav/ building-certification/resources/fees/current.aspx.

[4] GBCI, "LEED Project Registration and Certification Fees", LEED Online, http://www.gbci.org/main-nav/building-certification/resources/fees/current.aspx.

[5] USGBC, *LEED Reference Guide for Green Building Design and Construction* (USGBC, 2009), 221.

[6] Whole Building Design Guide, "Project Planning, Management and Delivery," http://www.wbdg.org/project/pm.php.

STAGE I: PROJECT DEFINITION AND GOAL SETTING

Introduction

The first stage of the LEED Project Management Process involves defining the project and establishing goals. It includes collecting information, conducting preliminary analyses, and creating a roadmap for achieving LEED certification. These activities should ideally be completed during the pre-schematic design phase. If, for some reason, the decision to pursue LEED certification is made at a later time, these activities must still be completed to ensure successful integration of LEED principles into the design.

The information that is required to complete activities in this stage of the process may be obtained from preliminary drawings, programming data, outline specifications, and information gathered from the owner or design team. The designated LEED Project Manager (PM) is primarily responsible for implementing the activities defined in this stage (refer to Chapter 2 for a detailed description of the LEED PM's responsibilities and skills desired).

This stage is composed of five primary steps that are further broken down into individual activities. Figure 3.1 illustrates the process flow for Stage I activities. The five steps that constitute this stage are:

Step 1: Collect and compile preliminary data
Step 2: Prepare a preliminary LEED assessment
Step 3: Register the project on LEED Online
Step 4: Organize a LEED Vision Workshop
Step 5: Prepare a LEED Project Workbook

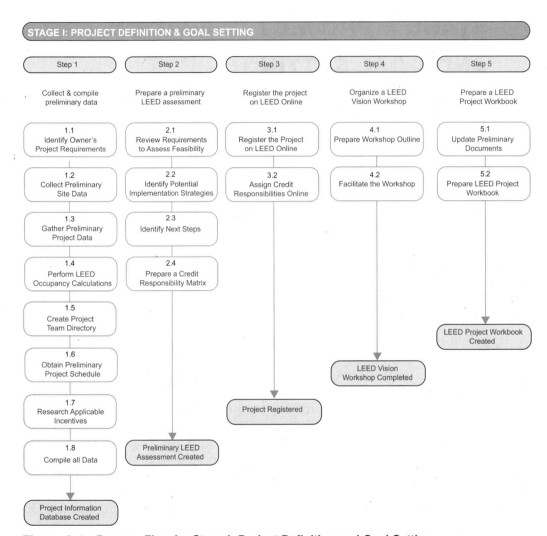

Figure 3.1   Process Flow for Stage I: Project Definition and Goal Setting.
Credit: Green Potential LLC (www.green-potential.com).

Step 1: Collect and Compile Preliminary Data

Objective	The main objective of this step is to collect and compile preliminary data to create a project information database that is available to and accessible by all team members. This is essential to ensure consistency of information across all LEED credits and will also help in preparing a preliminary LEED assessment.
Recommended Activities	1.1 Identify Owner's Project Requirements. 1.2 Collect Preliminary Site Data. 1.3 Gather Preliminary Project Data. 1.4 Perform LEED Occupancy Calculations. 1.5 Create Project Team Directory. 1.6 Obtain Preliminary Project Schedule. 1.7 Research Applicable Incentives. 1.8 Compile All Data.
Roles and Responsibilities	LEED PM: Collect and compile information. Owner: Provide requested information. Design Team: Provide requested information.
Duration	3 weeks
Outcome	Project information database created.

ACTIVITY 1.1—IDENTIFY OWNER'S PROJECT REQUIREMENTS

The first step is to identify and define the owner's goals and requirements for the project so that the design is aligned to meet the requirements. These requirements will become the *design drivers* that the project team can use, as they develop the program and design. The owner's project requirements (OPR) are also required as part of the formal documentation for the Fundamental Commissioning (EA P1) prerequisite. The Commissioning Authority (CxA) may assist the owner in developing the OPR. However, if the CxA has not yet been engaged, the LEED project manager (LEED PM) or the architect may provide a simple template or questionnaire that the owner may complete to communicate the goals and requirements. The key areas in which the owner's input should be obtained are described below:

Owner and user requirements   These define the purpose, program, and proposed use of the building. The owner may also include any goals related to future expansion, construction, and operational costs.

Environmental and sustainability goals   The environmental and sustainability goals include the owner's desired level of LEED certification and any additional goals, such as meeting any federal, state, or local green building standards.

Energy efficiency goals   These include the targeted energy efficiency goals with respect to the local energy code, ASHRAE, or LEED standards. While some owners may target basic level energy efficiency (10% better than ASHRAE 90.1-2007) required by the LEED-NC rating system, there are others who may target higher levels of energy efficiency for long-term operational savings or to benefit from any incentives.

Indoor environmental quality goals   The indoor environmental quality goals include the expected occupancy schedules; space environmental requirements in terms of lighting, temperature, acoustics, etc.; adjustability of system controls; and any accommodations for after-hours use. These requirements gain special importance in building types like hospitals, labs, or museums, which must adhere to strict indoor environmental requirements.

Equipment and system expectations   These define the owner's desired level of quality, automation, and maintenance requirements for equipment and systems used in the building. If desired, the owner may include information on efficiency targets or preferred manufacturers of building systems.

Building occupant and O&M personnel requirements   The building occupant and personnel requirements provide an understanding of how the facility will be operated and specify whether any training is necessary for the occupants to operate the building's systems.

Table 3.1 lists some sample questions that the LEED PM or CxA may ask the owner to define the *owner's project requirements.*

TABLE 3.1 SAMPLE QUESTIONNAIRE TO DEVELOP OWNER'S PROJECT REQUIREMENTS	
NO.	**QUESTIONS**
1	What are the functional requirements of this facility?
2	What are the *green* or LEED goals for the project?
3	What specific energy goals do you have for the project?
4	What conditions are important to your comfort in an ideal facility?
5	What activities generate pollutants in this facility?
6	What problems with previous projects should be avoided?
7	What must be accomplished for a successful project?

ACTIVITY 1.2—COLLECT PRELIMINARY SITE DATA

After the owner's project requirements have been identified, the next step is to collect preliminary site data. The *Sustainable Sites* category of the LEED-NC rating system provides guidance for selecting and developing building sites with minimal impact to the environment. Collecting and compiling site data is required to assess the opportunities that the site offers in achieving credits in this category. This section is written with the assumption that the site has already been selected by the owner or developer. If the site has not yet been selected, then the owners may use the LEED *Sustainable Sites* credit requirements as a guide for choosing an appropriate site. This activity may be completed by the designated LEED PM. The following information should be collected:

- Site location
- Site predevelopment conditions
- Site area data

Site Location

This is the physical address or location of the project site and is required to determine the neighborhood, transportation, and topographical characteristics of the site.

Site Predevelopment Conditions

The site's predevelopment conditions define the status of the site before the development work begins. The following information regarding the site's predevelopment conditions should be collected.

Greenfield or previously developed   Determine whether the site is a *greenfield* or has been previously developed. A *greenfield* site is essentially land that has not been built upon or developed. Building on such sites means disturbing the existing ecosystem of the place. It is therefore preferable to choose a previously developed site, as those sites have already been disturbed and can utilize existing transportation and utility infrastructure. The LEED rating system gives credit to projects that utilize previously developed sites. This information may be obtained from the owner or civil engineer.

Prime farmland   Determine whether the site is *prime farmland* as defined by the U.S. Department of Agriculture in the U.S. Code of Federal Regulations. According to this code, prime farmland is defined as "land that has the best combination of physical and chemical characteristics for producing food, feed, forage, fiber, and oilseed crops."[1] This information may be obtained from the local zoning classification data or the civil engineer.

Habitat for endangered species   Find out whether the site has been identified as a habitat for any species on the federal or state threatened or endangered lists maintained by the U.S. Fish and Wildlife Service. This information may be obtained from local zoning data or the U.S. Fish and Wildlife Service website.

Proximity to wetlands   Determine whether the site is within 100 feet of any wet-lands, as defined by the U.S. Code of Federal Regulations 40 CFR, or within setback distances prescribed in state or local regulations. The Code defines wetlands as "areas that are inundated or saturated by surface or groundwater at a frequency and duration sufficient to support a prevalence of vegetation typically adapted for life in saturated soil conditions."[2] Presence of wetlands on-site may be determined from local zoning plans, site surveys, and/or software applications such as Google Earth.

Proximity to water bodies   Determine whether the site is within 50 feet of a water body, defined as "seas, lakes, rivers, streams, and tributaries that support or could support fish, recreation, or industrial use,"[3] according to the Clean Water Act. This infor-mation may be obtained from site plans, site surveys, or software applications such as Google Earth.

Previously public parkland   Determine whether the site was public parkland prior to acquisition. It is recommended not to develop sites that are classified as public park-land unless land of equal or greater value as parkland is accepted in trade by the pub-lic land owner. This information can be determined from either local zoning data or directly obtained from the owner.

Brownfield or contaminated   Determine whether the site is a *brownfield* or has been documented as contaminated by means of a Phase II Environmental Site Assessment (ESA). According to the U.S. Environmental Protection Agency (EPA), a brownfield site is defined as "a real property, the expansion, redevelopment or reuse of which may be complicated by the presence of a hazardous substance, pollutant, or contaminant."[4] A brownfield site can be remediated for reuse and developed to preserve virgin land for future generations.

A Phase II ESA is "an investigation that collects original samples of soil, ground-water, or building materials to analyze for quantitative values of various contaminants. The most frequent substances tested are petroleum hydrocarbons, heavy metals, pes-ticides, solvents, asbestos, and mold. A Phase II ESA is usually conducted upon guid-ance received from a Phase I ESA."[5] Information on whether the site is a brownfield or contaminated as per Phase II ESA may be obtained from the owner, environmental consultant, or local zoning data.

Flood plain zone   Determine whether the elevation of the land is lower than 5 feet above the 100-year flood as defined by the Federal Emergency Management Agency (FEMA). This information may be obtained from a site survey, local city website, or directly from the FEMA website.

Site Area Data

The site area data will be required by the project team to complete a number of LEED credit calculations, and it is important that the numbers referenced by different team members are consistent. The following information should be obtained:

- Site area
- LEED site area
- Open space area (required by local zoning ordinance)
- Open space area (proposed or provided)
- LEED open space area

Site Area

Site area is defined as the area in square feet within the legal property boundary of the site. This area number may be obtained from a site plan or survey drawing.

LEED Site Area

The *LEED site area* is defined as the area within the *LEED project boundary* of the site. The *LEED project boundary* is the portion of the project site that will be submitted for LEED certification. Typically, for most single-building developments, the entire site is submitted for LEED certification requiring no separate definition of a LEED project boundary. In such cases, the LEED site area is the same as the overall site area. However, in some scenarios, the design team is allowed "to determine the limits of the project submitted for LEED certification differently from the overall site boundaries."[6] Projects that are part of a larger multiple building development or campus may utilize this option. Projects where the scope of work includes redoing the sidewalk and planting new trees outside of the legal site boundary may also utilize this option to define a LEED project boundary based on the scope of work. Figure 3.2 illustrates the difference

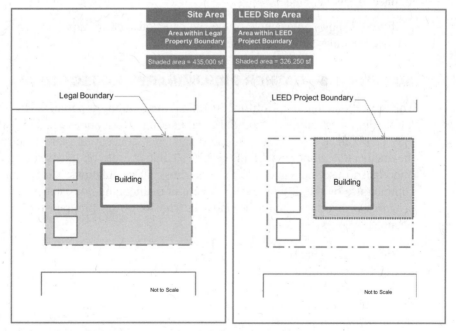

Figure 3.2   Distinction between site area and LEED site area.

between site area and LEED site area. Regardless of how the LEED project boundary is defined, the boundary and the area must be consistently used throughout all LEED calculations.

The LEED PM may determine the applicable site area in consultation with the owner, the architect, or the civil engineer and create an exhibit drawing that clearly identifies the LEED project boundary and site area.

Open Space Area (required by local zoning ordinance)

This is the minimum open space area required by the local zoning ordinance. The LEED PM may obtain this information from the local zoning ordinance or from the civil engineer. If there is no open space requirement, then that should be noted as well.

Open Space Area (proposed or provided)

This is the open space area that is proposed or has been provided for the project. This information can be obtained from a conceptual site plan, building program, or directly obtained from the architect or the civil engineer.

LEED Open Space Area

Open space area, for the purpose of LEED calculations, is defined as the area of the site that is vegetated and pervious and is calculated by subtracting the development footprint from the site area. Pedestrian-oriented hardscape and vegetated roof areas may be included in the LEED open space area calculations, if the project is located in high density or urban areas. The LEED PM may determine the proposed vegetated open space area from conceptual site plan drawings or obtain the number from the landscape architect or civil engineer.

Table 3.2 shows an example of the site data collected for a project. The LEED PM may use a similar table to compile all data.

ACTIVITY 1.3—GATHER PRELIMINARY PROJECT DATA

The project data includes information about the building program, such as the overall building area, building footprint area, parking information, etc. This information may be obtained from programming documents, conceptual plans or directly from the architect. A number of LEED credit calculations rely on this basic data, and it is important to use the same information across all calculations. Compiling them for common use by the design team will help in maintaining consistency. As the design progresses, this data should be updated. The following program data needs to be obtained:

- Building gross area
- Building conditioned area
- Building footprint area
- Parking spaces required
- Parking spaces provided

TABLE 3.2   SAMPLE PRELIMINARY SITE DATASHEET

SITE LOCATION: Green Avenue and W. 44th Street, Austin, Texas, USA

Site Predevelopment Conditions

ITEM	SITE DESCRIPTION	YES/NO	REMARKS
1	Greenfield or Previously Developed	Yes	Previously Developed
2	Prime Farmland	No	
3	Habitat for Endangered Species	No	
4	Within 100 Feet of Wetlands	No	
5	Within 50 Feet of Water Body	Yes	
6	Previously Public Parkland	No	
7	Brownfield or Contaminated	No	
8	Flood Plain Zone	No	Confirm with Civil Engineer

Site Area Data

ITEM	SITE INFORMATION	AREA (sf)	REMARKS
1	Site Area	600,000	
2	LEED Site Area	600,000	Same as Site Area
3	Open Space Area (required)	150,000	25% of Site Area
4	Open Space Area (provided)	200,000	
5	LEED Open Space Area	200,000	

Building gross area   This is total gross area in square feet (sf) of the building(s) being considered for LEED certification.

Building conditioned area   The conditioned area of the building is defined as the area (sf) of the building that is heated or cooled or both, for the comfort of the occupants.

Building footprint area   Building footprint is the area on a project site used by the building structure, defined by the perimeter of the building plan.

Parking spaces required   These are the number of parking spaces that are required to be provided according to local zoning ordinance or building code.

Parking spaces provided   These are the number of parking spaces that are planned to be provided for the project. A breakdown of the number of spaces that will be covered and those that will be uncovered should also be obtained.

TABLE 3.3 SAMPLE PRELIMINARY PROJECT DATASHEET		
ITEM	**PROJECT DATA**	**DESCRIPTION**
1	Building Area (square feet)	400,000
2	Conditioned Area (square feet)	300,000
3	Building Footprint Area (square feet)	100,000
4	Number of Parking Spaces (required)	200
5	Number of Parking Spaces (provided—covered)	165
6	Number of Parking Spaces (provided—uncovered)	40
7	Number of Parking Spaces (provided—total)	205

Table 3.3 illustrates preliminary data collected for a project. The LEED PM may use a similar table to compile the information gathered.

ACTIVITY 1.4—PERFORM LEED OCCUPANCY CALCULATIONS

The next step after collecting the site and program data is to calculate the total number of occupants in the building. This occupant count is required for a number of LEED credits and is calculated differently from the traditional building code occupancy calculations. The occupancy is based on the number and type of occupants expected to use the building, and this information may be obtained from the owner or building management. The following information needs to be collected:

1 Number of full-time staff (eight-hour work day)
2 Number of part-time staff and number of hours they work per day
3 Number of residents
4 Number of peak transients (students, volunteers, visitors, customers, etc.)

After collecting this information, the number of full-time equivalent (FTE) occupants for the project should be calculated. A full time occupant (eight-hour work day) has an FTE value of 1.0; a part-time occupant has an FTE value based on work hours per day divided by eight. Then, the total number of occupants may be calculated by adding the number of FTE occupants, peak transients, and residents. While some credit requirements are based on FTE numbers, others are based on total number of occupants. The LEED occupancy numbers will be required for calculations in Sustainable Sites credits: SS C4.2 and SS C4.3 and Water Efficiency credits: WE P1 and WE C3.

Table 3.4 shows an example of an occupancy calculation for a LEED project. The project is mixed-use and contains both residential and office components. The LEED PM may use a similar table to calculate the occupancy.

TABLE 3.4   SAMPLE LEED OCCUPANCY CALCULATION

FTE OCCUPANTS	NO. OF STAFF	NO. OF HOURS	FTE[§]
Full-Time Staff	100	8	100
Part-Time Staff	50	4	25
Total FTE Occupants			**125**
RESIDENTS	**NO. OF UNITS**	**NO. OF OCCUPANTS[§§]**	**TOTAL**
One Bedroom	100	2	200
Two Bedroom	75	3	225
Three Bedroom	50	4	200
Total Residents			**625**
Transient Occupants			**Daily Total[§§§]**
Total Visitors			15
Total Number of Occupants[§§§§]			**765**

[§] Full Time Equivalent (FTE) = [number of staff × (number of hours/8)]

[§§] Number of Occupants = number of units × number of occupants per unit (residential occupancy assumption according to LEED: assume 2 occupants in 1-bedroom unit, 3 occupants in 2-bedroom unit, 4 occupants in 3-bedroom unit, unless specific project type like student housing has different number of occupants per unit. If a project chooses to deviate from the LEED residential occupancy assumptions, then a suitable document, such as a lease agreement or letter from management, needs to be provided to validate the deviation.)

[§§§] Daily Total of Transient Occupants is an estimated number obtained from owner or building management.

[§§§§] Total Number of Occupants = sum of total FTE + total residents + total visitors

ACTIVITY 1.5—CREATE PROJECT TEAM DIRECTORY

After calculating the LEED occupancy numbers, the next step is to obtain contact information for all the team members involved with the project and compile it into a project directory. This directory will be helpful, later on, for assigning LEED responsibilities to each team member. Table 3.5 shows an example of a directory compiled for a project. The LEED PM may use a similar format to compile the team information.

ACTIVITY 1.6—OBTAIN PRELIMINARY PROJECT SCHEDULE

The next step is to obtain a preliminary design and construction schedule from the owner or the architect. This will be used to incorporate LEED milestones into the schedule (refer to Chapter 2 on how to integrate). At this stage, a simple timeline as shown in Table 3.6 should suffice.

TABLE 3.5   SAMPLE PROJECT TEAM DIRECTORY

DISCIPLINE	COMPANY	CONTACT PERSON	E-MAIL	PHONE
Owner	Company Name	John Smith	john@companyname.com	
Bldg. Mgmt.				
Architect				
Civil				
Structural				
MEP				
Landscape				
Contractor PM				
LEED				

ACTIVITY 1.7—RESEARCH *GREEN* INCENTIVES

The next recommended activity is to conduct a research of all available incentives for the project. It is important to do this, as the incentive requirements may influence the owner's project requirements as well as design decisions. This research may be performed by the owner or the LEED PM.

There are several federal, state, or local incentives that are currently driving the green building/LEED industry. These incentives vary from tax deductions and tax credits to cash incentives, rebates, and fast-track permitting. For instance, the State of Maryland provides an income tax credit to owners or tenants of green buildings. The credit equals eight percent of the allowable costs ($120 per square foot of the base building/$60 per square foot of the tenant space) for green buildings. Similarly, the Chicago Green Permit Program reduces the permitting process for developers and owners who build *green* to less than 30 business days and, in some cases, less than 15 days.[7] A

TABLE 3.6   SAMPLE PRELIMINARY PROJECT SCHEDULE

DATE	EVENT DESCRIPTION
April 15, 2009	Schematic Design Phase Begin
July 31, 2009	Schematic Design Phase Complete
December 10, 2009	Design Development Phase Complete
June 1, 2010	Construction Documents Complete
October 1, 2010	Bid and Negotiate Construction Contract
November 1, 2010	Pre-Construction Meeting
November 15, 2010	Construction Work Commencement
August 15, 2010	Construction Work Complete

TABLE 3.7   SAMPLE *GREEN* INCENTIVES RESEARCH DATA

NO.	CATEGORY	DESCRIPTION	PURSUE (Y/N)
1.	Level	Federal	N
	Incentive	Energy-Efficient Commercial Buildings Tax Deduction	
	Type	Corporate Deduction	
	Amount	$0.30–$1.80 per square foot, depending on technology and amount of energy reduction	
	Requirements	Install (1) interior lighting; (2) building envelope; and/or (3) heating, cooling, ventilation, or hot water systems that reduce the building's total energy and power cost by 50% or more in comparison to a building meeting minimum requirements set by ASHRAE Standard 90.1-2001.	
2.	Level	Local	Y
	Incentive	Austin Energy—Commercial New Construction Efficiency Rebates	
	Type	Utility Rebate Program	
	Amount	25% of cost of efficiency measures; Max $200,000 per site, per fiscal year	
	Requirements	Projects must exceed 2006 IECC energy code efficiency levels	

good resource to locate any incentives that may be available at both federal and state level is the website of the Database of State Incentives for Renewables and Efficiency (DSIRE)—http://www.dsireusa.org.

Table 3.7 is an example of research information on various incentives compiled for a project. In this case, the owner decided to pursue the local incentive and incorporated exceeding the 2006 IECC energy code as a key energy efficiency goal in the OPR. The design team also developed the design in accordance with the code requirements.

ACTIVITY 1.8—COMPILE ALL DATA

The final activity of this step is to compile all the information (Tables 3.1 to 3.7) into a common project database that is available to and accessible by all team members. As the design progresses, team members should inform the LEED PM about any changes to this information. The use of this common information database by all team members to perform credit calculations will help in avoiding inaccurate or inconsistent data across LEED credits. This will ensure that the documentation application and review process with GBCI occurs smoothly. This database will also help in completing the rest of the steps of this stage of the process.

Step 2: Prepare a Preliminary LEED Assessment

Objective	The objective of this step is to prepare a preliminary assessment of the project's status with respect to meeting LEED credit requirements. The assessment will provide direction on the strategies the team must use to integrate the credit requirements into the design and construction of the building.
Recommended Activities	2.1 Review Credit Requirements to Assess Feasibility. 2.2 Identify Potential Implementation Strategies. 2.3 Identify Next Steps. 2.4 Prepare a Credit Responsibility Matrix.
Roles and Responsibilities	LEED PM: Perform preliminary assessment.
Duration	3 weeks
Outcome	Preliminary LEED assessment completed. Target LEED score established.

ACTIVITY 2.1—REVIEW CREDIT REQUIREMENTS TO ASSESS FEASIBILITY

The purpose of the LEED credit framework is to guide the design team in making *green*, energy efficient, and sustainable choices. It is important to use this framework right from the early design stages to make informed design decisions. By doing this, LEED credit requirements will become integral to the design and not a superficial addition. The first step in this direction is to prepare a preliminary credit-by-credit assessment of the project, using the LEED checklist to determine which credits should be pursued, how those will affect design decisions, and what strategies will help in achieving them. The assessment will also help determine a target LEED score that the design team should strive to achieve. All the activities in this step are primarily for the designated LEED PM. However, input may be required from the owner or other design team members as well.

In order to perform this preliminary assessment, the first step is to review each credit's requirements and identify the appropriate *case* or *option* that will be applicable to the project. According to the *LEED Reference Guide, case* refers to the broader project criteria, such as whether the project is commercial or residential, and *option* refers to the various ways in which the credit requirement may be demonstrated. Then, based on the information available for the project, the LEED PM should assess each credit and determine its feasibility. The credits may be assigned to three categories: *Yes, No,* and *Maybe.* Those credits that are likely to be achieved will fall under the *Yes* category. Those credits that require further investigation, calculations, or input from the team will fall under the *Maybe* category. The credits in the *No* category will be those that are not applicable or viable for the project.

Example of a Credit Assessment

Let us assess the feasibility of a Sustainable Sites credit for a project: *SS C4.1, Alternative Transportation—Public Transportation Access.*

The intent of this credit, according to the *LEED Reference Guide,* is "to reduce pollution and land development impacts from automobile use."[8] There are two options available to meet the intent of this credit.

Option 1: Rail Station Proximity   "Locate the project within ½ mile walking distance (measured from a main building entrance) of an existing or planned and funded commuter rail, light rail, or subway station."[9]

Option 2: Bus Stop Proximity   "Locate the project within ¼ mile walking distance (measured from a main building entrance) of 1 or more stops for 2 or more public, campus, or private bus lines usable by building occupants."[10]

The first step is to identify whether there are any rail stations and/or bus stops in the vicinity of the site. This may be done by referring to the local transport authority's website or using software applications, such as Google Maps or Google Earth. Most of these tools also help measure distances.

If there are no rail stations or bus stops within the required distance, the credit should be assigned to the *No* category. The assessment may be stated as: "According to investigation of the site vicinity and local transportation network, the project does not meet the requirements of the credit, as there are no rail stations or bus stops within the required distances."

If there is a rail station within ½ mile or there are bus stops within ¼ mile, then assign the credit to the *Yes* category. The assessment may be stated as: "According to preliminary investigation of the site vicinity and local transportation network, there are 2 bus stops with 3 bus lines within ¼ mile of the project site. Therefore, the project meets the requirements of the credit via Option 2."

A third situation could be where there are no bus stops or rail stations within the required walking distances, but the owner is considering providing a shuttle service to the nearest transit center or is in discussion with the city's transit authority about the possibility of adding a stop to serve the occupants of the building. In this situation, the project may be able to achieve the credit by procuring all necessary approvals to demonstrate compliance with the credit. In this case, the credit can be assigned to the *Maybe* category.

ACTIVITY 2.2—IDENTIFY POTENTIAL IMPLEMENTATION STRATEGIES

The objective of this activity is to identify strategies that may be potentially implemented by the team to integrate LEED credit requirements into the design. These strategies should not be considered prescriptive but viewed as a starting point for further discussion among team members (the LEED Vision Workshop described in Step 4 will facilitate this discussion). Each strategy suggested will have an impact on the project design, schedule, and overall budget. Below are some questions that the LEED PM should consider while exploring these strategies.

Is the Strategy Applicable in the Context of the Project?

This question relates to the applicability of a strategy to the project scope, type of building, climate, and location. For instance, while recycling construction waste for implementing MR C2 credit may be fairly easy to execute, and perhaps cheaper in certain areas, some project locations may not have access to any construction waste recycling facilities. In such cases, the credit related to construction waste management may not be feasible, and there may be a need to investigate alternative recycling solutions.

What Are Some of the Intangible Benefits?

Both tangible and intangible benefits should be considered while evaluating strategies. It is common for teams to look at the tangible benefits and overlook the intangible benefits. Green buildings offer many intangible benefits, such as improved health, higher motivation, and increased occupant productivity, which could translate to savings for all parties involved. As mentioned in Chapter 1, a study conducted by the University of San Diego and the commercial real estate broker, CB Richard Ellis Group found that tenants in green buildings are more productive based on two measures: the average number of tenant sick days and the change in productivity. Respondents reported an average of 2.88 fewer sick days in their current green office versus their previous non-green office. About 55% of respondents indicated that employee productivity had improved. Based on an average tenant salary of $100,000, an office space of 250 square feet per worker, and 250 workdays a year, the decrease in sick days translated into a net impact of nearly $5.00 per square foot per year. The increase in productivity translated into a net impact of about $20 per square foot. The study also showed that green buildings have 3.5% lower vacancy rates and 13% higher rental rates than the market.[11]

Therefore, even though an immediate dollar value may not be apportioned to a strategy, it should be carefully evaluated for any intangible benefits that it might offer.

Are There Known Technology/Expertise/Skills to Implement the Solution?

There has been a rapid development of new *green* technologies, products, and materials over the last decade. A lot of technologies have undergone major improvements and modifications since the time they were first introduced into the market and sufficient expertise is becoming available to implement these new technologies. However, caution needs to be exercised before suggesting a completely new product or technology without sufficient research on the capabilities and performance of the product.

Are There Any Potential Synergies?

Another important aspect to consider is whether a strategy explores the synergies between various credits. Strategies should be evaluated for their potential in having a far-reaching impact across various credit categories, instead of having a minor impact in just one area. For instance, a single solution of harvesting rainwater has a far-reaching impact on a number of aspects, such as controlling stormwater quantity (SS C6.1), controlling stormwater quality (SS C6.2), reducing potable water use for landscape (WE C1), or reducing potable water for sewage conveyance (WE C2).

Is the Strategy Cost-Effective?

The answer to this question may not be straightforward. When evaluating a solution for its cost-effectiveness, look at both the initial costs and the long-term benefits associated with it. For instance, harvesting rainwater and reusing it for the purposes of landscape irrigation might be perceived as a capital intensive exercise, but the savings from reduced water bills will more than offset that initial expense in a matter of few years. A quick back-of-the-envelope calculation should be sufficient at this time (detailed cost-benefit analysis is described in Chapter 4).

ACTIVITY 2.3—IDENTIFY NEXT STEPS

The final component of the preliminary assessment is to identify the *next steps* or action items for the design team. These next steps will serve as a roadmap for further implementation of the credit requirements and become discussion items at the LEED Vision Workshop, which is described in Step 4. The next steps may be composed of tasks, such as performing research work, investigating design options, or performing detailed analyses to determine the feasibility of credits and finalize implementation strategies.

   Figure 3.3 shows an example of a preliminary LEED assessment completed for a project. All the aforementioned activities of assessing a credit (Activity 2.1), identifying potential implementation strategies (Activity 2.2), and identifying *next steps* (Activity 2.3) were compiled into a single spreadsheet.

ACTIVITY 2.4—PREPARE A CREDIT RESPONSIBILITY MATRIX

The final activity of this step is to prepare a responsibility matrix identifying the team members responsible for implementation of each credit. The LEED PM may use the general team roles and responsibilities discussed in detail in Chapter 2 as a guideline

Sustainable Sites		Max.	Yes	?	No
Prereq 1	**Construction Activity Pollution Prevention**	P	Y		
Requirements	Create and implement an erosion and sedimentation control plan for all construction activities associated with the project. The plan must conform to 2003 EPA Construction General Permit's NPDES program or local codes, whichever is more stringent				
Assessment	This is standard practice and requirement for sites over an acre. It is expected that erosion control plans will be provided by the civil engineer and contractor will implement the measures.				
Potential Implementation Strategies	Temporary Seeding, Silt Fence, Sediment Basin etc.				
Next Steps	Civil engineer to evaluate which code is more stringent; assess site conditions and propose control measures; confirm all requirements will be met				
Credit 1	**Site Selection**	1		1	
Requirements	Site should not be prime farmland, elevation is not lower than 5 feet above the 100-yr flood plain, is not identified as habitat for any species, is not within 100 feet of wetlands, is not previously undeveloped land within 50 feet of water body, was not previously a public parkland				
Assessment	Based on preliminary data provided, site meets all criteria except presence of wetlands				
Potential Implementation Strategies	Not applicable				
Next Steps	Investigate whether site meets all criteria or not				
Credit 2	**Development Density and Community Connectivity**	5		5	
Requirements	Option-2 Community Connectivity: located on previously developed site, within 1/2 mile of residential neighborhood with average density of 10 units per acre, is within 1/2 mile radius of at least 10 community services, has pedestrian access between the building and services				
Assessment	Site is previously developed and has 10 basic community services within 1/2 mile radius. Need to establish pedestrian access to those services and also investigate if there is a residential neighborhood with a density of 10 units/acre				
Potential Implementation Strategies	Design pathways and site entry points to facilitate pedestrian access to community services, during site planning				
Next Steps	Investigate densities of residential neighborhoods within 1/2 mile radius				
Credit 3	**Brownfield Redevelopment**	1			1

Activity 2.1 Activity 2.2 Activity 2.3

Figure 3.3 Example of a preliminary LEED assessment.
Credit: Green Potential LLC (www.green-potential.com).

in preparing this matrix. Since the implementation of a single credit might involve multiple parties, it is important to further classify the responsibilities based on who will be leading the implementation of the credit and who will be supporting that group. Figures 3.4 and 3.5 show an example of a responsibility matrix prepared for a project. If all members of the project team are known, then specific firm names, in lieu of the discipline they serve, may be used.

LEED NC 2009 Credit Responsibilty Matrix Project Name: Date: Legend: L = Lead (responsible for implementation and documentation) S = Support (provide supporting information, drawings, calculationse etc.)	Points possible	Owner	Bldg Manager	Architect	MEP Engineer	Civil	Landscape	Interior	General Contractor	CxA	LEED Consultant
Sustainable Sites	26										
Prereq 1 Construction Activity Pollution Prevention						L			S		S
Credit 1 Site Selection	1	S				S					L
Credit 2 Development Density and Community Connectivity	5			S							L
Credit 3 Brownfield Redevelopment	1	S				S					L
Credit 4.1 Alternative Transportation—Public Transportation Access	6										L
Credit 4.2 Alternative Transportation—Bicycle Storage and Changing Rooms	1			L							S
Credit 4.3 Alternative Transportation—Low-Emitting and Fuel-Efficient Vehicles	3		S	L							S
Credit 4.4 Alternative Transportation—Parking Capacity	2			L							S
Credit 5.1 Site Development—Protect or Restore Habitat	1						L				S
Credit 5.2 Site Development—Maximize Open Space	1						L				S
Credit 6.1 Stormwater Design—Quantity Control	1						L				S
Credit 6.2 Stormwater Design—Quality Control	1						L				S
Credit 7.1 Heat Island Effect—Non-roof	1			L							S
Credit 7.2 Heat Island Effect—Roof	1			L							S
Credit 8 Light Pollution Reduction	1			S	L						S
Water Efficiency	10										
Prereq 1 Water Use Reduction—20% Reduction				S	L						S
Credit 1 Water Efficient Landscaping	2 to 4						L				S
Credit 2 Innovative Wastewater Technologies	2			S	L						S
Credit 3 Water Use Reduction	2 to 4			S	L						S
Energy and Atmosphere	35										
Prereq 1 Fundamental Commissioning of Building Energy Systems					S				S	L	S
Prereq 2 Minimum Energy Performance					L						S
Prereq 3 Fundamental Refrigerant Management					L						S
Credit 1 Optimize Energy Performance	1 to 19				L						S
Credit 2 On-Site Renewable Energy	1 to 7			S	L						S
Credit 3 Enhanced Commissioning	2				S				S	L	S
Credit 4 Enhanced Refrigerant Management	2				L						S
Credit 5 Measurement and Verification	3				L				S		S
Credit 6 Green Power	2	S	S		S						L

Figure 3.4 Example of a typical credit responsibility matrix—part 1.

Credit: Green Potential LLC (www.green-potential.com).

LEED NC 2009 Credit Responsibilty Matrix Project Name: Date: Legend: L = Lead (responsible for implementation and documentation) S = Support (provide supporting information, drawings, calculationse etc.)		Points possible	Owner	Bldg Manager	Architect	MEP Engineer	Civil	Landscape	Interior	General Contractor	CxA	LEED Consultant
Materials and Resources		14										
Prereq 1	Storage and Collection of Recyclables			S	S							L
Credit 1.1	Building Reuse—Maintain Existing Walls, Floors, and Roof	1 to 3								L		S
Credit 1.2	Building Reuse—Maintain 50% of Interior Non-Structural Elements	1								L		S
Credit 2	Construction Waste Management	1 to 2								L		S
Credit 3	Materials Reuse	1 to 2			S					L		S
Credit 4	Recycled Content	1 to 2			S					L		S
Credit 5	Regional Materials	1 to 2			S					L		S
Credit 6	Rapidly Renewable Materials	1			S					L		S
Credit 7	Certified Wood	1			S					L		S
Indoor Environmental Quality		15										
Prereq 1	Minimum Indoor Air Quality Performance					L						S
Prereq 2	Environmental Tobacco Smoke (ETS) Control			S	L							S
Credit 1	Outdoor Air Delivery Monitoring	1				L						S
Credit 2	Increased Ventilation	1				L						S
Credit 3.1	Construction IAQ Management Plan—During Construction	1								L		S
Credit 3.2	Construction IAQ Management Plan—Before Occupancy	1	S							L		S
Credit 4.1	Low-Emitting Materials—Adhesives and Sealants	1							S	L		S
Credit 4.2	Low-Emitting Materials—Paints and Coatings	1							S	L		S
Credit 4.3	Low-Emitting Materials—Flooring Systems	1							S	L		S
Credit 4.4	Low-Emitting Materials—Composite Wood and Agrifiber Products	1							S	L		S
Credit 5	Indoor Chemical and Pollutant Source Control	1			S	L						S
Credit 6.1	Controllability of Systems—Lighting	1				L			S			S
Credit 6.2	Controllability of Systems—Thermal Comfort	1				L			S			S
Credit 7.1	Thermal Comfort—Design	1				L						S
Credit 7.2	Thermal Comfort—Verification	1				S						L
Credit 8.1	Daylight and Views—Daylight	1			S							L
Credit 8.2	Daylight and Views—Views	1			S							L
Innovation and Design Process		6										
Credit 1.1	Innovation in Design: Specific Title	1										L
Credit 1.2	Innovation in Design: Specific Title	1										L
Credit 1.3	Innovation in Design: Specific Title	1										L
Credit 1.4	Innovation in Design: Specific Title	1										L
Credit 1.5	Innovation in Design: Specific Title	1										L
Credit 2	LEED Accredited Professional	1										L
Regional Priority Credits		4										
Credit 1.1	Regional Priority: Specific Credit	1										L
Credit 1.2	Regional Priority: Specific Credit	1										L
Credit 1.3	Regional Priority: Specific Credit	1										L
Credit 1.4	Regional Priority: Specific Credit	1										L
Total		110										

Figure 3.5   Example of a typical credit responsibility matrix—part 2.

Credit: Green Potential LLC (www.green-potential.com).

Step 3: Register the Project on LEED Online

Objective	The objective of this step is to officially begin the process of LEED certification by registering the project with the Green Building Certification Institute (GBCI).
Recommended Activities	3.1 Register the Project on LEED Online. 3.2 Assign Credit Responsibilities on LEED Online.
Roles and Responsibilities	LEED PM: Register the Project. Owner/Developer: Provide information requested.
Duration	1 week
Outcome	Project registered on LEED Online.

ACTIVITY 3.1—REGISTER PROJECT ON *LEED ONLINE*

Project registration marks the official beginning of the process of LEED certification by the Green Building Certification Institute (GBCI). This is an important step in establishing contact with GBCI and gaining access to their electronic document submission system called LEED Online.

Project may be registered at *www.leedonline.com* by completing an application and paying the requested registration fees. It is required to have an account with the website prior to registering the project. The LEED PM should create a site username and password if he/she does not already have one. This is the same account information that is typically used to log into the USGBC or GBCI website. The information that is requested in the application is preliminary, project-related data and can be changed later on by the project administrator (by default the team member who registers the project becomes the administrator) at any point in the certification process. There are multiple payment options: check, credit card, or wire transfer. USGBC member organizations can avail themselves of a discount on both registration and certification fees. Table 3.8 lists the specific information that will be needed to complete the registration application.

What Is LEED Online?

LEED Online is the primary resource through which teams can manage the LEED registration and certification process. According to GBCI, with LEED Online, project managers can do the following:

TABLE 3.8 KEY INFORMATION REQUIRED FOR REGISTERING A PROJECT		
NO.	**INFORMATION REQUESTED**	**DESCRIPTION**
1	Project Title	This is the name of the project. It will appear in LEED Online and in the LEED Project Directory.
2	Project Address	This is the address where the project is or will be located.
3	Anticipated Construction Start Date	If not certain, estimated date should be entered.
4	Anticipated Construction End Date	If not certain, an estimated date should be entered.
5	Project Square Footage	This is the gross square feet of the entire project (all buildings going in for LEED certification).
6	Confidentiality	The LEED PM should indicate whether or not the project is confidential. If not confidential, then the information about the project could be shared and used by USGBC in case studies, exhibits, educational, and promotional materials, etc.
7	Anticipated Certification Level	An anticipated certification level based on the preliminary assessment may be entered.
8	Project Owner Information	Information about the owner of the project should be provided.

- Submit documentation to USGBC for review.
- Document compliance with LEED credit requirements.
- Coordinate resources among project team members.
- Submit technical inquiries and credit interpretation requests regarding LEED Credits.
- Track progress towards LEED certification.

The person who registers the project is designated as the LEED Project Administrator by default. The LEED project administrator has access to all areas of the project on LEED Online. He or she has capabilities of adding team members, removing them, assigning roles, attempting credits, and eventually submitting documentation to GBCI. It is preferable for the LEED PM to be the LEED project administrator or at least have similar level of access.

The new version of LEED Online has features and capabilities that are different from the previous versions. Some of the key differences listed on the GBCI website are described below:

Rating system decision   The registration phase of the new LEED Online now provides assistance to project teams in deciding the LEED rating system that is best suited to their project type.

Team member administration   The new LEED Online allows for project administrators to assign credits by team member name rather than by project role, as was the case previously.

Status indicators and timeline   The new version of LEED Online displays specific dates associated with each phase and step and includes target dates on which each review will be returned to the project team.

Mid-stream communication   During a LEED review, a mid-review clarification page allows the LEED reviewer to contact the project team through the system if any minor clarifications are needed to complete the review.

Data linkages   The new LEED Online shares data (like areas, FTE, etc.) within a project, auto-filling it in all appropriate places after the user enters it the first time. Override options are available in case the values truly need to differ.

Automatic data checks   The online system alerts users when there are errors in data between credits and gives them a final chance to correct the error before submitting the application.

ACTIVITY 3.2—ASSIGN CREDIT RESPONSIBILITIES ONLINE

After registration is complete, the project will move into the *Design Preliminary Application Phase* of the GBCI process. The LEED PM should use the preliminary assessment as a basis to attempt the credits that the project will likely pursue and assign them to the responsible team member. This can be done in the *LEED Scorecard* tab of the site where all the credits are listed and the project administrator can *attempt* credits as needed. The LEED PM should then assign the attempted credits to the responsible team members and invite them to join the project online. Invitations can be sent from the *Team Administration* portion of the site. Each team member will get an e-mail with instructions to join the project along with a project access code.

Step 4: Organize a LEED Vision Workshop

Objective	The objective of this step is to bring together all the various stakeholders in the project and facilitate discussion of various design possibilities using the LEED rating system framework.
Recommended Activities	4.1 Prepare Workshop Outline. 4.2 Facilitate the Workshop.
Responsibilities	LEED PM: Organize and facilitate workshop. Design Team: Participate in workshop.
Duration	3 weeks
Outcome	Green design strategies identified; LEED target score established; Team member responsibilities defined.

ACTIVITY 4.1—PREPARE WORKSHOP OUTLINE

The purpose of the LEED Vision Workshop is to provide a platform for all the team members to come together and discuss ideas, synergies, and strategies for the design of the project using the preliminary LEED assessment (prepared as part of Step 2) as a starting point. The LEED PM's role is to facilitate this workshop by allowing open communication and discussion between the various team members. The objectives of this workshop are to:

- Kick-off the LEED design process.
- Provide a platform for all the people involved with the project who can influence design decisions to meet and begin planning the project.
- Set *green* and energy performance goals and specifically a LEED goal for the project.
- Discuss ideas, issues, and concerns for the project to help avoid later redesign activities.
- Delegate LEED responsibilities to the various project team members.
- Initiate an integrated design process.
- Identify partners, available incentives, and potential collaborations that can provide expertise, funding, credibility, and support.
- Set a project schedule and budget that all team members feel comfortable following.
- Encourage enthusiasm for the project and provide a sense of project ownership to each team member.

The LEED PM should prepare an agenda for the workshop outlining the goals, duration, and items for discussion. The length of the workshop may range from a four to eight hour session, depending on the project size and the design phase it is in. Below is an example of an outline prepared for a similar workshop. In this agenda, time allotted for each activity has been specified in lieu of mentioning the actual time.

ACTIVITY 4.2—FACILITATE THE WORKSHOP

The LEED Vision Workshop may be broken down into two sessions:

Session I: Educational/Informational Session
Session II: Interactive/Discussion-Oriented Session

Sample Outline for LEED Vision Workshop

Project Name: YZ Corp Office
LEED Vision Workshop Agenda

Date: Friday, February 5, 2010   Time: 10:00 AM–2:00 PM Central Time
Location: Conference Room at Offices of XY & Partners, Architects
Address: 1000 Congress Avenue, Suite 200, Atlanta, GA

Goals of the Meeting:

1 Understand the LEED process—registration, rating, schedule.
2 Discuss preliminary LEED assessment.
3 Discuss owners project requirements (OPR).
4 Discuss design strategies and recommendations.
5 Discuss team member responsibilities.
6 Establish LEED score target.
7 Future steps—immediate action items, project schedule.

Agenda:

1 Introductions	10 minutes
2 LEED overview, OPR	60 minutes
3 Lunch	45 minutes
4 LEED checklist, credit assignment, schedule	80 minutes
5 Future steps and wrap up	15 minutes

Miscellaneous:

1 Please indicate any special diet requests.
2 Please confirm number of attendees from your firm by Jan. 29, 2010.

Both sessions may be equal in time duration, but it is preferable to have more time for the second session than the first one. On an average, for a total meeting time of four hours, one hour may be spent on the educational session and the remaining time on the interactive session. Figure 3.6 illustrates the two session framework of the workshop.

Session I: Educational/Informational Session

The first session will be primarily educational and informative in nature presenting the team members with an overview of the LEED process, key project information, and overview of LEED Online. Although the session is intended to be educational, there is sufficient scope for interaction in the form of questions asked by the team members. The following items should be presented as part of this session.

LEED process overview   The LEED PM should present an overview of the LEED process and how it will be managed. The overview should include the certification steps identified by the GBCI and the LEED milestones that the team should be aware of.

Figure 3.6 LEED Vision Workshop Framework.

Refer to Figure 2.7 in Chapter 2 for an illustration of the process and how it aligns with the architectural design phases.

Project information database   The LEED PM should present the project information database created as part of Step 1. The purpose of communicating this information is to ensure that all project team members use the same set of numbers and data to perform their calculations.

LEED Online overview   Gaining access to and using LEED Online is user-friendly and self-explanatory, but it will be beneficial to organize a live demonstration of the LEED Online documentation management system to ensure all team members are aware of the online tool.

Case study   Depending on the time allotted for this session, the LEED PM may present a case study of a similar project, explaining how the project was able to attain its LEED goals. Other items that may be included in this session are videos and presentations by other speakers that may motivate and encourage the team.

Session II: Interactive Session

The second session will be an interactive session where all the team members may participate and discuss ideas for further implementation. The following items should be discussed as part of this session.

Preliminary LEED assessment   The LEED PM should present the preliminary assessment that was prepared as part of Step 2 and facilitate discussion among the team members. As the facilitator, the LEED PM may introduce each credit, explain its requirements, and describe the preliminary assessment on the feasibility of the credit. After this, he/she should allow discussion among the team members to get their feedback. At the end of each credit's discussion, the LEED PM should make note of any changes in the credit status and determine the total points to establish a target rating that the project will pursue.

Credit responsibility matrix   The LEED PM should discuss the credit responsibility matrix prepared as part of Step 2 to ensure everyone is aware of their respective responsibilities. Any changes or concerns on the matrix should be noted, and a consensus on team member responsibilities should be obtained.

LEED schedule and communication protocols   Finally, the schedule and LEED milestones should be discussed and key deliverables should be identified (refer to Figure 2.11 in Chapter 2 for an illustration of how to integrate LEED milestones into the schedule). A schedule of meetings showing when the entire team would regroup to discuss items should be established. These could vary from biweekly phone calls to once a month face-to-face meetings.

The cornerstone of the LEED process is integrated design, which involves a collaborative effort from all team members. However, this may also lead to confusion on how information is communicated. The communication protocols between the various groups

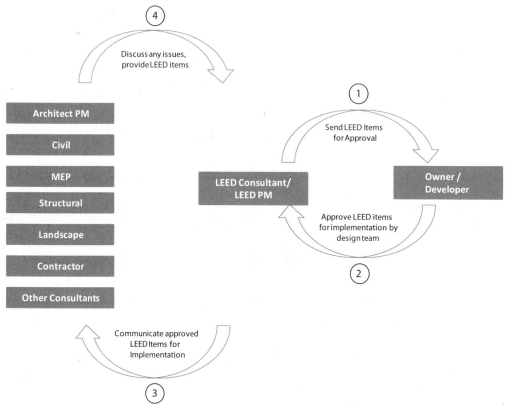

Figure 3.7   Example of a communication protocol.
Credit: Green Potential LLC (www.green-potential.com).

of people involved should be discussed. Key contact people for each group should be identified, and how information will flow from one group to another should be defined. Figure 3.7 shows an example of a communication protocol developed for a project. In this case, the LEED PM sent all LEED-related items to the owner for approval. After receiving the owner's approval, the information was passed to the design team. The design team then communicated any concerns to the LEED PM, who reviewed them and communicated to the owner. Overall, the LEED PM was responsible for coordinating the information between various groups of people.

After the workshop is over, the LEED PM should prepare notes identifying key discussion items and action items. The objective of these notes is not only to have a record of the proceedings but also to become a valuable document that will form the basis of discussion at all future meetings. Figure 3.8 shows an example of notes from a LEED Vision Workshop. The format shown here is an extension of the preliminary LEED assessment that was prepared as part of Step 2. Columns were added to the assessment spreadsheet to indicate the responsible team member along with items discussed at the meeting.

Sustainable Sites		Max	Yes	?	No	Responsibilities		Meeting Notes
		P	Y			Lead	Support	
Prereq 1	Construction Activity Pollution Prevention	P	Y			Civil	LEED Cons.	LEED Vision Workshop- 02/05/10
Requirements	Create and implement an erosion and sedimentation control plan for all construction activities associated with the project. The plan must conform to 2003 EPA Construction General Permit's NPDES program or local codes, whichever is more stringent							Civil Engineer confirmed that the ESC Plan will be included in drawings
Assessment	This is standard practice and requirement for sites over an acre. It is expected that erosion control plans will be provided by the civil engineer and contractor will implement the measures.							
Potential Implementation Strategies	Temporary Seeding, Silt Fence, Sediment Basin etc.							
Next Steps	Civil engineer to evaluate which code is more stringent; assess site conditions and propose control measures; confirm all requirements will be met							
Credit 1	Site Selection	1			1	LEED Cons.	Owner, Civil	LEED Vision Workshop- 02/05/10
Requirements	Site should not be prime farmland, elevation is not lower than 5 feet above the 100-yr flood plain, is not identified as habitat for any species, is not within 100 feet of wetlands, is not previously undeveloped land within 50 feet of water body, was not previously a public parkland							It was realized that the project doesn't meet the criteria set forth by the credit because of the presence of wetlands; this credit will not be feasible
Assessment	Based on preliminary data provided, site meets all criteria except presence of wetlands							
Potential Implementation Strategies	Not applicable							
Next Steps	Investigate whether site meets all criterial or not							
Credit 2	Development Density and Community Connectivity	5		5		LEED Cons.	Owner, Civil	LEED Vision Workshop- 02/05/10
Requirements	Option-2 Community Connectivity: located on previously developed site, within 1/2 mile of residential neighborhood with average density of 10 units per acre, is within 1/2 mile radius of at least 10 community services, has pedestrian access between the building and services							It was confirmed by owners that the site does have pedestrian access to community services and there is a residential neighborhood with the required density, so this credit can be achieved. LEED Consultant will prepare the required documents
Assessment	Site is previously developed and has 10 basic community services within 1/2 mile radius. Need to establish pedestrian access to those services and also investigate if there is a residential neighborhood with a density of 10 units/acre							
Potential Implementation Strategies	Design pathways and site entry points to facilitate pedestrian access to community services, during site planning							
Next Steps	Investigate densities of residential neighborhoods within 1/2 mile radius							
Credit 3	Brownfield Redevelopment	1			1	Lead	Support	LEED Vision Workshop- 02/05/10

Figure 3.8 Example of LEED Vision Workshop notes.
Credit: Green Potential LLC (www.green-potential.com).

95

Step 5: Prepare a LEED Project Workbook

Objective	This step marks the culmination of all the steps and activities in Stage I of the LEED project management process. The objective of this step is to create a ready-reference LEED Project Workbook that the team can refer to on an ongoing basis as they develop the design.
Recommended Activities	5.1 Update Preliminary Documents. 5.2 Prepare LEED Project Workbook.
Roles and Responsibilities	LEED PM: Prepare LEED Project Workbook. Design Team: Refer to and implement items in Project Workbook.
Duration	3 weeks
Outcome	LEED Project Workbook compiled.

ACTIVITY 5.1—UPDATE PRELIMINARY DOCUMENTS

Update preliminary LEED assessment   The LEED PM should make changes to the preliminary LEED assessment based on the comments and discussion at the LEED Vision Workshop. The next steps in the assessment checklist will serve as a guideline for implementation by the team.

Finalize LEED schedule   The LEED PM should review the previously prepared LEED schedule and make appropriate changes to the milestones, dates, and planned meeting schedule based on discussions at the workshop.

Update responsibility matrix   The LEED PM should update the responsibility matrix based on discussions from the workshop. This matrix may be combined with the preliminary assessment checklist, or it may be kept as a separate document.

ACTIVITY 5.2—PREPARE LEED PROJECT WORKBOOK

The LEED Project Workbook is a compilation of documents that were prepared as part of activities completed in all the previous steps. The purpose of compiling all information into a single binder is to provide the team with key information regarding the project that they can refer to as they proceed with the design. It will serve as a project-specific reference manual for the team. The LEED Project Workbook may include:

1 Project team directory

2 Preliminary LEED assessment with LEED target score

3 Schedule with LEED milestones

4 Responsibility matrix

5 Owner's project requirements (OPR)

6 Applicable incentives list

7 Preliminary site data with LEED project boundary exhibit

8 Preliminary project data

9 LEED occupancy count (including total FTEs)

All the information was prepared as part of Steps 1 to 4. However, before compiling all documents, the LEED PM should ensure that any changes that may have been discussed at the workshop are incorporated and that the design team receives the most updated version of all the documents. This LEED Project Workbook may be made available electronically to all team members.

Summary

In summary, the first stage of the LEED project management process aligns with the pre-schematic design or early schematic design phase of the architectural design process. It is during this stage that basic project information is collected and target LEED goals for the project are established. LEED certification process and specific team member responsibilities are also defined. The average time to complete the various steps and activities that comprise this stage is estimated at nine weeks, as is shown in the process schedule in Figure 3.9. However, this duration may vary from project to project. Project managers may use the checklist provided at the end of this chapter (Figure 3.10) to keep track of activities that have been completed.

Notes

[1] USGBC, *LEED Reference Guide for Green Building Design and Construction* (USGBC, 2009), 19.

[2] Ibid.

[3] Ibid.

[4] Ibid., 37.

[5] Ibid., 37.

[6] USGBC, *LEED Reference Guide for Green Building Design and Construction* (USGBC, 2009), 2.

[7] AIA, *Local Leaders in Sustainability: Green Incentives* (AIA, 2007)

[8] USGBC, *LEED Reference Guide for Green Building Design and Construction* (USGBC, 2009), 41.

[9] Ibid., 41.

[10] Ibid., 41.

[11] Norm G. Miller, Dave Pogue, *Do Green Buildings Make Dollars and Sense?* (Burnham-Moores Center for Real Estate University of San Diego, CB Richard Ellis, 2009).

		Duration*	Month 1				Month 2				Month 3			
Stage I	Project Definition	12 weeks	WK1	WK2	WK3	WK4	WK5	WK6	WK7	WK8	WK9	WK10	WK11	WK12
Step 1	**Collect and compile preliminary data**	**3 weeks**												
1.1	Identify Owner's Project Requirements	2 weeks												
1.2	Collect Preliminary Site Data	2 weeks												
1.3	Gather Preliminary Project Data	3 weeks												
1.4	Perform LEED Occupancy Calculations	1 week												
1.5	Create Project Team Directory	1 week												
1.6	Obtain Preliminary Project Schedule	1 week												
1.7	Research Applicable Incentives	2 weeks												
1.8	Compile all Data	1 week												
Step 2	**Prepare a preliminary LEED assessment**	**3 weeks**												
2.1	Review Credit Requirements to Assess Feasibility	3 weeks												
2.2	Identify Potential Implementation Strategies	2 weeks												
2.3	Identify Next Steps	2 weeks												
2.4	Prepare a Credit Responsibility Matrix	1 week												
Step 3	**Register the project on LEED-Online**	**1 week**												
3.1	Register the Project on LEED Online	1 week												
3.2	Assign Credit Responsibilities Online	1 week												
Step 4	**Organize a LEED Vision Workshop**	**3 weeks**												
4.1	Prepare Workshop Outline	2 weeks												
4.2	Facilitate the Workshop	2 weeks												
Step 5	**Prepare a LEED Project Workbook**	**3 weeks**												
5.1	Update Preliminary Documents	2 weeks												
5.2	Prepare LEED Project Workbook	1 week												

* Duration shown is approximate - will vary based on project schedules

Process Step Duration
Activity Duration
Milestone

Figure 3.9 Process Schedule for Stage I.
Credit: Green Potential LLC (www.green-potential.com).

Project Name:

Stage I	Project Definition & Goal Setting	Due Date	Not started	In Progress	Complete	Remarks
Step 1	**Collect and compile preliminary data**	**4/26/2010**				
1.1	Identify Owner's Project Requirements	4/5/2010		x		
1.2	Collect Preliminary Site Data	4/5/2010		x		
1.3	Gather Preliminary Project Data	4/5/2010		x		
1.4	Perform LEED Occupancy Calculations	4/5/2010		x		
1.5	Create Project Team Directory	4/17/2010	x			
1.6	Obtain Preliminary Project Schedule	4/17/2010	x			
1.7	Research Applicable Incentives	4/17/2010	x			
1.8	Compile all Data	4/26/2010	x			
Step 2	**Prepare a preliminary LEED assessment**	**5/17/2010**				Milestone
2.1	Review Credit Requirements to Assess Feasibility	5/17/2010	x			
2.2	Identify Potential Implementation Strategies	5/10/2010	x			
2.3	Identify Next Steps	5/17/2010	x			
2.4	Prepare a Credit Responsibility Matrix	5/17/2010	x			
Step 3	**Register the project on LEED-Online**	**5/24/2010**				
3.1	Register the Project on LEED Online	5/24/2010	x			
3.2	Assign Credit Responsibilities Online	5/24/2010	x			
Step 4	**Organize a LEED Vision Workshop**	**6/7/2010**				Milestone
4.1	Prepare Workshop Outline	5/31/2010	x			
4.2	Facilitate the Workshop	6/7/2010	x			
Step 5	**Prepare a LEED Project Workbook**	**7/28/2010**				
5.1	Update Preliminary Documents	7/21/2010	x			
5.2	Prepare LEED Project Workbook	7/28/2010	x			

Figure 3.10 Process Checklist for Stage I.
Credit: Green Potential LLC (www.green-potential.com).

STAGE II: DESIGN PHASE INTEGRATION

Introduction

The second stage of the LEED project management process involves the integration of green building principles, strategies, and products into the design, using the LEED credit system as a guiding framework. This stage entails activities that coincide with the design phases of the building process—the schematic design (SD), design development (DD), and construction documents (CD). The activities are iterative in nature where feedback from multiple types of analyses (energy simulations, stormwater calculations, cost-benefit analyses, etc.) is used to make informed design decisions. The requirements of construction-related LEED credits are also defined in order to facilitate easy implementation by the construction team. The final product deliverable at the end of this stage is a complete documentation package that will be submitted to GBCI for Design Review. The key assumption here is that the project team is pursuing the split design and construction review path offered by GBCI (refer to Chapter 2 for details of this GBCI process).

Stage II is composed of a total of ten steps, and each step, along with recommended activities, has been explained in detail. There are four steps in the schematic design phase, four in the design development phase, and two in the construction documents phase. The ten steps that constitute this stage are as follows:

Schematic Design (SD) LEED Phase

Step 1: Perform credit analyses

Step 2: Research and identify *green* products

Step 3: Integrate *green* requirements into design

Step 4: Review schematic design drawings

Design Development (DD) LEED Phase

Step 5: Update credit analyses and drawings

Step 6: Prepare for construction phase

Step 7: Develop building management policies

Step 8: Review design development drawings

Construction Documents (CD) LEED Phase

Step 9: Prepare documentation for *design credits*

Step 10: Submit to GBCI for Design Review

Schematic Design (SD) LEED Phase

The goal of the design team during the SD phase is to review all credit requirements, and complete calculations and analyses to determine the direction of the design. The LEED PM's role is to provide research support, offer advice on selection of materials, perform cost-benefit analyses, assist in credit calculations, and review drawings at the end of the SD phase. Figure 4.1 illustrates the process flow for activities in the SD LEED phase of Stage II.

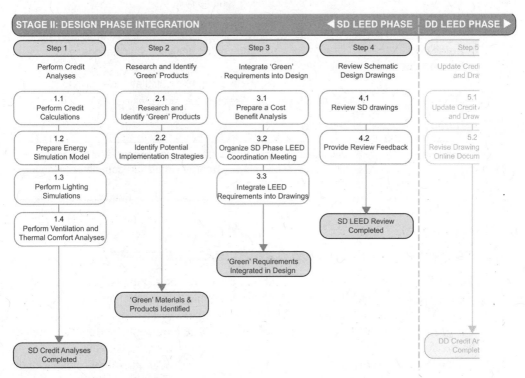

Figure 4.1 **Process flow for Stage II: SD LEED Phase.**

Credit: Green Potential LLC (www.green-potential.com).

SD LEED Phase
Step 1: Perform Credit Analyses

Objective	The objective of this step is to perform various preliminary analyses to inform early design decisions. These analyses vary from completing calculations and conducting research to preparing energy simulation models.
Recommended Activities	1.1 Perform Credit Calculations. 1.2 Prepare Energy Simulation Model. 1.3 Perform Lighting Simulations. 1.4 Perform Ventilation and Thermal Comfort Analyses.
Roles and Responsibilities	Design Team: Perform credit analyses, calculations, and simulations. LEED PM: Provide research and calculations support. Owner/Developer: Assist in decision making.
Duration	5 weeks
Outcome	Site, parking and transportation, stormwater, and water use calculations completed; Energy simulation completed; Lighting simulation completed.

ACTIVITY 1.1—PERFORM CREDIT CALCULATIONS

The LEED credits reference various industry standards and require calculations to determine if a project meets the credit requirements. While some are simplistic calculations, there are others that require detailed engineering expertise. The goal for the team in the SD phase should be to review the respective credit requirements, perform preliminary level calculations, and use them as a tool to make decisions related to the design of the building's envelope, materials, and systems. Figure 4.2 provides an example of some of the early design decisions that may be impacted by LEED credit analyses.

The preliminary LEED assessment that was created after the LEED Vision Workshop may be used as a starting point by team members as they proceed with the design. Most of the information required to complete these preliminary calculations should be available to the team after the Stage I activities. It is the responsibility of the LEED PM to ensure that the team members are aware of their credit responsibilities. The following are some key calculations that will impact the design and should be completed by the team during the SD phase. Note that these do not represent all calculations for every credit, but describe the basic calculations/analyses that will impact design decisions during the SD phase. Refer to Part 2 for credit-by-credit implementation activities.

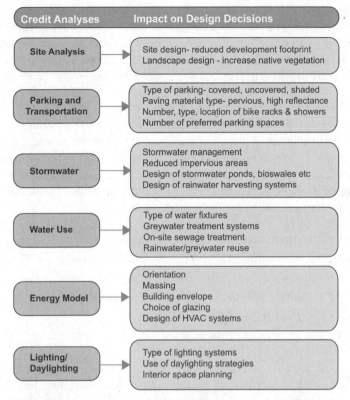

Figure 4.2   Impact of credit analyses on early design decisions.

- Site development calculations
- Parking and transportation calculations
- Stormwater calculations
- Water use calculations

Site Development Calculations

The *Sustainable Sites* category of the LEED rating system offers guidance on developing the site in a manner that minimizes environmental impact and disturbance to the natural, ecological balance of the site. Strategies include setting limits of disturbance, maximizing open space, reducing development footprint, planting native vegetation, and minimizing urban heat island effect. During the SD phase, the architect, the civil engineer, and the landscape architect should design the development footprint with the requirements of these credits in mind. The specific credits that might impact site and landscape design are *SS C5.1, Site Development—Protect or Restore Habitat* and *SS C5.2, Site Development—Maximize Open Space.*

SS C5.1 requires limiting site disturbance for greenfield sites or restoring the site with native vegetation for previously developed sites. SS C5.2 requires maximizing vegetated open space on-site. Usually these two credits fall under the civil engineering and landscape disciplines. The team members who have been designated with the responsibility of implementing these credits should perform the required calculations to determine design strategies that will help in meeting these credit requirements. At the SD stage, these calculations will be preliminary and are only to establish basic parameters for the design team. Detailed activities to implement these credits during the various stages of the design are provided in Part 2 of this book under *Sustainable Sites* credit implementation.

Parking and Transportation Calculations

The next set of calculations that need to be completed are related to parking and accessibility to alternative modes of transport. Based on information gathered during Stage I and discussions at the LEED Vision Workshop, the responsible team member should complete preliminary calculations and determine requirements that need to be incorporated in the design. LEED credits that might impact the design of the parking and transportation options for the project are as follows:

SS C4.1, Alternative Transportation—Public Transportation Access

SS C4.2, Alternative Transportation—Bicycle Storage and Changing Rooms

SS C4.3, Alternative Transportation—Low-Emitting and Fuel-Efficient Vehicles

SS C4.4, Alternative Transportation—Parking Capacity

SS C7.1, Heat Island Effect—Non-Roof

SS C7.2, Heat Island Effect—Roof

Credit SS C4.1 requires having access to public transportation to reduce the environmental impact caused by individual vehicular emissions. This will be largely determined by the location of the site and the availability of public transport in that location. However, it is also important that the bus stops be within a quarter mile or rail stations be within a half mile of main building entrances. Designers should carefully study the various pedestrian routes possible to bus stops and/or rail stations and plan the building location, entrances, and pedestrian pathways accordingly.

Credit SS C4.2 requires the incorporation of bike racks to promote bicycle ridership among occupants. The team member responsible for the credit should calculate the number of bike racks and showers required, research suitable bike rack solutions, identify locations, and include them in the design.

Credits SS C4.3 and SS C4.4 require providing preferred parking spaces for fuel-efficient vehicles and/or carpools or providing alternative-fuel stations. Designers must review the requirements of the options applicable for the project and plan their parking design accordingly.

Credit SS C7.1 requires providing shaded and/or covered parking spaces or high reflectance paving for uncovered parking spaces to reduce the urban heat island effect caused by large uncovered, high-absorption paved areas. The civil engineer, the architect, and the landscape architect must work together to determine how the parking may be designed to mitigate the heat island effect caused by large, uncovered parking areas. They should consider providing a covered parking garage in lieu of large, paved parking areas.

Credit SS C7.2 encourages use of high reflectance roofs and/or vegetated roofs to further reduce the urban heat island effect. The team must evaluate all opportunities and make a decision of whether to provide vegetated roofs or have high reflectance roofs or a combination of both. Deciding this early-on will be beneficial, as the type of roof, the building structure, and the project budget will be impacted.

Most of these credits fall under the responsibility of the architect, the civil engineer, and the landscape architect. The LEED PM may assist in the selection of materials and calculations required for the credits. Detailed activities to implement these credits during the various stages of the design are provided in Part 2 of this book under *Sustainable Sites* credit implementation.

Stormwater Calculations

The LEED rating system gives credit for the proper management of stormwater, both with respect to the quantity and quality that is discharged from a site after development. Stormwater that is transported through urban gutters, pipes, and sewers to receiving waters is often polluted with contaminants, like atmospheric deposits, vehicle-fluid leaks, mechanical equipment wastes, etc. This polluted stormwater further contaminates downstream waters, thus degrading aquatic habitats.[1] A number of cities in states, such as California, Texas, and Washington, have stringent regulations for stormwater management, and these calculations are mandatory for obtaining permits.

The two credits that outline the requirements and implementation strategies for stormwater management are SS C6.1, *Stormwater Quantity Control* and *SS C6.2, Stormwater Quality Control.* The intent of these credits is to "limit disruption of natural hydrology by reducing impervious cover, increasing on-site infiltration, reducing or eliminating pollution from stormwater runoff, and eliminating contaminants."[2]

Credit SS C6.1 requires the implementation of stormwater management strategies that help in reducing the post-development discharge rate and quantity from the predevelopment discharge rate. This can be achieved by using strategies, such as pervious paving, harvesting stormwater for reuse in irrigation, designing infiltration swales, and installing vegetated roofs. During SD phase, the civil engineer should perform preliminary stormwater quantity calculations in coordination with the landscape architect to determine the measures needed to control stormwater runoff and quantity. The results of this analysis will help in the careful integration of any stormwater control strategies within the overall site and landscape design.

Credit SS C6.2 requires the implementation of a stormwater management plan that is able to "capture and treat stormwater runoff from 90% of the average annual rainfall using best management practices (BMPs) that are able to remove 80% of the average total suspended solids in the stormwater."[3] The civil engineer should perform an assessment of the site conditions and propose the various BMPs that will enable the project to control the quality of stormwater released from the site.

Coordination with the plumbing engineer and the landscape architect might be required to complete this analysis, especially if any of the stormwater is planned to be reused either within the building or for irrigation purposes. Calculations of potable water use (which will be talked about in the next section) might be affected based on the decisions made for stormwater quality and quantity control.

Water Use Calculations

The increasing consumption of potable water by buildings has resulted in a strain on supplies and a drop in water levels of underground aquifers. The LEED-NC rating system encourages strategies that reduce the total potable water use in buildings for sewage conveyance and irrigation. The intent behind this set of credits and prerequisites is to "increase water efficiency within buildings to reduce the burden on municipal water supply and wastewater systems."[4] The efficient use of water will also help owners to reduce operational costs associated with high water utility bills.

During the SD stage, the plumbing engineer along with the landscape architect and the civil engineer must perform a comprehensive analysis of water usage to determine strategies that will help in reducing potable water utilization. Some examples of strategies to supplement water use requirements are reusing collected rainwater and treated greywater, or using municipally supplied non-potable water. The specific credits that might impact the building water systems design are as follows:

WE P1, Water Use Reduction

WE C1, Water Efficient Landscaping

WE C2, Innovative Wastewater Technologies

WE C3, Water Use Reduction

Prerequisite WE P1 and credit WE C3 require the reduction of potable water use from a baseline calculated according to the Energy Policy Act (EPAct) 1992. This may be achieved by using high-efficiency water closets, waterless urinals, as well as low flow faucets and showerheads. The plumbing engineers must perform calculations with multiple scenarios, estimating the total annual water saved compared to the baseline.

Credit WE C1 requires the reduction of potable water used for landscape purposes by using strategies, such as installing high-efficiency irrigation systems, grouping plants with similar needs, planting species that are native and have adapted to the region, reusing rainwater, etc. The landscape architect may use this credit as a tool to choose strategies that will be most water conserving.

Credit WE C2 requires further reduction of potable water use for building sewage conveyance and encourages strategies, such as using treated greywater or rainwater for sewage conveyance or treating wastewater on-site and reusing it for irrigation purposes. The plumbing engineer must perform calculations and consider various options that will help minimize potable water use in the building.

Both the water use analysis described here and the stormwater analysis described in the previous section require collaboration between the civil engineer, the landscape architect, and the plumbing engineer to arrive at solutions that address the reduction of water use in a comprehensive and holistic manner. Detailed activities to implement these credits during the various stages of the design are provided in Part 2 of this book under *Water Efficiency* credit implementation.

ACTIVITY 1.2—PREPARE ENERGY SIMULATION MODEL

In simple terms, energy simulation is the process of simulating a *proposed* building using computer programs to predict the annual energy costs of the building in operation.

The performance of the building is dependent on a number of factors, like construction details, orientation of walls and windows, occupancy patterns, climate, operating schedules, the lighting and HVAC systems, etc. Due to the complexity of the calculations and the multiple variables involved, software packages have been developed that carry out complex equations to describe energy usage of buildings. The best-known hourly simulation software package is DOE-2 (developed by the Simulation Research Group at Lawrence Berkeley National Laboratory), which can accept and produce a lot of data. eQuest and VisualDOE are relatively new programs that are far more user-friendly and popular among practitioners.[5]

Modeling the energy performance of a building early on in the design process plays a significant role in a green project. The energy model can provide the team with valuable information that they can use to make better decisions regarding the building's envelope (insulation, glazing, etc.), systems (HVAC), and lighting. In addition, the energy model can also help determine if the proposed building will be able to meet the LEED prerequisite *EA P2, Minimum Energy Performance*, which is required to achieve certification. Further, the model may be used to test various energy efficiency options that can improve the performance of the building, which is required for LEED credit *EA C1, Optimizing Energy Performance*. Note that there are multiple options to demonstrate compliance with the prerequisite and the credit. Not all projects are required to perform a building energy simulation. However, a majority of buildings (especially those over 50,000 square feet) are required to conduct a simulation to demonstrate compliance. It is recommended that design teams use the energy simulation option to take full advantage of the synergies between various building systems. This section discusses the requirements of only the energy simulation options of EA P2 and EA C1.

Prerequisite EA P2 requires buildings to "demonstrate a 10% improvement in the *proposed* building performance rating for new buildings or 5% improvement in the *proposed* building performance rating for major renovations to existing buildings, compared with the *baseline* building performance rating calculated according to the method in Appendix G of ANSI/ASHRAE/IESNA Standard 90.1-2007, using a computer simulation model for the whole building project."[6]

Credit EA C1 requires demonstrating "a percentage improvement (new buildings—12% to 48%, major renovations—8% to 44%) in the *proposed* building's performance rating compared with the *baseline* building performance rating calculated according to the method in Appendix G of ANSI/ASHRAE/IESNA Standard 90.1-2007."[7] A total of up to 19 points can be achieved through this credit by showing varying levels of percentage improvement in the building performance rating.

Note that according to the performance rating method (see Appendix G of ASHRAE 90.1-2007), the *proposed* building *performance* is defined as "the annual energy cost calculated for a proposed design."[8] Similarly, the *baseline* building *performance* is defined as the "annual energy cost for a building intended for use as a baseline for rating the proposed design."[9] The annual energy costs can be determined using local utility rates for purchased energy, such as electricity, gas, oil, etc., or can be obtained from state average prices that are published annually by the U.S. Department of Energy's Energy Information Administration (EIA).

The mechanical engineer or energy simulation expert should create an energy model to determine if the proposed building will be able to meet the minimum energy efficiency level set forth by the prerequisite, and also test various options to suggest measures for optimizing energy performance.

What Information Is Needed to Prepare the First Energy Model?

It is not necessary that the initial energy model developed during SD phase of the project have all the details of the building systems or design. It can be developed with basic information based on the size and shape of the building, the intended use of the building, its location, and the type of HVAC system. Although a simplistic model such as this may not be sufficient to predict the absolute energy consumption of the building, it can give sufficient guidance to compare design options. As the design evolves, more detail can be added to the energy model, so that it can be used to answer more comprehensive design questions.

Table 4.1 identifies the preliminary information required to develop a schematic level energy model. The energy simulation expert must obtain this information from the project design team and make reasonable assumptions for information that is not available at the time.

Which Energy Simulation Program to Use?

The choice of the energy simulation program is dependent on a number of factors, such as the purpose of the model and the level of detail that is required. However, the most important factor to consider before selecting a particular program is the level of familiarity of the energy simulation expert with a particular software program. While a novice modeler may be more comfortable using simpler, user-friendly tools like eQUEST, advanced users might want to use detailed energy analysis programs like TRACE and Hourly Analysis Program (HAP). Most of these programs deliver similar results. It is recommended that the modeler choose the program that he/she is

TABLE 4.1 BASIC INFORMATION TO DEVELOP SCHEMATIC LEVEL ENERGY MODEL

NO.	INFORMATION REQUIRED
1	Size and rough shape of building
2	Proposed exterior wall and roof construction with insulation values
3	Approximate interior layout (for zones)
4	Exterior elevation to determine wall and window area
5	Occupancy type for all areas, i.e. office, conference room, etc.
6	Lighting types and quantity per area
7	Basic mechanical system description

conversant with to ensure that the inputs are entered accurately. Table 4.2 shows a list of the most commonly used software programs.

What to Expect from the First Energy Model Results?

The purpose of the energy analysis created in the SD phase is to answer basic design questions (see Figure 4.3) and determine whether the planned orientation, construction, glazing, window size, or HVAC system type need to be modified to make the building more energy efficient. Once the energy model has been created, the energy analyst must communicate the results of the analysis to the design team and recommend any design options or energy efficiency measures that will help improve the efficiency of the building. The LEED PM may then organize an *energy model charrette* to bring the

TABLE 4.2 COMMONLY USED ENERGY SIMULATION PROGRAMS	
PROGRAM	**DESCRIPTION**
eQUEST	Microsoft Windows–based graphical user interface (GUI), based on the DOE-2.2 calculation engine. Most commonly used software with default templates for a range of building types. Available for free download at www.doe2.com.
Energy-10	Developed jointly by the National Renewable Energy Laboratory, Lawrence Berkeley National Laboratory, Berkeley Solar Group, Sustainable Building Industries Council, and the U.S. Department of Energy. Intended for performing quick energy simulations. Program comes with the *Designing Low-Energy Buildings* manual.
DOE-2	Been in existence since 1978. There is no user interface, and program inputs are made with a standard text-editing program. Extensive documentation of program inputs is available in printed form.
VisualDOE 2.6	A Microsoft Windows—based GUI for DOE-2.1E that can greatly reduce input time. Has an extensive library of predefined schedules, construction assemblies, and glass types. Has the ability to import an AutoCAD *.dxf file for developing zone inputs.
PowerDOE	Feature-rich GUI based on the DOE-2.2 calculation engine. Allows non-rectangular walls and provides renderings.
EnergyPro	Based on DOE-2.1E and used for documenting compliance with Title 24 requirements. It is also good for setting up non-compliance models due to its extensive library of building materials and HVAC equipment.
Trane Trace 600	HVAC load calculation program that also has many energy analysis capabilities. Provides engineering checks of most program inputs for reasonableness.
Carrier Hourly Analysis Program	An hourly calculation HVAC load calculation program with a Windows-based interface that can also be used for energy analysis.

Questions Schematic Energy Model is Expected to Answer

▶ Is the proposed wall construction type and insulation efficient?

▶ Do the window sizes or glazing type need to be changed?

▶ What will be the energy impact of adding more windows to a building?

▶ If an efficient lighting system is installed, how much smaller can the cooling system be made?

▶ How much energy will be saved by specifying high-efficiency HVAC components?

▶ Will the proposed design meet or exceed the baseline defined by ASHRAE 90.1-2007?

▶ What can be done to improve the energy efficiency of the proposed building?

Figure 4.3 Typical questions expected to be answered by schematic energy model.

design team members, owner, and the energy analyst together to discuss the findings and any potential impact to the design.

ACTIVITY 1.3—PERFORM LIGHTING ANALYSES

The *type* and *nature* of lighting used in a building play a significant role in the energy performance and indoor environmental quality of buildings. While the *type* of lighting fixtures (such as incandescent vs. compact fluorescent) is critical for reducing the building energy consumption, the *nature* of light (artificial vs. natural) plays a key role in the indoor experience and productivity of the building occupants. This section discusses the two lighting analyses that address the *nature* of light according to the LEED credit framework—one is light pollution reduction analysis and the second is daylighting analysis. The two credits that specifically reference these analyses are *SS C8, Light Pollution Reduction* and *EQ C8, Daylighting and Views*. Please note that the analysis of the *type* of interior lighting fixtures are not part of this section but form part of the energy analysis covered in the previous section—Activity 1.2.

Light Pollution Reduction Analysis—Site Illumination Model

Credit SS C8 requires minimizing "light trespass from the building and site, reducing sky-glow to increase night sky access, improving nighttime visibility through glare reduction, and reducing development impact from lighting on nocturnal environments."[10] The design team must use the guidance provided by this credit to design the site, façade, landscape, and interior lighting in a way that reduces night sky pollution.

The credit comprises two parts. The first part provides requirements for interior lighting and the second part describes exterior lighting requirements. The interior lighting requirements can be met by two options. The first option is to provide automatic turn-off of interior luminaires with direct line of sight to openings in the envelope. The second option is to provide automatic shading of openings in the envelope with a direct line of sight to luminaires during non-business hours. This part does not require computer-generated simulation but needs a decision by the design team on whether automatic control devices can be incorporated into the design.

In the case of exterior lighting, a detailed analysis is needed to design the site lighting according to the credit requirements. The first step is to calculate the exterior lighting power densities (LPDs) in accordance with ANSI/ASHRAE/IESNA 90.1-2007 Section 9 and determine whether the LPDs are less than those allowable for the project. The second step is to develop a site illumination model or photometric site plan to show that illumination at the site boundary is within the allowed range. There are a number of professional lighting simulation software programs that are available to perform this analysis, such as Radiance, Dialux, and Ecotect.

The lighting consultant should perform a site photometric analysis prior to making decisions on the site lighting design, the type of light fixtures, and the electrical requirements of the project. The consultant must also coordinate with the landscape architect and the architect for design of landscape and façade lighting fixtures. The LEED PM should review the analysis prepared by the lighting consultant to ensure that the proposed lighting design is in accordance with LEED credit requirements.

Performing this analysis during the SD phase is valuable, as it will help in making necessary changes early on without significant cost impact. For instance, a site lighting model prepared for a multifamily project during the SD phase showed that the proposed light fixtures did not have appropriate shielding. The lighting consultant suggested a different lighting product with sufficient shielding. However, the owner was concerned about the high cost of the proposed fixture and was hesitant to pursue this approach. Upon running the lighting simulation, the model showed that the *number* of fixtures could be reduced to meet the desired lighting levels. The high price of the product with shielding was offset by the reduction in their total quantity. As a result there was no net increase in the lighting budget.

Daylighting Simulation and View Analysis

Daylighting refers to the practice of using natural light as a source of lighting interior building spaces instead of relying completely on artificial or electric lighting. Studies show that a well-designed daylit building can reduce lighting energy use by 50% to 80%.[11] In addition, the use of daylight allows occupants to connect to the outdoors and also has a positive impact on their productivity and well-being.[12] The LEED rating system encourages incorporating daylighting strategies under the Indoor Environmental Quality (EQ) category for credits, *EQ C8.1, Daylighting and Views—Daylighting* and *EQ C8.2, Daylighting and Views—Views.*

Credit EQ C8.1 requires projects to achieve minimum light levels in regularly occupied spaces of the building. The *LEED Reference Guide* defines regularly occupied spaces as "areas where people sit or stand as they work. In residential buildings these spaces include all living and family rooms and exclude bathrooms, closets, and storage or utility areas."[13] Note that there are multiple options to demonstrate compliance with this credit, such as prescriptive design strategies or physical light measurements. Not all projects are required to use a computerized simulation model to demonstrate compliance. However, it is recommended that the project pursue the *simulation* option to develop an integrated design strategy. The amount of daylight that is received (in the interior of a space) is dependent on a number of factors, such as the type of windows, glazing, the shading devices, interior surfaces, exterior obstructions, and location. By means of a computer-generated model, the light levels within a space can be determined by taking into account all these factors. Further, integrating daylighting strategies requires careful considerations with respect to heat gain, glare control, visual quality, and variations in daylight availability. The daylight simulation model will help in understanding the interplay of all these factors and help project teams create a comprehensive daylighting strategy.

There are a number of software programs available that can be used to perform daylighting simulations to get both performance feedback and visual information to determine how well the design functions. Historically, computer software tools were complex to use, and only researchers involved with product development used them. But now there are many tools available that are user-friendly and give valuable outputs. Some of the more frequently used programs are Radiance, DaySim, Ecotect, ADELINE, and SkyCalc.

Developing a daylight simulation model during the SD phase will help the project team make design decisions related to the building envelope, fenestrations, and also interior layouts. This simulation could be performed by a daylighting consultant, the LEED consultant, or the architect of the project, depending on their level of expertise. The team members responsible for the daylighting simulation should convey the results of their analysis to the LEED PM and design team. The LEED PM should review the results and coordinate with the design team on any significant changes.

Architects and interior designers should also carefully plan interior spaces to maximize the amount of daylight received in regularly occupied spaces. For instance, in an office building the open office spaces may be planned towards the perimeter, while non-regularly occupied spaces could be planned in the core space of the building.

Credit EQ 8.2 requires regularly occupied spaces to have access to vision glazing (between 30 inches to 90 inches) with an unobstructed view at a height of 42 inches from the floor. The interior spaces should be designed keeping these requirements in mind, especially with respect to partition types and their heights. Detailed activities for implementation of these credits have been provided in Part 2 of this book under— *Indoor Environmental Quality* credit implementation.

ACTIVITY 1.4—PERFORM VENTILATION AND THERMAL COMFORT ANALYSES

Americans and people all over the world spend a majority of their time indoors, in sealed buildings that are controlled through the means of mechanical HVAC systems. The quality of the indoor environment should, therefore, be of prime concern to building designers, owners, and operators. Several studies document the impact of indoor environment on occupant health, productivity, and well-being. The LEED rating system encourages use of strategies, such as providing better indoor air quality, reducing sources of indoor pollution, and providing optimal thermal comfort conditions to enhance indoor environments. Two important factors that contribute to the indoor environmental quality are: indoor air quality (IAQ) and thermal comfort. During the SD phase, the mechanical engineers, architects and owners must work collaboratively to design the building in a manner that improves both indoor air quality and thermal comfort for its occupants.

Indoor Air Quality

Indoor air quality, or rather, acceptable indoor air quality is defined by ASHRAE 62.1-2007 as "air in which there are no known contaminants at harmful concentrations as determined by cognizant authorities and with which a substantial majority (80 percent or more) of the people exposed do not express dissatisfaction."[14] The standard specifies minimum ventilation rates and IAQ levels and prescribes ventilation system designs that minimize contaminants. There are a number of studies that document the benefits of proper ventilation in eliminating airborne pollutants. One study found as much as a 50 percent higher incidence of upper respiratory problems in military recruits housed in new buildings compared to those living in older, less airtight buildings. Studies such as these clearly highlight the fact that mechanically ventilated buildings have to be designed, operated, and maintained correctly, if they are not to cause an increase in health problems for occupants.[15]

The LEED rating system has specific prerequisites and credits that address indoor air quality. The general intent of these credits is to provide better ventilation and minimize sources of pollution to improve the air quality of interior spaces. They are as follows:

EQ P1, Minimum Indoor Air Quality Performance

EQ P2, Environmental Smoke Control

EQ C1, Outdoor Air Delivery Monitoring

EQ C2, Increased Ventilation

EQ C3.1, EQ C3.2, Construction Indoor Air Quality—during Construction and before Occupancy

EQ C4.1 to EQ C4.4, Low-Emitting Materials

EQ C5, Indoor Chemical and Pollutant Source Control

Ventilation

Teams must evaluate various options to determine the most appropriate ventilation strategy for the building. The decision on whether to use mechanical or natural ventilation is dependent on several factors, such as type of building, occupancy, quality of outside air, noise pollution, etc. In order to determine whether natural ventilation is an effective strategy, the *LEED Reference Guide* recommends the use of the flow diagram process provided by the Chartered Institute of Building Services Engineers (CIBSE) Applications Manual, *Natural Ventilation in Non-domestic Buildings.*

EQ P1 requires mechanically ventilated spaces to meet the requirements of Sections 4 through 7 of ASHRAE 62.1-2007. Naturally ventilated spaces are required to comply with Section 5.1 of the same standard. The design team must use the requirements identified in the standard as a basis for designing the building ventilation system and controls.

For mechanical ventilation, the ASHRAE standard describes two methods by which minimum ventilation rates may be determined—either using the ventilation rate procedure or the IAQ procedure. The ventilation rate procedure prescribes outdoor air quality levels acceptable for ventilation, treatment measures for contaminated outdoor air, and ventilation rates for different types of spaces. The IAQ procedure uses a performance-based approach where the ventilation system maintains contaminant levels at or below determined limits to achieve acceptable indoor air quality. The LEED prerequisite EQ P1 and credit EQ C1 both reference the ventilation rate procedure for mechanical ventilation design. The requirements are outlined in Section 6.2 of ASHRAE 62.1-2007. The mechanical engineer should perform calculations to determine the amount of outdoor air needed to ventilate interior spaces. The calculations take into account several factors, such as the number of occupants and their activities and the type of occupancy.

Natural ventilation requirements are outlined in Section 5.1 of the standard, which provides specific guidance on size and location of ventilation openings (requires that the *openable* area be a minimum of 4% of the usable floor area). The mechanical engineer along with the architect should perform calculations to determine suitable opening sizes. In addition, credit EQ C2 requires the implementation of strategies that can help increase the ventilation rates to at least 30% above the ASHRAE standard. Calculations and analyses should be performed to determine strategies that will help increase the ventilation rates.

In some climates, the increase of ventilation rates to meet the requirements of credit EQ C2 may lead to higher HVAC energy costs and potentially greater HVAC capacity needs. If ventilation rates are increased, then additional energy may be required to heat and cool outdoor air. However, the high energy costs may be offset by using strategies such as heat recovery wheels that can precondition the intake air and minimize the extent to which additional energy may be required. The mechanical engineer should work with the energy simulation expert to find a solution that allows increase in ventilation rates with minimal impact to building energy use.

The additional energy cost that may be incurred because of increasing ventilation rates may be offset by the dollars saved because of reduced sick days of occupants (due to better air quality). The *Design Briefs: Indoor Air Quality* by Energy Design Resources illustrates this with a study that determined the overall savings of low ventilation (i.e., 5 cubic feet per minute [cfm] per person) range from negative 1 cent per square foot each year (an increase in electricity costs) in very mild climates to positive 14 cents per square foot each year in more extreme climates. This translated to an energy savings of about $1,425 annually for a small office building. The study further calculated the savings associated with reduced absenteeism to be $25,000 per year. If the building is owner-occupied, then it is much more profitable for the owner to invest in the additional energy cost rather than suffer losses due to employee absenteeism.[16]

Credit EQ C1 requires monitoring the ventilation systems to ensure that the design minimum requirements are maintained. The purpose of monitoring is also to notify the occupants and facility managers of any variations in the outdoor air levels. This credit does not require any calculations; but appropriate monitoring devices should be incorporated into the ventilation design of the building during early design phases.

Source Control

Prerequisite EQ P2 and Credits EQ C3.1, C3.2, EQ C4.1–C4.4, and EQ C5 highlight ways by which the various sources of indoor air pollution—contaminants, harmful VOCs, etc.—may be contained. These credits do not require any specific calculations or analyses. However, their requirements will impact the building design, filtration media, exhaust systems, and interior finishes. During SD phase, the design team must review the requirements and investigate options prior to making decisions regarding the building interiors.

Thermal Comfort

The second important factor contributing to an optimal indoor environment is the level of thermal comfort of building occupants. Although comfort is subjective and will vary from occupant to occupant, the ASHRAE 55-2004 Standard prescribes thermal comfort criteria for the design of building envelope and HVAC systems. It takes into account several factors, such as temperature, thermal radiation, humidity, air speed, activity level, and clothing of occupants. The LEED rating system references this standard and provides guidance on improving thermal comfort conditions for occupants through various credits. The specific credits are: *EQ C7.1, Thermal Comfort Design, EQ C7.2, Thermal Comfort Survey,* and *EQ C6.2, Controllability of Systems—Thermal Comfort.*

Credit EQ C7.1 requires that the building's HVAC systems and envelope be designed according to the ASHRAE 55-2004 standard. During the SD phase, the mechanical engineer and the architect must collaboratively review the requirements of the standard and incorporate strategies that will help achieve optimal thermal comfort conditions. The thermal comfort criteria offered by the standard should be used as a basis of design for the building systems.

Credit EQ C7.2 requires implementation of a thermal comfort survey after occupancy. This credit does not have any impact in the SD phase and has been discussed in the DD phase as part of the building management policies (Activity 7.1).

Credit EQ C6.2 encourages individual control of thermal comfort by means of individual thermostats, diffusers, etc. During SD phase, the team should evaluate various strategies that will facilitate this individual control. Strategies such as underfloor air distribution systems make it easier to provide thermal comfort control compared to traditional ventilation systems. The Center for the Built Environment (CBE) at the University of California, Berkeley defines underfloor air distribution (UFAD) as "a method of delivering space conditioning in offices and other commercial buildings. This technology uses the open space (underfloor plenum) between the structural concrete slab and the underside of a raised access floor system to deliver conditioned air directly into the occupied zone of the building. Air can be delivered through a variety of supply outlets located at floor level (most common), or as part of the furniture and partitions. UFAD systems have several potential advantages over traditional overhead systems, including improved thermal comfort, improved indoor air quality, and reduced energy use. By combining a building's heating, ventilating, and air-conditioning (HVAC) system with all major power, voice, and data cabling into one easily accessible service plenum under the raised floor, significant improvements can be realized in terms of increased flexibility and reduced costs associated with reconfiguring building services."[17]

In summary, during the SD phase, it is important for the mechanical engineer to investigate various ventilation design strategies that can help improve indoor air quality and maintain optimal thermal comfort conditions for the building occupants. The design of these systems will also have an impact on other aspects of the building, and therefore, it is important to perform these analyses early on.

SD LEED Phase
Step 2: Research and Identify
Green Products

Objective	The objective of this step is to research and identify materials, products, systems, and solutions that will help in conserving natural resources, and also improve the overall efficiency of the building.
Recommended Activities	2.1 Research and Identify *Green* Products.
Roles and Responsibilities	Design Team: Research available products. LEED PM: Support the design team with *green* product research. Owner/Developer: Decision making.
Duration	8 weeks
Outcome	Product and material options researched and identified.

ACTIVITY 2.1—RESEARCH AND IDENTIFY *GREEN* PRODUCTS

A green projecct warrants the selection of materials and products that have minimal environmental impact, enhance the quality of indoor spaces, and also contribute towards better energy performance. The *LEED Reference Guide* states that "during the life cycle of a material, its extraction, processing, transportation, use, and disposal can have negative health and environmental consequences, polluting water and air, destroying native habitats, and depleting natural resources."[18] Research on materials, products, and other innovative technologies should be conducted during the early design phases. In recent years, the *green* product industry has exploded with new materials released almost on a daily basis. It is crucial to discern the best performing materials/products and select those that are most appropriate. The design team must carefully consider the sustainability criteria along with other factors, like durability, budget, and aesthetics, that are typically considered for material selection. If feasible, the general contractor should be consulted regarding the availability and costs associated with procurement of such materials. The LEED PM should provide research support and assist the design team in selecting materials based on *green* criteria.

What Materials or Products to Choose?
Project teams must consider the relative environmental, social, and health benefits of the available choices when specifying materials and products. Below are some types of products, materials, and equipment that would qualify a product as *green*. These criteria are based on various LEED credits and will also help satisfy many credit

requirements. Most manufacturers now provide information on their websites stating the LEED point potential of their products. Figure 4.4 presents an overview of the various types of materials and products that should be considered for a LEED project.

Salvaged Products

These include salvaged, refurbished, or reused materials. The *LEED Reference Guide* defines salvaged or reused materials as "construction materials recovered from existing buildings or construction sites. Common salvaged materials include structural beams, flooring, doors, cabinetry, brick, and decorative items."[19]

'Green' Materials & Products	Examples	Credits
Salvaged Materials	Salvaged brick, pavers, lumber	MR C3
Materials with Recycled Content	Steel, gypsum ceilings/boards, carpet, insulation	MR C4
Regionally Sourced Materials	Harvested, extracted, manufactured within 500 miles of project site	MR C5
Rapidly Renewable Materials	Cotton insulation, bamboo flooring, linoleum flooring, wheatboard cabinets	MR C6
Certified Wood Products	Forest Stewardship Council (FSC) certified wood	MR C7
Low Emitting Materials	Low or no VOC paints, coatings, adhesives, sealants	EQ C4
Efficient Systems & Equipment	ENERGY STAR rated, high efficiency lighting, energy recovery systems	EA C1
Renewable Energy Equipment	Solar panels, wind energy, geothermal	EA C2
High Efficiency Water Fixtures	High efficiency or dual flush toilets, waterless urinals, low flow faucets	WE C3
High Reflectance Products	High reflectance paving and roofing products	SS C7
Stormwater Control Products	Rainwater harvesting, stormwater quality control products	SS C6

Figure 4.4 Examples of *green* materials and products.

The benefit of using salvaged materials and products is to reduce demand for virgin materials and reduce waste, thereby minimizing the impact associated with the extraction and processing of virgin resources. The impact of producing new construction products and materials is significant because buildings account for a large portion of natural resource consumption, including 40% of raw stone, gravel, and sand as well as 25% of virgin wood.[20] Using salvaged or reused materials can contribute toward two LEED credit points for *MR C3, Materials Reuse.*

Products with Recycled Content

These include products and materials with recycled content. Many commonly used products, such as metals, concrete, masonry, gypsum wallboard, carpet, tile, and insulation contain recycled content. There are two types of materials available—one that is made with postconsumer recycled content and the other that is made with pre-consumer recycled content.

Postconsumer recycled content is defined as the "percentage of material in a product that was derived from consumer waste—household, commercial, industrial—or institutional end-users and that can no longer be used for their original purpose. Some examples include construction and demolition debris, materials collected through recycling programs, discarded products like furniture and cabinetry, and landscaping waste like leaves, grass clippings, etc."[21] Preconsumer recycled content is defined as "the percentage of material in a product that is recycled from manufacturing waste. Examples include planer shavings, sawdust, trimmed materials, and obsolete inventories."[22]

Using products with recycled content can contribute up to two points for LEED credit *MR C4, Recycled Content.* More weight is given to products with postconsumer recycled content compared to products with pre-consumer recycled content because of the increased environmental benefit over the life cycle of the postconsumer product.

Regional Materials

These include products that have been extracted, harvested, and manufactured within close proximity to the project location, preferably within 500 miles of the project site. Use of regionally sourced products promotes local economies and reduces the environmental impacts resulting from transportation. The availability of regionally manufactured materials depends on the project location. While it may be easier to procure a majority of products within a 500 mile radius for some project locations, it may not be feasible in some areas. The use of regionally sourced products can help the project achieve up to two points for credit *MR C5, Regional Materials.*

Rapidly Renewable Products

These include products that are made with rapidly renewable materials. Rapidly renewable materials are made from plants that are typically harvested within a 10-year or shorter cycle. Sourcing rapidly renewable materials reduces the use of raw materials whose extraction and processing have greater environmental impacts. Common examples of products that are considered rapidly renewable are cork, bamboo, natural rubber, wheat, and cotton. More products are becoming available now in the market at cost competitive rates. Use of rapidly renewable products can help the project achieve one point for credit *MR C6, Rapidly Renewable Materials.*

Certified Wood Products

These include wood or wood-based products that have been certified in accordance with the Forest Stewardship Council's (FSC) principles and criteria. The use of these products encourages responsible forest management and contributes to the long-term health and integrity of forest ecosystems. Irresponsible forest practices can include forest destruction, wildlife habitat loss, soil erosion, water and air pollution, and waste generation. Using certified wood products can help the project achieve one point for credit *MR C7, Certified Wood.*

Low-Emitting Materials

Low emitting materials emit low or no volatile organic compounds (VOCs). VOCs have a negative impact on the indoor air quality and can adversely impact the health and well-being of building occupants. They also have a negative impact on the earth's atmosphere, contributing to smog generation and air pollution. VOC content on all interior adhesives, sealants, paints, and coatings should be lower than that prescribed by the standards referenced by LEED. In addition, carpets that meet Carpet and Rug Institute's (CRI) requirements and flooring systems that meet the FloorScore Standard should be used. Also, composite wood and agrifiber products that do not have any added urea-formaldehyde should be considered. Use of low-emitting products can contribute up to four points under the IEQ credits *EQ C4, Low Emitting Materials.*

Efficient Envelope Components

Building envelope components that contribute toward lowering the heating and cooling loads, and thereby reducing the overall energy used by the building should be considered. Examples of products include structurally insulated panels (SIPs), high performance glazing, windows, and insulation.[23] Though the percentage contribution of these products toward building energy reduction cannot be isolated, their use will contribute towards achieving credit *EA C1, Optimize Energy Performance.*

Energy-Conserving Equipment

Equipment that helps conserve energy and manage loads, thereby contributing to the overall reduction in the building's energy use, should be considered. Examples of such products are Energy Star rated appliances, high-efficacy lighting and lighting control equipment, sensors, etc. These products can also contribute towards achieving credit EA C1.

Renewable Energy Systems

Equipment and products that will enable the use of renewable energy on-site for supplementing the energy needs of the building must be considered. Solar water heaters, photovoltaic systems, and wind turbines are examples of products that fall under this category. Using on-site renewable energy systems can help the project achieve up to five points through the credit *EA C2, Renewable Energy* and also contribute toward optimizing the building's energy performance to obtain points under EA C1 credit.

High Efficiency Plumbing Fixtures

These include plumbing fixtures—toilets, showerheads, and faucets—that are considered high efficiency and are at least 20% more efficient than the Federal Energy Policy Act (EPAct) 1992 Standard. The U.S. Environmental Protection Agency (EPA) has introduced a new *WaterSense* label to identify high efficiency plumbing fixtures. The use of water-conserving fixtures can help the project achieve up to three points for *WE C3, Water Use Reduction.*

High Efficiency Irrigation Systems

These include irrigation systems that are water-efficient and conserve potable water. The project team should also consider using dedicated water meters for landscaping and rain or moisture sensors to further control the amount of water used for landscaping purposes. Note that the benefit of an efficient irrigation system can be better realized if the landscape plantings are also water efficient. The choice of the irrigation system can play a significant role in reducing water use and help the project achieve points for *WE C1, Water Efficient Landscaping.*

Stormwater Control Products

These include porous paving products and vegetated roofing systems, which can help in reducing stormwater runoff, thereby reducing surface water pollution. Using products that help reduce stormwater quantity or treat stormwater quality can help the project achieve up to two points under site credits SS C6.1 and SS C6.2. Reuse of stormwater for irrigation or for sewage conveyance (toilet flushing, etc.) will also help in reducing potable water consumption by the building, thereby contributing toward the *Water Efficiency* credits: WE C1 and WE C2.

High Reflectance Products

These include hardscape materials on-site and roofing products with high solar reflectance. These products help reduce the heat island effect that is created by the radiation of heat absorbed by dark roof, parking, or hardscape surfaces. Using materials and products with high reflectance can help the project achieve up to two points through credits SS C7.1 and SS C7.2.

SD LEED Phase
Step 3: Integrate *Green* Requirements into Design

Objective	The next step is to integrate the results of the various credit analyses and research into the design of the building. In order to discern which strategies the team should pursue, it is recommended that a cost benefit analysis be done to determine the strategies that are most cost effective for the owners in the long term.
Recommended Activities	3.1 Prepare a Cost Benefit Analysis. 3.2 Organize a SD Phase LEED Coordination Meeting. 3.3 Integrate LEED Credit Requirements into Drawings.
Roles and Responsibilities	Cost Estimator/General Contractor/LEED PM: Perform cost benefit analysis. Owner: Decision making.
Duration	12 weeks
Outcome	Cost-Benefit Analysis completed. SD Phase LEED Coordination Meeting organized. *Green* strategies, materials, products, and technologies identified. *Green* requirements integrated with design.

ACTIVITY 3.1—PREPARE A COST-BENEFIT ANALYSIS

The choice of strategy, material, product, and design options is primarily driven by the additional initial costs (*green* premium) and the associated short- and long-term benefits. This area is of utmost concern to owners, developers, and investors who seek to maximize the net return on their investment. It is obvious that the true benefit of *green* design is realized by an integrated synergistic approach to credit points. However, it is widely recognized that a credit-by-credit analysis of costs and benefits helps owners with decision making. It is not a straightforward exercise to assign a cost per *green* item, and it is difficult to assign a dollar value to the intangible benefits derived from *green* buildings. There are several sophisticated tools available that help with conducting this *green* building analysis. The LEED PM should collaborate with the cost estimator and the owners to prepare a cost-benefit analysis that takes into account both operational savings and intangible benefits such as increased productivity. The goal of the analysis is to provide the owners with a summary of the additional first-costs along with benefits, so that they may choose strategies appropriately.

Figure 4.5 shows an example of a preliminary cost-benefit analysis performed for a typical office building. *Green* features that were considered additional to the base building design were identified along with their estimated costs. Annual benefits from

PROJECT INFORMATION ASSUMPTIONS		SUMMARY OF RESULTS	
Project Type	Office	Total Green Premium	$453,000
Gross Area (sf)	100,000	Total Green Premium/sf	$4.53
Rentable Area (sf)	80,000	Total NPV/sf	$48.57
Total occupants	400		
Number of FTEs	320		
Location	United States		

Green/Energy Efficiency Features (in addition to base building design)	Green Premium	1 year payback scenario	10 year payback scenario	20 year payback scenario
	[$]	[$]	[$]	[$]
Sustainable Sites				
A) Bike racks (20)	$2,000	Benefits included in item # H&I below		
B) Shower/changing rooms (2)	$48,000	Benefits included in item # H&I below		
Water Efficiency				
C) High Efficiency Plumbing Fixtures	$3,000	$430	$3,762	$6,536
Energy and Atmosphere				
D) Energy Efficiency Measures	$300,000	$45,360	$397,015	$689,783
Materials and Resources				
E) Rapidly Renewable Materials	$50,000	Benefits included in item # H&I below		
Indoor Environmental Quality				
F) CO$_2$ monitors	$50,000	Benefits included in item # G below		
Sub-Total (Operational only)	**$453,000**	**$45,790**	**$400,777**	**$696,319**

		1 year	10 year	20 year
G) Total Productivity Savings		$424,320	$3,713,873	$6,452,573
H) Total Marketing Benefit		$80,000	$700,202	$1,216,548
I) Total CO$_2$ Emissions Savings		$8,670	$75,882	$131,839
Total Benefits (Operational+Intangible)	**$453,000**	**$558,780**	**$4,890,734**	**$8,497,280**

ASSUMPTIONS
1) Benefit of high-efficiency fixtures= [Gallons of potable water saved annually x cost of water/gallon]; water use savings calculated using LEED credit WE C3
2) Benefit of energy efficiency measures obtained from energy cost savings estimated via energy model/EA C1
3) Productivity Savings = savings from reduced sick days and average increase in productivity Assumptions: Average Salary of employee: $50,000/year; average working days: 250; average reduced sick days: 2.88 and average increase in productivity: 1.50%; Net impact at 250 sf per worker: $5.30/sf
4) Marketing Benefit = Higher rent value of $1.00/sf annually times rentable area
5) CO$_2$ Emission savings = tons of CO$_2$ saved because of reduction in vehicles, electricity and energy consumption; Assumption= Cost of Carbon= $10/ton

Figure 4.5   Sample cost-benefit analysis.
Credit: Green Potential LLC (www.green-potential.com).

each feature were identified and assigned to one or more of the following categories—operational savings (reduced energy costs and water costs), productivity savings (increased productivity and reduced sick days), marketing benefit (high rent value per square foot [sf]) and CO_2 emissions savings (future carbon tax). Payback scenarios for 10 years and 20 years were calculated with an assumed depreciation rate of 3%. Another metric used to calculate the viability of the investment was net present value (NPV). NPV represents the present value of an investment's future financial benefits sans any initial capital costs. The investment should be made only for positive NPV projects. In this example, the NPV of the *green* premium portion of the project was calculated using a discount rate assumption of 5%. The positive NPV and less than 10-year payback period were favorable to the owners, and the proposed features were incorporated into the design. Note that this example does not reflect all the additional features that might typically be required for a LEED project but identifies only a few, for illustrational purposes.

ACTIVITY 3.2—ORGANIZE AN SD-PHASE LEED COORDINATION MEETING

After the preliminary research, calculations, and analyses have been completed by the design team, the LEED PM should organize a coordination meeting to discuss the results of all the analyses, address any issues, and track progress towards achieving the target LEED goal. This meeting may be conducted through a conference call, webinar, or face-to-face meeting. The following paragraphs outline the key agenda items for this coordination meeting.

Discuss analyses and research   The project team should discuss the findings of the energy model, lighting analyses, and other credit calculations to identify the various design measures that need to be integrated into the design. The cost-benefit analysis should be used as a basis to determine the most cost-effective measures based on long-term benefit for the owner. The LEED PM should establish consensus between the team members and discuss any issues or concerns on the implementation of any measure.

Reassess LEED score   The LEED PM should review the LEED assessment that was prepared during the pre-SD phase and evaluate the status of the credits with the design team. Based on the findings of the analyses, the team may realize that some targeted credits may no longer be feasible. The design team should look for other credit opportunities to ensure that the project is still able to achieve its target LEED goals. The LEED PM should discuss any concerns about credits that are crucial to meeting these goals.

Identify next steps   The LEED PM should identify the next steps for the team with respect to integrating the chosen measures into the design. By the end of this meeting, the team should have clear direction on the credits the team is pursuing, their impact on the design development, and the specific *green* measures that need to be integrated into the design.

ACTIVITY 3.3—INTEGRATE LEED CREDIT REQUIREMENTS INTO DRAWINGS

The team's next activity is to proceed with developing the design drawings, using the items discussed at the coordination meeting as guidance. The LEED PM should prepare a detailed documentation checklist for the design team's use so that they are aware of the credit requirements. Refer to the credit-specific implementation presented in Part 2 for a credit-by-credit documentation checklist.

SD LEED Phase
Step 4: Review Schematic
Design Drawings

Objective	The final step in this phase involves the review of SD drawings by the LEED PM to ensure that credit requirements have been appropriately demonstrated. The completion of this review is a significant milestone and will provide a basis for further development of the design drawings.
Recommended Activities	4.1 Review SD drawings. 4.2 Provide Review Feedback.
Roles and Responsibilities	Design Team: Provide SD drawings for review. LEED PM: Review drawings and compile comments.
Duration	3 weeks
Outcome	SD drawings reviewed. Review comments compiled and distributed to team for implementation.

ACTIVITY 4.1—REVIEW SD DRAWINGS

The final step that marks the culmination of the SD LEED phase is the review of the SD drawings by the LEED PM. The purpose of this review is to ensure that the design reflects the necessary *green* strategies that will help the project team demonstrate compliance with the targeted LEED credit requirements. The level of information reflected in the SD drawings will vary from project to project. At this stage, the drawings may not represent all information for all LEED credits. The objective of the review is to ensure that the project is moving in the correct direction and that the design reflects the intent of achieving the targeted credits. The review would also be a means of providing the design team with feedback on missing information that should be considered as the design is developed. Below are some examples of items that should be reflected in the drawings at this stage of the design. Not all items will be applicable to all projects; they will vary depending on the credits that the team is pursuing.

Architectural Drawings
- Building orientation determined based on schematic energy model results
- Building envelope design (wall section, windows, glazing, etc.) determined based on schematic energy model results

- Bike rack locations identified
- Shower and changing rooms incorporated
- Location of alternative-fuel fueling stations identified
- Parking numbers and layout determined— shaded, covered, surface
- Potential preferred parking spaces identified
- Spaces for storage and collection of recyclables identified
- Opportunities for daylighting and access to views reflected in interior layout
- Ventilation strategies—natural ventilation, underfloor air distribution, etc.— reflected in design
- Rooms needing isolation identified

Civil Drawings

- Erosion and sedimentation control measures identified
- Limits of construction disturbance identified
- Stormwater quality and quantity control strategies reflected in design
- Development footprint minimized

Landscape Drawings

- Native and adapted vegetation incorporated
- Vegetated open-space area maximized
- High-efficiency irrigation system planned
- Rainwater harvesting system planned

MEP Drawings

- Site lighting design based on site illumination model results
- High-efficiency plumbing fixtures selected based on water-use analysis
- HVAC system designed based on schematic energy model results
- Ventilation system designed in accordance with ASHRAE 62.1-2007
- Individual lighting and thermal comfort controls incorporated

ACTIVITY 4.2—PROVIDE REVIEW FEEDBACK

After reviewing the SD drawings, the next step is to provide feedback that the design team may use to develop the drawings. The LEED PM may compile the comments in a spreadsheet format as shown in Figure 4.6. This sheet will become a *working check-list* that will be used again for the drawing reviews at the completion of the DD and CD phases. This checklist will help keep track of missing information in the drawings and identify items pending implementation. The LEED PM may schedule a conference call or meeting to discuss the comments.

Project Name		
Date on Drawings		
Reviewed By		
Architectural Drawings Review		
Drawing No.	**Description**	**Credit**
Dwg # A151	Parking Garage Plan	SS C4.3
SD Review Comments [02/15/10]	Designate (65) preferred parking spaces for low emitting and fuel efficient vehicles. Provide signage and/or striping marked as 'FEV' to identify them. These will be spots that will be closest to the building entrances and can be located at all levels of the garage.	
Dwg # A200	First Floor Plan	MR P1
SD Review Comments [02/15/10]	Trash room to show area for storage of recyclables. Coordinate with recycling hauler to determine requirements (one container with all recyclables or separate containers) Minimum recycling required for paper, plastics, glass, metal and cardboard	
Dwg # A205	Roof Plan	SS C7.2
SD Review Comments [02/15/10]	Ensure roof specified has a Solar Reflectance Index (SRI) of at least 78. Indicate roof specification along with SRI on drawings	
Dwg # A802	Details	EQ C5
SD Review Comments [02/15/10]	Show complete isolation (deck to deck partition) of janitors rooms, heavy printing/copying/fax rooms and any other rooms where there may be hazardous chemicals	
MEP Drawings Review		
Drawing No.	**Description**	**Credit**
Dwg # P102	Plumbing Plan	WE C3
SD Review Comments [02/15/10]	Change fixture type to match flow rates discussed at coordination meeting based on water use calculations	
Dwg # M202	Mechanical Plans	EA C4
SD Review Comments [02/15/10]	Include list of refrigerants being used on drawings, perform refrigerant calculations to demonstrate credit compliance;	
Dwg # A202	Mechanical plans	EA C1
SD Review Comments [02/15/10]	Incorporate C02 monitors & outdoor airflow measurement devices as per credit requirements; show on drawings	
Landscape Drawings Review		

Figure 4.6   Example of an SD drawing review feedback sheet.
Credit: Green Potential LLC (www.green-potential.com).

Design Development (DD) LEED Phase

The second phase in Stage II is the DD LEED phase, which includes activities, such as updating calculations and energy analyses and refining the design. Key decisions regarding materials and construction implementation procedures are made during this phase. A review of the DD drawings and revision of the LEED scorecard to ensure that the project is on track to achieve the desired credits is also completed. This phase comprises four steps, which are described in detail in this section. Figure 4.7 presents an overview of these steps and activities.

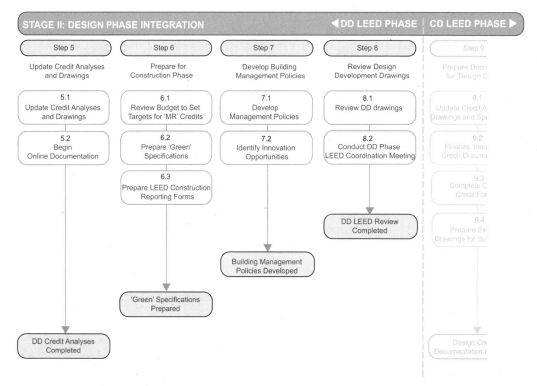

Figure 4.7   Process flow for Stage II: DD LEED Phase.

Credit: Green Potential LLC (www.green-potential.com).

DD LEED Phase
Step 5: Update Credit Analyses and Drawings

Objective	The main objective of this step is to update the credit analyses, refine the design, and continue to integrate *green* measures into the design.
Recommended Activities	5.1 Update Credit Analyses and Drawings. 5.2 Begin Online Documentation.
Responsibilities	Project Team: Continue development of design, update credit analyses, and begin online documentation.
Duration	12 weeks
Outcome	Credit analyses—stormwater, water calculations, energy model, lighting simulation—updated. Drawings updated. Online documentation initiated.

ACTIVITY 5.1—UPDATE CREDIT ANALYSES AND DRAWINGS

During the DD phase, the design team should continue to develop the design with the LEED requirements and goals in mind. The results from the preliminary credit analyses should be used to further optimize the building design. The team should regard the credit analyses as decision support tools that provide feedback on the design direction. Figure 4.8 describes the feedback-oriented nature of the LEED project process and how the drawing review and credit analyses help develop the design.

The involvement of the MEP design team is most significant during this phase to ensure that the HVAC systems are designed and detailed to optimize the building's energy performance. The design team must update the analyses done in the SD phase—energy simulation, lighting simulation, water use, stormwater, etc.—as the design progresses, so that both the analyses and the design are coordinated. Failure to update the credit analyses is risky for the design team, as the final design may deviate so much from the initial analyses' assumptions that making changes to either at that time may be costly and/or time-consuming. A similar situation arose for a LEED project where the energy model was not reviewed after the initial schematic design. The design was developed and just before submitting the documents to GBCI, it was realized that the input assumptions in the energy model were different from those shown in the construction drawings. Specifically, the Solar Heat Gain Coefficient of the glazing and the wall insulation values in the model were different from those listed in the drawings.

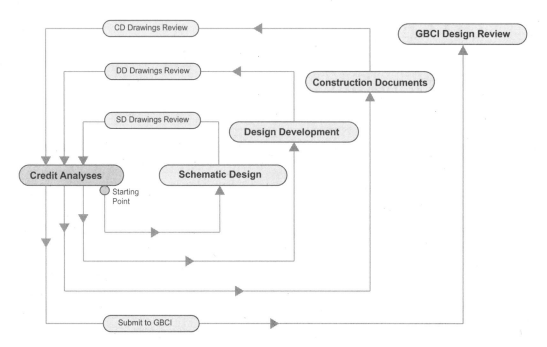

Figure 4.8 **Credit analyses as feedback tools.**
Credit: Green Potential LLC (www.green-potential.com).

Since the vendors for the insulation and windows had already been finalized, it was hard to change either at this stage. The energy model was redone with the proposed glazing and insulation specifications, and it was seen that the building was performing below what was earlier anticipated. As a result of this, the project lost a few credit points and more importantly, the owner lost out on a significant energy savings opportunity. Therefore, it is critical for the design team to be aware of how their design decisions might impact the performance of the building, and they must update their analyses and calculations as the design progresses. The LEED PM must be vigilant in keeping track of design changes and their potential impact on the LEED score.

ACTIVITY 5.2—BEGIN ONLINE DOCUMENTATION

The entire certification process is managed by GBCI through the online portal called LEED Online. Refer to Chapter 2 for details on the certification process and LEED Online. It is recommended that by the end of the DD phase, all team members join the project space on LEED Online. Team members must complete their respective online credit forms and all required calculations. Although this information might change as the design progresses into the CD phase, the purpose of this exercise is to ensure that team members become aware of the documentation requirements, and gain familiarity with using the credit forms. This will also give the LEED PM an opportunity to review the information.

DD LEED Phase
Step 6: Prepare for Construction Phase

Objective	The main objective of this step is to prepare for the successful implementation of LEED credit requirements during the construction phase. This is done by preparing specifications and providing tools that the construction team may use for implementation during construction.
Recommended Activities	6.1 Review Budget to Set Targets for MR Credits. 6.2 Prepare *Green* Specifications. 6.3 Prepare LEED Construction Reporting Forms.
Roles and Responsibilities	Architectural/Design Team: Prepare *green* specifications. General Contractor/Cost Estimator: Set goals for MR credits. LEED PM: Assist in preparation of specifications and reviewing budget to set goals for MR credits; prepare LEED construction reporting forms.
Duration	5 weeks
Outcome	Green Specifications prepared. Targets established for MR credits. Construction reporting forms prepared.

ACTIVITY 6.1—REVIEW BUDGET TO SET TARGETS FOR MR CREDITS

The LEED rating system encourages projects to use materials and construction practices that will help reduce the overall demand on virgin resources and minimize the environmental impact associated with production of *new* materials. Specific credits that address this issue are *MR C3, Materials Reuse, MR C4, Recycled Content, MR C5, Regional Materials, MR C6, Rapidly Renewable Materials,* and *MR C7, Certified Wood.* These credits reward use of salvaged materials, materials with recycled content, materials sourced regionally, rapidly renewable materials, and certified wood products, respectively. In order to achieve the credits, projects must demonstrate utilizing a certain percentage (based on cost) of such materials in their building. As soon as a budget is available, preliminary calculations should be performed to identify materials that will allow the project to meet the required credit percentages. Planning and setting goals at this stage will ensure that materials are procured accordingly.

According to the *LEED Reference Guide,* "material costs include all expenses to deliver the material to the project site and should account for all taxes and transportation costs incurred by the contractor but exclude any cost for labor and equipment once the material has been delivered to the site."[24] Since this information may not be available at the DD phase, teams may perform calculations assuming the default value for materials to be 45% of the total cost. This assumption of the default value is allowed by the LEED rating system. The default values are only for estimate purposes and can be updated later with the actual material costs, when they become available during construction.

This activity may require the collaborative effort of the architect, the general contractor, the cost estimator, and the LEED PM. The following describes the steps to create this preliminary estimate for materials:

1 A line item budget should be obtained only for materials that fall under the following CSI Master Format 2004 Divisions:

Divisions 3 to 10

Division 31: Section 31.60.00—Foundations

Division 32: Sections 32.10.00—Paving, 32.30.00—Site Improvements, and 32.90.00—Planting

Only materials that are permanently installed in the project should be included. Furniture may be included but only if it is consistently included in MR C3 to MR C7. Mechanical, electrical, and plumbing components, as well as specialty items such as elevators should not be included.

2 Next, the line item costs should be multiplied by .45 to estimate the default value of the materials (default value allowed by LEED is 45% of total costs). This is based on the assumption that the actual material costs of the items are not available at this stage.

3 The materials that are anticipated to meet requirements of MR C3 through C7 should be identified based on information available from manufacturer and suppliers. A separate column for each credit category should be created, and the dollar-value equivalent of the default value for credits MR C3, C5, C6, and C7 should be entered. For credit MR C4, postconsumer and pre-consumer recycled-content percentages should be entered to calculate the total recycled-content dollar value, which equals the sum of postconsumer, recycled-content value plus half of pre-consumer, recycled-content value.

4 Thereafter, the total percentages of the material cost that will qualify toward the MR C3–C7 credits should be calculated to see if the required percentage thresholds could be met. If not, then other *high-cost* materials that may help meet the credit percentage thresholds should be identified.

5 Finally, a list of materials/products that will be required to satisfy the various criteria—recycled content, regional materials, etc.—should be prepared. If there are additional materials that will help the project achieve greater percentages, thus, enabling achievement of exemplary performance points, then those may be listed as *optional materials*. The LEED PM should discuss the identified materials with the team to ensure that they are specified in the design. He/she should also obtain confirmation from the general contractor that these materials can be procured and installed.

The contractor can use this preliminary estimate spreadsheet as a basis for procuring the appropriate materials. Once the actual material costs are available, the default value assumption can be updated with actual costs. Figure 4.9 shows an example of a preliminary budget spreadsheet that was created to set percentage targets for MR credits. Specific materials that would help the project meet its targets were identified.

ACTIVITY 6.2—PREPARE *GREEN* SPECIFICATIONS

The next activity is to develop specifications to communicate the various *green* procedures that need to be followed and implemented by the construction team. Ideally,

Preliminary Budget Estimates for LEED MR Credits			MR Credit 3	MR Credit 4			MR Credit 5	MR Credit 6	MR Credit 7
	Total Construction Estimate from Preliminary Budget	Material Cost (Default value= 45% of estimated cost)		Recycled Content					
CSI MasterFormat 2004 Divisions (Applicable towards LEED Credits MR C3 to MR C7)			Materials Reuse	Post-Consumer	Pre-Consumer	Value	Regional Materials	Rapidly Renewable Materials	Certified Wood
	[$]	[$]	[$]	[%]	[%]	[$]	[$]	[$]	[$]
03 00 00 CONCRETE									
03 30 00 Architectural Concrete	$3,000,000	$1,350,000					$1,350,000		
Other sections									
04 00 00 MASONRY									
04 21 13 Brick	$1,000,000	$450,000	$450,000						
Other sections									
05 00 00 METALS									
05 12 00 Steel	$2,000,000	$900,000	$0	90		$810,000			
Other sections									
06 00 00 WOOD, PLASTICS & COMPOSITES									
06 11 00 Wood Framing	$500,000	$225,000							
Other sections									
07 00 00 THERMAL/MOISTURE PROTECTION									
07 12 13 Built Up Asphalt Waterproofing	$350,000	$157,500							
Other sections									
08 00 00 OPENINGS									
08 11 00 Metal Doors and Frames	$450,000	$202,500		75		$151,875			
Other sections									
09 00 00 FINISHES									
09 20 00 Plaster and Gypsum Board Assemblies	$600,000	$270,000							
Other sections									
10 00 00 SPECIALTIES									
10 11 00 Visual Display Surfaces	$250,000	$112,500							
Other sections									
12 00 00 FURNISHINGS (optional)									
All sections									
31 00 00 EARTHWORK									
31 60 00 Foundations	$1,000,000	$450,000							
32 00 00 EXTERIOR IMPROVEMENTS									
32 10 00 Paving	$100,000	$45,000							
32 30 00 Site Improvements	$500,000	$225,000	$100,000						
32 90 00 Planting	$300,000	$135,000							
TOTAL	$10,050,000	$4,522,500	$550,000			$961,875	$1,350,000	$0	$0

Credit	Description	Percentages	Remarks
MR C3	**Materials Reuse**	12.16%	2 points projected
MR C4	**Recycled Content**	21.27%	2 points projected
MR C5	**Regional Materials**	29.85%	2 points projected
MR C6	**Rapidly Renewable Materials**	0.00%	0 points projected
MR C7	**Certified Wood**	0.00%	0 points projected

Figure 4.9 Example of a preliminary budget target for MR credits.
Credit: Green Potential LLC (www.green-potential.com).

the general contractor should be involved right from the beginning of the project and give feedback to the design team on costs and procurement. However, this may not be possible in many project scenarios, such as those where contractors bid on projects after the entire construction document set has been created. Therefore, it is important for the design team to provide specifications that clearly identify the *green* product requirements, the additional construction measures to be implemented, and the document submittals required from the construction team. This will help the construction team to execute the project, as the choice of products (already determined to have LEED value) would be clearly identified in the specifications.

The design team must develop the specifications, covering all aspects of the LEED credits that the team is pursuing. Specification writing programs, such as ARCOM's MasterSpec, are available in the market to facilitate this. Design teams may use such programs as a starting point to insert LEED requirements into the specification sections but must also be careful to exclude any unnecessary or additional requirements that the specification-writing program may generate. It is critical to keep the information simple and relevant to what is required from a LEED credit perspective. The specification writer from the design team should discern what is *required* from what is *optional*. The LEED PM may assist in developing the specification sections.

Green Specifications Sections

The following are key sections that should be added to the Division 1: General Requirements (as per CSI MasterFormat 2004) of the specification manual:

Sustainable Design Requirements (Section 01 81 13)

The purpose of this section is to communicate the general responsibilities of the construction team with respect to meeting LEED requirements. It should include the LEED checklist, product requirements, as well as required submittals and their duration. This section may also be titled as *LEED Requirements*. The specification writer should ensure that only the targeted credits along with their requirements are listed in the checklist (these are credits in the *Yes* category). This will avoid confusion for the construction team that may occur if credits in the *Maybe* category are also included. The intent is to provide clear direction on what *must* be done rather than what *could* be done.

Construction Waste Management (Section 01 74 19)

The purpose of this section is to define the construction and demolition waste management procedures for the construction team. This section should be included if the project team is pursuing the credit *MR C2, Construction Waste Management,* which requires diverting construction waste from landfills by either recycling, reuse, or donation. The section should clearly identify the project's target recycling rate (50% or 90%) and communicate specific procedures that need to be implemented on-site to achieve that rate. The section should also include documentation requirements and a timeline for their submission. It is recommended that a sample/draft construction waste management plan and a waste reduction reporting form be included. These will help the contractor understand the requirements and report information efficiently.

Indoor Air Quality Requirements (Section 01 81 19)

The purpose of this section is to identify measures that need to be implemented during construction and prior to occupancy in order to manage the air quality of the interior spaces. This section should be included if the team is pursuing credits *EQ C3.1, Indoor Air Quality Management—During Construction* and *EQ C3.2, Indoor Air Quality Management—Before Occupancy.* A key documentation requirement for these credits is the inclusion of photographs showing the measures that have been implemented. The section should also include documentation requirements and a timeline for their submission. It is recommended that a sample/draft indoor air quality management plan and an indoor air quality inspection checklist be included.

Testing for Indoor Air Quality (Section 01 81 09)

The purpose of this section is to identify the requirements and procedures for testing the air quality of the interior spaces. This section should be included if the team is pursuing the credit EQ C3.2 through the option of testing indoor air quality. (The other option is to provide a building flush-out whose requirements should be specified in the Section 01 81 19.) The owner and the design team must make a decision on which option will be pursued to ensure that the appropriate option is reflected in the specification sections. Third-party agencies may be engaged by the owner to perform this air-quality testing. The section should communicate the requirements, number of samples to be taken, contaminants to be measured, and reporting requirements.

General Commissioning Requirements (Section 01 91 00)

The purpose of this Section is to identify the commissioning requirements for the project. This section is required to meet the prerequisite *EA P1, Fundamental Commissioning.* Since this is a prerequisite, all LEED projects must include this section. If the team is pursuing credit *EA C3, Enhanced Commissioning,* then additional commissioning requirements may be included. The designated Commissioning Authority should work with the design team/specification writer in developing this section. The section should clearly identify the construction team's responsibilities in implementing commissioning requirements.

Sustainable Design Reporting (Section 01 33 29)

The purpose of this section is to provide instructions to the construction team on the required LEED credit-related documents and the frequency at which they should be submitted. These documents serve as an important progress monitoring tool for the architect and/or LEED PM to ensure that the project is on track to achieve the targeted LEED credits. The frequency of submitting the progress reports may be tied to the payment draws to ensure that the paperwork is not lost. Document submittals may be requested at pre-construction (for approval), during construction (to monitor progress), and finally at project completion.

Sustainable Design Closeout Documentation (Section 01 78 53)

The purpose of this section is to identify the final documents that the contractor must provide to the design team or LEED PM prior to construction closeout. Depending on the credits the team is pursuing, these should include the final construction waste and indoor air quality management plans, waste reduction reports, inspection checklists, photographs, product data, and manufacturers cut sheets, as requested in the specifications.

Language for Other Specification Sections

In addition to the above individual sections, the remaining sections of the specifications must also reflect the requirements of the LEED project. The procedures, products, execution, and submittals required in all sections must echo the LEED credit requirements. For instance, the limits of disturbance for a *greenfield* site are defined by the credit requirements of SS C5.1. If the team is pursuing this credit, then disturbance limits should be specified under *use of premises* in the Section 01 10 00—Summary. The specification writer must carefully integrate LEED items that need to be executed during construction into the specifications. Below are some key parts that might need modifications.

Products   The product requirements must reflect the sustainability criteria based on the targeted LEED credits. For instance, if the team is pursuing MR C4, then product requirements should clearly state the recycled content requirements; or if the team is pursuing credit EQ C4, then product requirements should clearly state the allowable volatile organic compounds (VOCs) limits. Specific products may be identified based on the materials research (Activity 2.1) and the preliminary budget projections (Activity 6.1). A choice of acceptable manufacturers must be provided as well.

Submittals   For each section, the submittal requirements should be clearly identified to ensure that contractor and subcontractors are aware of the documentation expectations.

ACIVITY 6.3—PREPARE LEED CONSTRUCTION REPORTING FORMS

The next activity after integrating the LEED requirements into the specifications is to prepare forms that the contractor and subcontractors may use to document the information requested in the specifications. The LEED PM should prepare these forms in consultation with the architect and the general contractor. If the construction team has prior experience with documenting the LEED credit requirements, then the LEED PM need not prepare these forms. The following are key forms to be prepared.

Waste reduction reporting form   This form should be prepared so that the contractor may report the amount of waste that was diverted or landfilled to demonstrate compliance with the requirements outlined in Section 01 74 19—Construction Waste Management/LEED credit MR C2. Figure 4.10 provides an example of a waste reduction reporting form.

Indoor air quality inspection reporting form   This is a checklist that the contractor's superintendent or designated LEED champion on-site may use to verify that indoor

Project Name:					
Waste Reduction Reporting Form					
Time Period: __/__/__ to __/__/__					
DEMOLITION WASTE	**Diverted Waste**	**Landfilled**	**Commingled Waste Diverted Off-site**	**Off-Site % Diverted**	**Hauler or Location**
Material	(cu. yd./tons)	(cu. yd./tons)	(cu. yd./tons)	%	(Name/Location)
Wood					
Metal Recycled	286				
Inert Materials for Fill	9844				
Concrete Recycled					
Unsorted Debris/Solid Waste		170			
Subtotal Demolition Waste	**10130**	**170**			
CONSTRUCTION WASTE	**Diverted Waste**	**Landfilled**	**Commingled Waste Diverted Off-site**	**Off-Site % Diverted**	**Hauler or Location**
Material	(cu. yd./tons)	(cu. yd./tons)			(Name/Location)
Wood	239				
Metal Recycled	154				
Inert Materials for Fill	30				
Concrete Recycled	225				
Unsorted Debris/Solid Waste		222			
Subtotal Construction Waste	**648**	**222**			
Total Combined Waste	**10778**	**392**			
Total Waste Diverted (%)	**96.49%**				
Signature					
Name					
Company					
This form must be accompanied by Weight Tickets and Receipts					
Diverted Waste includes w aste that w as recyled, salvaged and/or donated					

Figure 4.10 Sample waste reduction report.

Credit: Green Potential LLC (www.green-potential.com).

air quality management measures have been implemented according to the requirements provided in Section 01 81 19—Indoor Air Quality Requirements/LEED credit EQ C3. Figure 4.11 provides an example of an indoor air quality management checklist.

Materials data (MR credits) reporting forms These forms may be used to document information required to complete credits MR C3 to MR C7. There are two

No.	Description of Measure	Yes	No	Remarks

Project Name:

Indoor Air Quality Inspection Checklist

Time Period: __/__/__ to __/__/__

No.	Description of Measure	Yes	No	Remarks
HVAC Protection				
1	MERV 8 filters at any used intakes or returns	x		
2	Intakes and returns sealed with plastic near construction and demolition	x		
3	Mechanical rooms void of construction and waste materials	x		
4	Filters periodically inspected for dust loading			
5	Filters, ducts and air handlers periodically inspected for leakage and repaired	x		
6	Coils, fans, and air handler chambers inspected for cleanliness before startup and cleaned if necessary			
7	All Air handling equipment, ducts, and accessories wrapped and protected or otherwise kept clean during transportation and storage			
Source Control				
1	Low emission chemicals and low VOC products as specified			
2	Vehicle traffic controlled to minimize contaminants			
3	Equipment turned off when not in use			
4	Exhaust pollution sources to outside			
5	Keep containers of wet products or waste materials closed or sealed			
6	Carpets aired out in warehouses for at least three days before installation			
7	Furniture aired out in warehouse for at least seven days before entering building			
Pathway Interruption				
1	Exhaust fans installed to adequately remove dust and contaminants			
2	Provide dust curtains or temporary enclosures			
Housekeeping				
1	All leftover food properly disposed off			
2	Walk off mats installed at entrances and cleaned daily since the time finishes were applied to enclosed building			
3	All workers used proper personal protective equipment per OSHA standards during any operation involving chemicals or dust			
4	No food or drink taken on carpet during or after installation			
5	Dust minimized with wetting agents or sweeping compounds. All sweeping done with dust reducing wax based sweeping compounds			
6	Any accumulated water promptly removed			
7	Porous materials protected			
Scheduling				
1	Schedule high pollution activities prior to installing absorptive materials			
2	Contaminating operations grouped together when possible and exhausted to outside			

Figure 4.11   Sample indoor air quality management report.

Credit: Green Potential LLC (www.green-potential.com).

Project Name:

Materials Data (MR Credits) Reporting Form - General Contractor

General Contractor to fill this form after obtaining information from Subcontractors

Time Period: __/__/__ to __/__/__

CSI MasterFormat 2004 Divisions (Applicable towards LEED Credits MR C3 to MR C7)	Total Bid Cost for Item	Material Cost (Default value= 45% of estimated cost)	Actual Material Cost (from subcontractors) 02/01/2010 to 03/01/2010	MR Credit 3 Materials Reuse	MR Credit 4 Recycled Content Post-Consumer	MR Credit 4 Recycled Content Pre-Consumer	MR Credit 4 Recycled Content Value	MR Credit 5 Regional Materials	MR Credit 6 Rapidly Renewable Materials	MR Credit 7 Certified Wood
	[$]	[$]	[$]	[$]	[%]	[%]	[$]	[$]	[$]	[$]
03 00 00 CONCRETE										
03 30 00 Architectural Concrete	$3,000,000	$1,350,000	$500,000					$500,000		
Other sections										
04 00 00 MASONRY										
04 21 13 Brick	$1,000,000	$450,000	$100,000	$100,000						
Other sections										
05 00 00 METALS										
05 12 00 Steel	$2,000,000	$900,000	$300,000		90		$270,000	$300,000		
Other sections										
06 00 00 WOOD, PLASTICS & COMPOSITES										
06 11 00 Wood Framing	$500,000	$225,000								
Other sections										
07 00 00 THERMAL/MOISTURE PROTECTION										
07 12 13 Built Up Asphalt Waterproofing	$350,000	$157,500								
Other sections										
08 00 00 OPENINGS										
08 11 00 Metal Doors and Frames	$450,000	$202,500								
Other sections										
09 00 00 FINISHES										
09 20 00 Plaster and Gypsum Board Assemblies	$600,000	$270,000								
Other sections										
10 00 00 SPECIALTIES										
10 11 00 Visual Display Surfaces	$250,000	$112,500								
Other sections										
12 00 00 FURNISHINGS (optional)										
All sections										
31 00 00 EARTHWORK										
31 60 00 Foundations	$1,000,000	$450,000								
32 00 00 EXTERIOR IMPROVEMENTS										
32 10 00 Paving	$100,000	$45,000								
32 30 00 Site Improvements	$500,000	$225,000								
32 90 00 Planting	$300,000	$135,000								
TOTAL	**$10,050,000**	**$4,522,500**	**$900,000**	**$100,000**			**$270,000**	**$800,000**	**$0**	**$0**

Credit	Description	Percentages	Remarks	
MR C3	Materials Reuse	2.21%	2 points projected for project	[Percentage calculated over total default value cost]
MR C4	Recycled Content	5.97%	2 points projected for project	[Percentage calculated over total default value cost]
MR C5	Regional Materials	17.69%	2 points projected for project	[Percentage calculated over total default value cost]
MR C6	Rapidly Renewable Materials	0.00%	0 points projected for project	[Percentage calculated over total default value cost]
MR C7	Certified Wood	0.00%	0 points projected for project	[Percentage calculated over total default value cost]

Figure 4.12 Sample materials data (MR credits) report—general contractor.
Credit: Green Potential LLC (www.green-potential.com).

forms required. The first form (example shown in Figure 4.12) is for use by the general contractor to compile all the information gathered from the subcontractors. The second form is for use by the subcontractors to report information (example shown in Figure 4.13).

Project Name:					
Materials Data (MR Credits) Reporting Form - Subcontractors					

1. Fill this form if your material/product falls under any of these Divisions or Sections of CSI Master Format 2004

    Divisions 3 to 10

    Division 31 – Section 31.60.00 Foundations

    Division 32 – Sections 32.10.00 Paving, 32.30.00 Site Improvements, and 32.90.00 Planting

2. Include Material Costs Only. Material Costs include all expenses to deliver the material to the project site; includes taxes and transportation costs; excludes cost of labor & equipment once the material has been delivered to site

3. If any section is not applicable, then indicate with NA

Time Period: __/__/__ to __/__/__

No.	CSI Division	Material/Product Name	Manufacturer's Name	Cut Sheets/ Manuf Letters provided? Y/N	Material Cost ($)
1	05 12 00	Steel	Nucor	Y	$100,000

No.	Material/Product Name	Reused or Salvaged Material	Recycled Content		Distance from Project Site to		Rapidly Renewable Materials	Certified Wood		If Assembly, then Weight of component
		Y/N	% Post consumer	% Pre consumer	Extraction location (miles)	Manufacturing location (miles)	Y/N	Y/N	COC Certificate Number	lbs
1	Steel	N	78	16	450	450	NA	NA	-	NA

Figure 4.13   Sample materials data (MR credits) report—subcontractors.
Credit: Green Potential LLC (www.green-potential.com).

Volatile organic compounds (VOC) data (EQ credit) reporting forms   These forms may be used to document information required to complete credits EQ C4.1 to EQ C4.4. There are two forms that should be included. The first form (example shown in Figure 4.14) is for the contractor to compile all the information gathered from subcontractors. The second form is for the subcontractors to report information pertaining to their products (example shown in Figure 4.15).

Project Name:							
Sample VOC Data (EQ credits) Reporting Form - General Contractor							
General Contractor to complete this form after obtaining information from subcontractors							
Time Period: __/__/__ to __/__/__							
		EQ C4.1- EQ C4.2		EQ C4.3		EQ C4.4	
Manufacturer's Name	Product Name/Model	VOC Limit allowed (g/L)	VOC Level (g/L)	CRI Certified? Y/N	FloorScore Compliant? Y/N	No added ureaforma-ldehyde? Y/N	Source attached? Y/N
Adhesives/Sealants							
Firestone	I.S.O Twin Pack Insulation Ac	70.00	0.00	NA	NA	NA	Y
3M	Fire Barrier Water Tight Seala	250.00	14.00	NA	NA	NA	Y
Paints/Coatings							
Benjamin Moore	Eco Spec Interior Latex Flat 2	50.00	0.00	NA	NA	NA	Y
Carpet							
Atlas	Chartwell	NA	NA	Y	NA	NA	Y
Carpet Cushion							
Hardsurface Flooring							
Composite Wood Products							
Sierra Pine	Medex MDF Panel	NA	NA	NA	NA	Y	Y

Figure 4.14   Sample VOC data (EQ credits) report—general contractor.

Credit: Green Potential LLC (www.green-potential.com).

Project Name:				
VOC Data (EQ credits) Reporting Form - Subcontractors				
Project Name:				
1. Fill this form as applicable for all products used in the interior of the building (anywhere within moisture barrier)				
2. If any section is not applicable, then indicate with NA				
No.	Adhesives/ Sealant Name	Manufacturer's Name	Cut Sheets/ Manufacturer's Letters provided? Y/N	VOC Level (g/L)
1	I.S.O Twin Pack Insulation Adhesive	Firestone	Y	0.00
2	Fire Barrier Water Tight Sealant 3000 W	3M	Y	14.00
No.	Paints/Coatings Name	Manufacturer's Name	Cut Sheets/ Manufacturer's Letters provided?	VOC Level (g/L)
1	Eco Spec Interior Latex Flat 219	Benjamin Moore	Y	0.00
No.	Carpet & Carpet Cushion Name	Manufacturer's Name	Cut Sheets/ Manufacturer's Letters provided? Y/N	CRI Certified (Y/N)
1	Chartwell	Atlas	Y	Y
No.	Hard Surface Flooring	Manufacturer's Name	Cut Sheets/ Manufacturer's Letters provided? Y/N	FloorScore compliant (Y/N)
No.	Composite Wood Product Name	Manufacturer's Name	Cut Sheets/ Manufacturer's Letters provided?	No Added Ureaformaldehyde (Y/N)
1	Medex MDF Panel	Sierra Pine	Y	Y

Figure 4.15 Sample VOC data (EQ credits) report—subcontractors.

Credit: Green Potential LLC (www.green-potential.com).

Photo-log template This is a template (example shown in Figure 4.16) that the contractor may use to document photographs taken during construction. Note that photographs are required to demonstrate implementation of erosion and sedimentation control measures (SS P1), construction waste management (MR C2), and indoor air quality management (EQ C3.1).

Project Name:	
Photo - Log	

[GC to Insert Photo Here]

Photo No.	1/1
Photo Description	HVAC Protection: Ducts sealed in Plastic
Date Taken	

Figure 4.16 Sample Photo-Log template.
Photo Courtesy: Harvard Office of Sustainability and Shawmut Construction.

DD LEED Phase
Step 7: Develop Building
Management Policies

Objective	The purpose of this step is to create policies related to the building operations and maintenance that are required to demonstrate implementation of various LEED credit requirements.
Recommended Activities	7.1 Develop Management Policies. 7.2 Identify Innovation Opportunities.
Responsibilities	LEED PM: Develop management policies. Building Management: Assist in preparation of policies. Design Team: Investigate innovation opportunities.
Duration	3 weeks
Outcome	LEED-related building management policies developed. Innovation opportunities identified.

ACTIVITY 7.1—DEVELOP MANAGEMENT POLICIES

The next activity during the DD phase is to develop policies that will help the building management demonstrate their intent to implement LEED requirements after occupancy. The LEED PM should work with the building owners and/or building management to develop these policies. The policies may be compiled into a single document or integrated with existing policy manuals that the management may have. A copy of the policies should be made available to the building occupants at the time of lease or sale. The key building management programs or policies that need to be developed are:

- Preferred parking policy
- Recycling policy
- No smoking policy
- Thermal comfort survey

Preferred parking policy   The purpose of this policy is to demonstrate the building management's intention of providing preferred parking to drivers of fuel-efficient vehicles and/or carpools. Preferred parking refers to either designated spots closest to the building entrance or parking at a discounted price. This policy will be required if the team is pursuing credits SS C4.3 and/or SS C4.4. The policy should also include a list of qualifying vehicles and the logistics of how it will be enforced. An exhibit plan showing the location of the preferred parking spaces may be included.

Recycling policy   The purpose of the recycling policy is to encourage occupants to recycle materials by providing space(s) for storage and collection of recyclables. At a minimum, there should be provision to recycle paper, cardboard, glass, plastics, and metals. This is required to satisfy the requirements of prerequisite MR P1. The policy should clearly indicate the logistics of transferring the recyclables from the occupants' space to the point source of collection in the building. For instance in an office building, occupants may be requested to recycle paper in bins provided at their individual workstations and asked to recycle other items in the appropriate containers in the break rooms on each floor. Thereafter, custodial staff would transfer the recyclables to the central/common storage and collection area of the building from where the waste hauling company would pick up the recyclable items.

No smoking policy   This is required to demonstrate compliance with prerequisite EQ P2. Depending on the option being pursued, a no smoking policy for the building and the site should be developed. If there are designated areas for smoking, clear information should be provided on where smoking is allowed, within and/or outside the building.

Thermal comfort survey   This survey is required to meet the requirements of EQ C7.2. The building management must agree to conduct a survey of the building occupants after 6 to 18 months of occupancy and then implement a plan for corrective action if more than 20% of the occupants are dissatisfied with their thermal comfort. Either facility operators or outside consultants may be engaged at the time to conduct the survey. Some sample surveys are available at the websites of the Center for the Built Environment (www.cbesurvey.org) and The Usable Buildings Trust (www.usablebuilding.co.uk).

ACTIVITY 7.2—IDENTIFY INNOVATION OPPORTUNITIES

As the design progresses, the design team, owner, building management, and the LEED PM should explore and identify opportunities for innovation in design, construction, and operations of the building. The LEED rating system encourages teams to demonstrate exceptional or innovative performance and awards up to 5 points in the *Innovation in Design* (ID) category. These credits may be obtained in two ways. The first option is to demonstrate exemplary performance beyond the thresholds set forth by a particular credit. The second option is to demonstrate innovation in areas that are not addressed by the credits in the rating system. In order to encourage more teams to explore these opportunities, the USGBC has published a catalog of ID credits that were achieved by previous projects. This catalog is available on the USGBC website. Team members may view the established precedents in the catalog as a starting point to identify innovation opportunities and develop their own ID credits.

It is recommended that the owner/management consider opportunities for innovation in the operations and maintenance of the building to carry forward the *green* design and construction principles into the building operations as well. Some examples of such opportunities are *green* cleaning, *green* education, and preventive maintenance. The benefits of each option should be evaluated against the cost and ease of implementation by the building management. Programs such as *green* cleaning help enhance the indoor air quality of the building; *green* education might help the owners demonstrate their commitment to sustainability and create a positive brand value; and preventive maintenance can help the management save money by extending the expected life of the building's equipment through regular maintenance instead of having high replacement costs because of equipment failure due to lack of proper maintenance.

The LEED PM may investigate various innovation opportunities and present options to the team, including the building management. Some of these options may involve implementation by third-party organizations. Therefore, it is important to identify specific innovation programs during the DD phase, so that a contract with the third-party may be finalized by the end of the CD phase.

DD LEED Phase
Step 8: Review Design Development Drawings

Objective	This step marks the culmination of this phase, where the DD drawings prepared by the team are reviewed by the LEED PM to ensure previous feedback comments have been addressed and the design is on track to achieve its targeted LEED goals. After the review, a coordination meeting may be organized to discuss progress and address any pending issues.
Recommended Activities	8.1 Review DD Drawings. 8.2 Conduct a DD Phase LEED Coordination Meeting.
Responsibilities	LEED PM: Review DD drawings and compile feedback comments.
Duration	3 weeks
Outcome	DD LEED review of drawings completed.

ACTIVITY 8.1—REVIEW DD DRAWINGS

At the end of the DD phase, the LEED PM should review the drawings prepared by the team to determine whether the comments from the previous SD LEED review have been addressed, identify any issues or concerns in achieving any credits, and provide feedback on any additional items that should be included in the drawings. The comments may be compiled in the same spreadsheet that was created during the SD LEED review (Figure 4.6). Figure 4.17 shows an example of a DD phase drawing review.

Since the level of detail in the drawings will be more than that during the SD phase, the review and comments should also be more specific and detailed. Below are some examples of items that should be reflected in the drawings at this stage of the design. Not all items will be applicable to all projects; they will vary depending on the credits that the team is pursuing.

Architectural Drawings
- LEED related general notes and construction notes added
- Bike rack locations finalized, type of bike racks identified
- Shower and changing rooms detailed
- Location of alternative-fuel fueling stations finalized
- Parking numbers and structure finalized
- Preferred parking spaces finalized

Project Name		
Date on Drawings		
Reviewed By		
Architectural Drawings Review		
Drawing No.	**Description**	**Credit**
Dwg # A151	**Parking Garage Plan**	**SS C4.3**
SD Review Comments [02/15/10]	Designate (65) preferred parking spaces for low emitting and fuel efficient vehicles. Provide signage and/or striping marked as 'FEV' to identify them. These will be spots that will be closest to the building entrances and can be located at all levels of the garage.	
DD Review Comments [05/15/10]	Number of preferred parking spots shown are 50- add 15 more spaces	
CD Review Comments [12/15/10]		
Dwg # A200	**First Floor Plan**	**MR P1**
SD Review Comments [02/15/10]	Trash room to show area for storage of recyclables. Coordinate with recycling hauler to determine requirements (one container with all recyclables or separate containers) Minimum recycling required for paper, plastics, glass, metal and cardboard	
DD Review Comments [05/15/10]	SD Review comments addressed- item resolved	
CD Review Comments [12/15/10]		
Dwg # A205	**Roof Plan**	**SS C7.2**
SD Review Comments [02/15/10]	Ensure roof specified has a Solar Reflective Index (SRI) of at least 78. Indicate roof specification along with SRI on drawings	
DD Review Comments [05/15/10]	Roof Specification not provided- please provide	
CD Review Comments [12/15/10]		
Dwg # A802	**Details**	**EQ C5**
SD Review Comments [02/15/10]	Show complete isolation (deck to deck partition) of janitors rooms, heavy printing/copying/fax rooms and any other rooms where there may be hazardous chemicals	
DD Review Comments [05/15/10]	Rooms with isolation have been identified; need to incorporate construction details	
CD Review Comments [12/15/10]		
MEP Drawings Review		

Figure 4.17   Example of DD drawings review.
Credit: Green Potential LLC (www.green-potential.com).

- Signage locations and details included
- High SRI roof specified
- High reflectance paving specified
- Building envelope components—wall sections, window schedules, glazing, etc.—finalized and coordinated with DD energy model
- Spaces for storage and collection of recyclables finalized
- MR and EQ credit requirements reflected in the material and product specifications on drawings
- Weatherstripping for doors provided for smoke control
- Daylighting strategies—window sizes, locations, type, glare control, etc.—reflected according to daylighting simulation analysis
- Deck-to-deck separation for smoking rooms and rooms with hazardous materials included
- Details for entryway systems to trap dirt shown
- Operable window data in accordance with ASHRAE 62.1-2007 shown

Civil Drawings

- Erosion and sedimentation control plan developed and detailed
- Drawings showing limits of construction disturbance developed
- Stormwater quality and quantity control strategies developed and detailed
- Development footprint coordinated with landscape drawings

Landscape Drawings

- Landscape plan developed and detailed
- Native and adapted planting species list provided
- Details on high-efficiency irrigation system included
- Rainwater harvesting system developed and detailed
- Vegetated open-space areas coordinated with civil drawings

MEP Drawings

- Site lighting plan developed and details provided
- Plumbing fixture schedules provided and coordinated with water use analysis
- HVAC system design developed and detailed according to optimized DD energy model
- Ventilation system design developed and detailed as per ASHRAE 62.1-2007
- Individual lighting and thermal comfort controls shown
- Outdoor air flow measurement devices and CO_2 monitors included
- Exhaust systems designed to isolate rooms with hazardous materials

ACTIVITY 8.2—CONDUCT A DD PHASE LEED COORDINATION MEETING

After the review of the DD drawings, the LEED PM should organize a coordination meeting to discuss comments and address any issues. The LEED PM should reassess the LEED score and present an updated LEED scorecard to the design team to keep everyone abreast of any changes. By the end of the DD phase and this meeting, all LEED-related decision-making should be completed by the project design team. This will ensure that there are minimal changes to the design during the CD phase.

Construction Documents (CD) LEED Phase

The third and final phase of Stage II involves completion and submittal of all credit documentation to GBCI for Design Review. This is based on the assumption that project teams will elect to pursue the Split Design and Construction Review path to take advantage of the feedback received from GBCI. However, even if the project team chooses the Combined Design and Construction Review path, it is recommended that all activities be completed up until the final submission to GBCI. This will ensure that all documentation is completed and uploaded to LEED Online while the information is fresh in the mind of the design team members. Typically, it has been seen that it becomes difficult to track and compile all the design information after construction is complete because of the long time gap. There are two steps in this phase and Figure 4.18 describes each of these steps along with recommended activities.

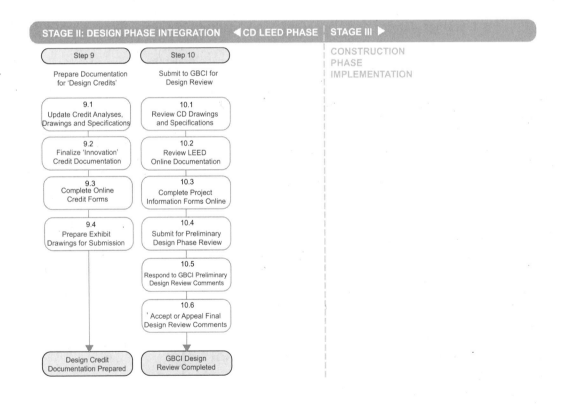

Figure 4.18 **Process flow for Stage II: CD LEED Phase.**
Credit: Green Potential LLC (www.green-potential.com).

CD LEED Phase
Step 9: Prepare Documentation
for *Design Credits*

Objective	The purpose of this step is to compile all information and prepare documentation for submission to GBCI. At this stage only documents related to the *design credits* need to be prepared. All previously prepared calculations, energy models, and lighting simulations should be updated with the most current information as well.
Recommended Activities	9.1 Update Credit Analyses, Drawings, and Specifications.
	9.2 Finalize *Innovation* Credit Documentation.
	9.3 Complete Online Credit Forms.
	9.4 Prepare Exhibit Drawings for Submission.
Responsibilities	Design Team: Finalize documentation.
	LEED PM: Compile documentation; prepare exhibit documentation.
Duration	8 weeks
Outcome	Design credit documentation prepared.

ACTIVITY 9.1—UPDATE CREDIT ANALYSES, DRAWINGS, AND SPECIFICATIONS

During the CD phase, the design team must revisit and revise the initial calculations and analyses to reflect the most current information. The design team should also prepare the final construction drawings and specifications, taking the credit analyses and previous LEED review comments into consideration.

Update Credit Analyses

While not all credit calculations and analyses done during the SD and DD phases may need to be revised, it is important to go through all of them and ensure that the values and information assumed in those calculations is accurate and aligned with the most current design. Some analyses that will need careful coordination and review at this stage are as follows:

Energy simulation analysis   The energy simulation expert must review the input information for both the *baseline* and *proposed* buildings and ensure that it is according to the current design. If the design has changed significantly in terms of areas, building envelope, or systems, all those need to be updated in the energy model. In other words, the energy model must represent the proposed building as accurately as possible. Errors in basic assumptions and discrepancies between the design and the energy model are common mistakes that are found during the GBCI Design Review. The LEED PM should review the efforts of the energy modeler as a quality assurance measure.

Lighting analyses   The lighting consultant for the project must incorporate the latest information regarding the lighting systems, controls, room areas, and building envelope—window sizes, types, glazing specifications, etc.—into the daylighting simulation model that was prepared during the SD and DD phases. Further, the site photometric analysis should be updated to reflect any changes in the exterior lighting fixtures.

Stormwater and water calculations   The civil, landscape, and plumbing engineers must update the calculations pertaining to the stormwater and water use, based on the final decisions made regarding the project's stormwater management systems, irrigation systems, and plumbing fixtures. If captured rainwater is reused for either irrigation or indoor water use, then the engineers must coordinate their calculations to reflect similar amounts of rainwater.

Other credit calculations   In general, all team members must review their respective credit calculations, as applicable, and either update the calculations or let the LEED PM know if there are no changes since the last time the calculations were performed.

Revise Drawings

During the CD phase, as the team finalizes all construction details, they must also incorporate comments from previous LEED reviews. The reason for preparing the construction drawings for submission to GBCI is to submit drawings that are as complete as possible so that changes are not required at a later stage. The drawings must be coordinated with the credit analyses to ensure that the information presented is consistent.

Finalize Specifications

The last item to be updated or finalized during the CD phase is the specification book. The specifications prepared during the DD phase should be updated and finalized to reflect any changes in the design. The LEED PM should review the specifications to ensure that all LEED credit requirements have been appropriately represented and provide any final comments concerning additional changes or revisions for the design team to make.

ACTIVITY 9.2—FINALIZE *INNOVATION* CREDITS DOCUMENTATION

The next step is to finalize the ID credits and develop documentation to demonstrate the innovation ideas. The process of identifying the innovation areas was initiated during the DD phase (Step 7). During the CD phase, the LEED PM should work with the team, the owners, and the building management to finalize those initial ideas. The LEED PM should also review all the credit calculations to determine if any exemplary performance points are available. Once the ideas have been identified, drawings/documents should be prepared to demonstrate the innovation idea. Typically, documentation is prepared by the team member responsible for the area in which the innovation is achieved. For instance, if the project intends to submit an innovative stormwater management strategy, then the civil engineer or the landscape architect may prepare the documents and/or drawings to describe the innovative strategy. The LEED PM may assist the team members in preparing documentation.

ACTIVITY 9.3—COMPLETE ONLINE CREDIT FORMS

The design team members should complete the credit forms on LEED Online by the end of the CD phase. While not all credits require narratives, it is recommended that team members provide at least a brief description on their design and implementation approach to a credit. This narrative will help the GBCI reviewer better understand the process behind the credit implementation.

ACTIVITY 9.4—PREPARE EXHIBIT DRAWINGS FOR SUBMISSION

The construction documents may be submitted as-is without any special reference to LEED credits. However, it is recommended that the relevant aspects of the credits be highlighted for the benefit of the GBCI review team. Note that these drawings will be *exhibit* drawings and not actually part of the construction set. The physical size of the drawings may be the same as the original CD drawings or may be reduced to a tabloid size. Portable document format (PDF) or design web format (DWF) drawings may be used to highlight the LEED aspects. For instance, for credit SS C4.2, the site plan may be highlighted with a circle or arrow to show the location of the bike racks. This activity can be completed by the respective design team members and/or the LEED PM before uploading the documents to LEED Online.

CD LEED Phase
Step 10: Submit to GBCI
for Design Review

Objective	This is the final step of both the construction documents LEED phase as well as the entire Stage II—Design Integration. The main objective of this step is to conduct a last review of the CD drawings and specifications to ensure that all credit requirements have been demonstrated suitably.
Recommended Activities	10.1 Review CD Drawings and Specifications. 10.2 Review LEED Online Documentation. 10.3 Complete Project Information Forms Online. 10.4 Submit for Preliminary Design Phase Review. 10.5 Respond to GBCI Preliminary Design Review Comments. 10.6 Accept or Appeal Final Design Review Comments.
Roles and Responsibilities	LEED PM: Prepare and compile final submittal information and submit to GBCI.
Duration	4 weeks
Outcome	Design credits submitted to GBCI for design review.

ACTIVITY 10.1—REVIEW CD DRAWINGS AND SPECIFICATIONS

This is the third and final review of the construction documents and specifications by the LEED PM before submitting them to GBCI for Design Review. Although there may not be many comments at this stage because of the prior reviews, it is still important to verify one last time that all previous comments have been addressed and that any additional new information is aligned with the overall LEED goals. A similar exercise should be completed with the specifications manual as well.

ACTIVITY 10.2—REVIEW LEED ONLINE DOCUMENTATION

The next important task for the LEED PM is to review all the credit forms, documents, drawings and narratives that have been uploaded to LEED Online by the team members. This is an added quality assurance measure that will allow for any final changes before all the documentation is submitted to GBCI.

What to Look for While Reviewing Online Documentation?

Consistency of information The LEED PM must look for any inconsistencies in information across the various credits. The credit forms in the new version of LEED Online (v3) have been designed to reflect the same information across all forms. But care should be taken to ensure that the information presented in the credit forms is the same as that mentioned in the drawings or documents uploaded.

Missing or inaccurate calculations Any missing items or inaccurate calculations in the credit forms should be identified. Special attention must be paid to the water-efficiency, and energy-model-related calculations.

Missing supporting documents or narratives The LEED PM should look for any documents that are *required* but have not been uploaded by a team member or narratives that have not been included in the credit forms.

Design credit tally Finally, the projected design credit point total should be reviewed to ensure that it aligns with the target LEED goal.

Compile Comments and Conduct Final Coordination

After the final review of the construction drawings, specifications, and online documentation, the LEED PM should compile a list of comments and organize a final coordination meeting to discuss any last-minute changes. Due to the consistent reviews during the progress of the design, the volume of changes that will be required at this stage should be minimal.

ACTIVITY 10.3—COMPLETE PROJECT INFORMATION FORMS ON LEED ONLINE

Next, the LEED PM should complete the information requested in the Project Information forms. The following Project Information forms need to be completed before submitting the documents for GBCI Design Review.

Minimum program requirements LEED-NC v3 requires projects to adhere to minimum program requirements (MPRs) to be eligible for certification. This form lists the seven MPRs and requires acknowledgement from the owner that the MPRs have been satisfied. Table 4.3 identifies the MPRs as defined by GBCI on their website.

Project summary details This form requires entry of basic project information related to the site and building, such as building gross areas, number of stories, site areas, etc. The LEED PM may obtain this information from the project information database that was created during the pre-SD phase and update it as needed.

Occupant and usage data This form requires data regarding the occupancy and type of building—full time equivalent (FTE) occupants, visitors, residents, students, etc. This information may be obtained from the LEED occupancy calculations done during the pre-SD phase.

TABLE 4.3 LEED-NC V3 MINIMUM PROGRAM REQUIREMENTS[25]	
NO.	**MINIMUM PROJECT REQUIREMENTS (MPRS)**
1	Must comply with environmental laws.
2	Must be a complete, permanent building or space.
3	Must use a reasonable site boundary.
4	Must comply with minimum floor area requirements of at least 1000 square feet.
5	Must comply with minimum occupancy rates serving at least 1 full-time equivalent (FTE) occupant.
6	Must commit to sharing whole-building energy and water usage data for at least five years.
7	Must comply with a minimum building area to site area ratio of at least 2%.

Schedule and overview documents   This form requires information regarding the approximate schedule of the project and when each phase of the project was completed—pre-schematic, SD, DD, and so on. The LEED PM may obtain this information from the architect. The form also requires representative drawings to provide a big picture overview of the project.

ACTIVITY 10.4—SUBMIT CREDITS FOR PRELIMINARY DESIGN PHASE REVIEW

The submittal to GBCI for a Design Review is composed of two reviews—one is a preliminary Design Review and the second one is a final Design Review (refer to Chapter 2 for details on the GBCI process). The preliminary Design Review gives the team a chance to provide clarifications to any questions that the GBCI review team may have. At the end of the preliminary Design Review, each reviewed prerequisite and credit will be designated by GBCI as *anticipated*, *pending*, or *denied*, and will be accompanied with technical advice from the review team. Project information forms will be designated as *approved* or *not approved*, and accompanied with technical advice as well.

The first step before submitting the credits is to verify that all credits have been satisfactorily completed by the team members. Once all the *attempted* design credits are marked as *complete*, the online instructions screen will display a description of the next steps. It will indicate whether or not all the attempted design credits are complete and ready to be submitted for preliminary Design Review. Please note that even if there is one attempted credit that is still *in progress*, the option to continue to the Design Review submission will not appear. If there is a credit that is incomplete or was attempted in the beginning of the project but is no longer being attempted, then that credit must be *unattempted* to allow the successful submission of the rest of the credits. Payment options for the certification fee will be provided. The LEED PM should choose the appropriate option and submit the design credits for the preliminary Design Review to GBCI.

ACTIVITY 10.5—RESPOND TO GBCI'S PRELIMINARY DESIGN REVIEW COMMENTS

After the preliminary Design Review is completed, GBCI will send notification via e-mail to the project administrator stating that the preliminary Design Review is complete and available to view on LEED Online. Comments will be provided on each attempted credit, and it will be indicated whether that credit is *anticipated, denied,* or *pending* and needs clarification. All project teams have the option to accept the results of the preliminary Design Review as final. However, at the discretion of the project team, a response to the preliminary Design Review may be submitted via LEED Online. The team is allowed 25 business days to respond with clarifications. The LEED PM should coordinate the project team's response to GBCI by contacting the team members responsible for those credits that require clarifications. The response of the design team must include revised documentation for any prerequisite or project information form marked as *pending* or *denied.* Alternatively, project teams have the option of withdrawing such credits from the application.

ACTIVITY 10.6—ACCEPT OR APPEAL FINAL DESIGN REVIEW COMMENTS

If the project team chooses to accept the results of the preliminary Design Review as final, then this activity will not be required. However, if the team elects to respond to the preliminary Design Review, then GBCI will review the clarifications provided by the project team and issue a final Design Review designating each prerequisite and credit as *anticipated* or *denied* along with technical advice from the review team. The project team has the option of either accepting the final Design Review, or if they are not satisfied with GBCI's decision, then they can appeal the decision by paying a fee of $500 per credit. This appeal should be made within 25 business days of receiving the final review decision from GBCI.

    If no appeals are made, then the LEED PM should accept the Design Review comments and begin the construction application process.

Summary

The integration of LEED requirements during the various phases of design takes the most time among all the stages of the LEED project management process. A number of decisions addressing various components and disciplines need to be made concurrently as early as possible in the design. The role of the design team during this stage is to use the LEED credit framework and analyses as decision support tools. The role of the LEED PM during this stage is to provide research support, design direction, rating system clarifications, and constant oversight/review of the design drawings and specifications. The schedule for implementing these steps depends on the design duration of the project and will vary from project to project. Figures 4.19, 4.20, and 4.21 provide a typical process schedule for the activities for each phase in Stage II. Project

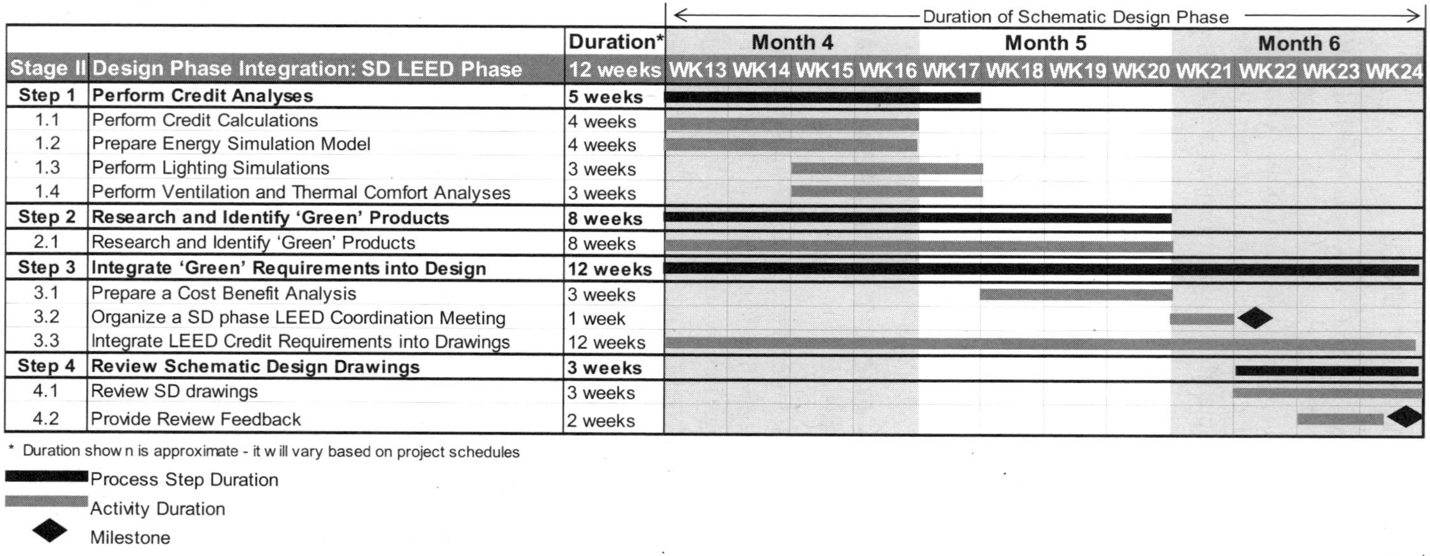

	Stage II Design Phase Integration: SD LEED Phase	Duration*
		12 weeks
Step 1	**Perform Credit Analyses**	**5 weeks**
1.1	Perform Credit Calculations	4 weeks
1.2	Prepare Energy Simulation Model	4 weeks
1.3	Perform Lighting Simulations	3 weeks
1.4	Perform Ventilation and Thermal Comfort Analyses	3 weeks
Step 2	**Research and Identify 'Green' Products**	**8 weeks**
2.1	Research and Identify 'Green' Products	8 weeks
Step 3	**Integrate 'Green' Requirements into Design**	**12 weeks**
3.1	Prepare a Cost Benefit Analysis	3 weeks
3.2	Organize a SD phase LEED Coordination Meeting	1 week
3.3	Integrate LEED Credit Requirements into Drawings	12 weeks
Step 4	**Review Schematic Design Drawings**	**3 weeks**
4.1	Review SD drawings	3 weeks
4.2	Provide Review Feedback	2 weeks

Duration of Schematic Design Phase

Month 4: WK13 WK14 WK15 WK16 WK17 — Month 5: WK18 WK19 WK20 WK21 — Month 6: WK22 WK23 WK24

* Duration shown is approximate - it will vary based on project schedules

Process Step Duration

Activity Duration

◆ Milestone

Figure 4.19 Process Schedule for Stage II: SD LEED Phase.
Credit: Green Potential LLC (www.green-potential.com).

Stage II Design Phase Integration: DD LEED Phase	Duration*	Month 7 WK25 WK26 WK27 WK28	Month 8 WK29 WK30 WK31 WK32	Month 9 WK33 WK34 WK35 WK36
Step 5 **Update Credit Analyses and Drawings**	**12 weeks**			
5.1 Update Credit Analyses and Drawings	12 weeks			
5.2 Begin Online Documentation	3 weeks			
Step 6 **Prepare for Construction Phase**	**5 weeks**			
6.1 Review Budget to Set Targets for 'MR' Credits	2 weeks			
6.2 Prepare 'Green' Specifications	3 weeks			
6.3 Prepare LEED Construction Reporting Forms	3 weeks			
Step 7 **Develop Building Management Policies**	**3 weeks**			
7.1 Develop Management Policies	1 week			
7.2 Identify Innovation Opportunities	7 weeks			
Step 8 **Review Design Development Drawings**	**3 weeks**			
8.1 Review DD drawings	3 weeks			
8.2 Conduct a DD phase LEED Coordination Meeting	1 week			

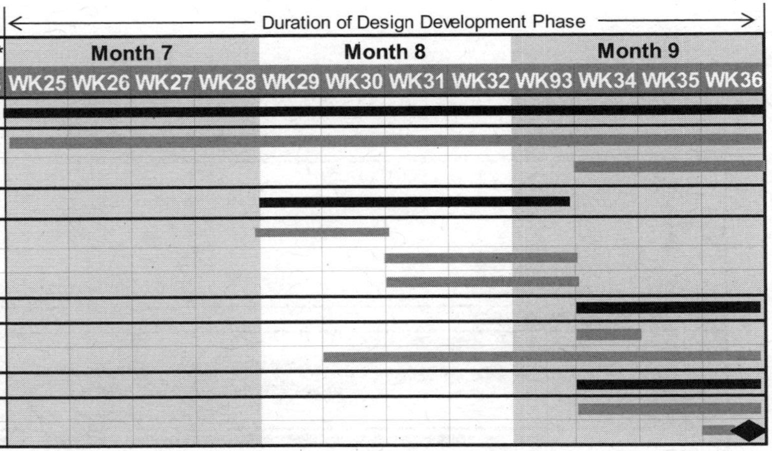

Duration of Design Development Phase

* Duration shown is approximate - it will vary based on project schedules

Process Step Duration
Activity Duration
◆ Milestone

Figure 4.20   Process Schedule for Stage II: DD LEED Phase.
Credit: Green Potential LLC (www.green-potential.com).

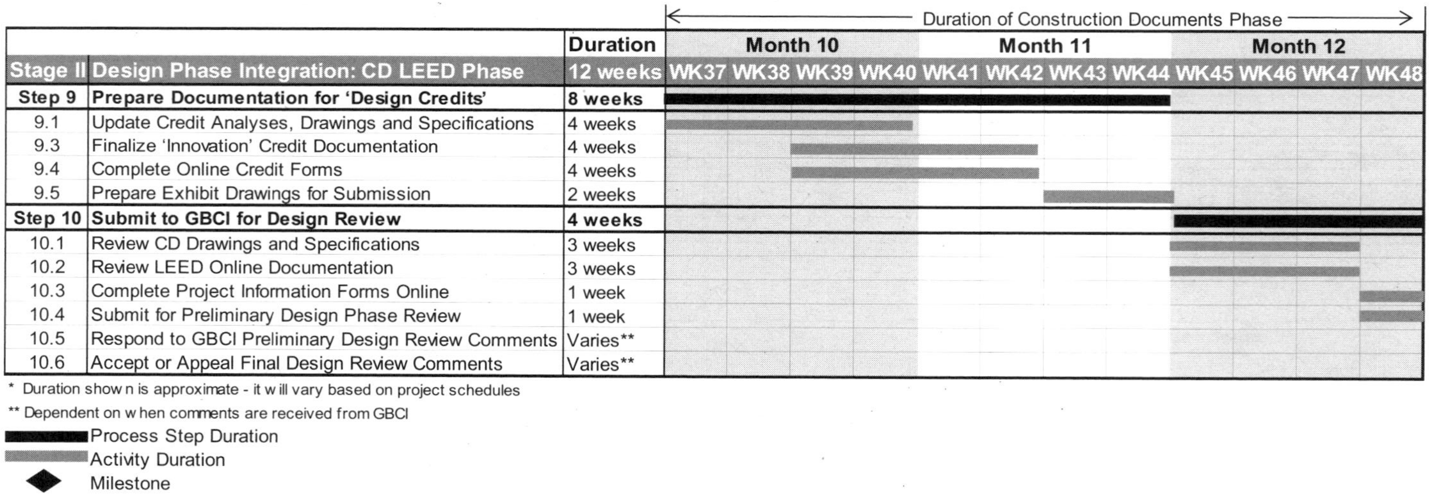

		Duration	Month 10				Month 11				Month 12			
			WK37	WK38	WK39	WK40	WK41	WK42	WK43	WK44	WK45	WK46	WK47	WK48
Stage II Design Phase Integration: CD LEED Phase		12 weeks												
Step 9	**Prepare Documentation for 'Design Credits'**	8 weeks												
9.1	Update Credit Analyses, Drawings and Specifications	4 weeks												
9.3	Finalize 'Innovation' Credit Documentation	4 weeks												
9.4	Complete Online Credit Forms	4 weeks												
9.5	Prepare Exhibit Drawings for Submission	2 weeks												
Step 10	**Submit to GBCI for Design Review**	4 weeks												
10.1	Review CD Drawings and Specifications	3 weeks												
10.2	Review LEED Online Documentation	3 weeks												
10.3	Complete Project Information Forms Online	1 week												
10.4	Submit for Preliminary Design Phase Review	1 week												
10.5	Respond to GBCI Preliminary Design Review Comments	Varies**												
10.6	Accept or Appeal Final Design Review Comments	Varies**												

— Duration of Construction Documents Phase —

* Duration shown is approximate - it will vary based on project schedules

** Dependent on when comments are received from GBCI

█ Process Step Duration

█ Activity Duration

◆ Milestone

Figure 4.21 Process Schedule for Stage II: CD LEED Phase.

Credit: Green Potential LLC (www.green-potential.com).

Managers may need to adapt the durations of each activity to their project. Figure 4.22 presents a checklist that project managers may use to mark the completion of specific activities and steps in this stage.

Stage II Design Phase Integration		Due Date	Not Started	In Progress	Complete	Remarks
Project Name:						
SD LEED Phase						
Step 1	**Perform Credit Analyses**	**10/26/2010**				
1.1	Perform Credit Calculations	10/19/2010		x		
1.2	Prepare Energy Simulation Model	10/26/2010		x		
1.3	Perform Lighting Simulations	10/26/2010		x		
1.4	Perform Ventilation and Thermal Comfort Analyses	10/26/2010		x		
Step 2	**Research and Identify 'Green' Products**	**10/26/2010**		x		
2.1	Research and Identify 'Green' Products	10/26/2010	x			
Step 3	**Integrate 'Green' Requirements into Design**	**10/26/2010**	x			
3.1	Prepare a Cost Benefit Analysis					
3.2	Organize a SD phase LEED Coordination Meeting					
3.3	Integrate LEED Credit Requirements into Drawings					
Step 4	**Review Schematic Design Drawings**					
4.1	Review SD drawings					
4.2	Provide Review Feedback					Milestone
DD LEED Phase						
Step 5	**Update Credit Analyses and Drawings**					
5.1	Update Credit Analyses and Drawings					
5.2	Begin Online Documentation					
Step 6	**Prepare for Construction Phase**					
6.1	Review Budget to Set Targets for 'MR' Credits					
6.2	Prepare 'Green' Specifications					
6.3	Prepare LEED Construction Reporting Forms					
Step 7	**Develop Building Management Policies**					
7.1	Develop Management Policies					
7.2	Identify Innovation Opportunities					
Step 8	**Review Design Development Drawings**					
8.1	Review DD drawings					Milestone
8.2	Conduct a DD phase LEED Coordination Meeting					
CD LEED Phase						
Step 9	**Prepare Documentation for 'Design Credits'**					
9.1	Update Credit Analyses, Drawings and Specifications					
9.3	Finalize 'Innovation' Credit Documentation					
9.4	Complete Online Credit Forms					
9.5	Prepare Exhibit Drawings for Submission					
Step 10	**Submit to GBCI for Design Review**					
10.1	Review CD Drawings and Specifications					
10.2	Review LEED Online Documentation					
10.3	Complete Project Information Forms Online					
10.4	Submit for Preliminary Design Phase Review					
10.5	Respond to GBCI Preliminary Design Review Comments					
10.6	Accept or Appeal Final Design Review Comments					

Figure 4.22   Process Checklist for Stage II.
Credit: Green Potential LLC (www.green-potential.com).

Notes

[1] USGBC, *LEED Reference Guide for Green Building Design and Construction* (USGBC, 2009), 91.

[2] Ibid.

[3] USGBC, *LEED Reference Guide for Green Building Design and Construction* (USGBC, 2009), 165.

[4] Ibid., 101.

[5] Energy Design Resources, *Design Brief: Building Simulation* (Energy Design Resources, Jan. 01, 2002), http://www.energydesignresources.com/Resources/Publications/DesignBriefs/tabid/74/articleType/ArticleView/articleId/3/Design-Briefs-Building-Simulation.aspx (accessed February 2010).

[6] USGBC, *LEED Reference Guide for Green Building Design and Construction* (USGBC, 2009), 237.

[7] Ibid., 257.

[8] Ibid., 263.

[9] Ibid.

[10] Ibid., 129.

[11] Public Technology Inc., *Sustainable Building Technical Manual* (Public Technology Inc., U.S. Department of Energy, 1996), 90.

[12] Norm G. Miller, Dave Pogue, *Do Green Buildings Make Dollars and Sense?* (Burnham-Moores Center for Real Estate, University of San Diego, CB Richard Ellis, 2009)

[13] USGBC, *LEED Reference Guide for Green Building Design and Construction* (USGBC, 2009), 564.

[14] Energy Design Resources, *Design Briefs: Indoor Air Quality* (Energy Design Resources, 2003), http://www.energydesignresources.com/Resources/Publications/DesignBriefs/tabid/74/articleType/ArticleView/articleId/123/Design-Briefs-Indoor-Air-Quality.aspx (accessed April 2010).

[15] Ibid.

[16] Energy Design Resources, *Design Briefs: Indoor Air Quality* (Energy Design Resources, 2003), http://www.energydesignresources.com/Resources/Publications/DesignBriefs/tabid/74/articleType/ArticleView/articleId/123/Design-Briefs-Indoor-Air-Quality.aspx (accessed April 2010).

[17] Center for the Built Environment, "Underfloor Air Technology: Overview," (The Regents of the University of California), http://www.cbe.berkeley.edu/underfloorair/default.htm.

[18] USGBC, *LEED Reference Guide for Green Building Design and Construction* (USGBC, 2009), 335.

[19] Ibid., 368.

[20] Ibid., 364.

[21] Ibid., 377.

[22] Ibid.

[23] Alex Wilson, "Building Materials: What Makes a Product Green?, *Environmental Building News*, Vol. 9, No. 1 (2006), http://www.buildinggreen.com/auth/article.cfm/2000/1/1/Building-Materials What-Makes-a-Product-Green/.

[24] USGBC, *LEED Reference Guide for Green Building Design and Construction* (USGBC, 2009), 372.

[25] GBCI, "LEED 2009 Minimum Program Requirements," http://www.gbci.org/DisplayPage.aspx?CMSPageID=130.

5

STAGE III: CONSTRUCTION PHASE IMPLEMENTATION

Introduction

The final stage of the LEED project management process involves implementing *green* construction practices and procedures on-site. The activities in this stage align with the construction phase of the building and include procuring *green* materials, implementing construction measures, commissioning systems, and recording information. This chapter outlines a set of steps that will help the construction team understand their LEED responsibilities and execute the project efficiently. The steps may seem to be very simplistic to the reader but these are listed to give the reader a framework for systematic and timely completion of the project.

Construction can begin independently of the GBCI review process without waiting for feedback from GBCI on the Design Review. This stage entails organizing a pre-construction LEED kick-off meeting, monitoring construction progress, and finally, submitting documents to GBCI for Construction Review. The responsibilities for these activities primarily lie with the general contractor (GC), the subcontractors, the owner, and the LEED PM. Because construction schedules vary from project to project, the steps described in this chapter may last for different periods of time from what is shown. This stage has five primary steps, and each step has been further described with recommended activities. Figure 5.1 presents the process flow for these steps. The five steps that comprise this stage are as follows:

Step 1: Organize a Pre-Construction LEED Kick-Off Meeting

Step 2: Review Progress of Credit Implementation

Step 3: Prepare Documents for GBCI Construction Review

Step 4: Review Final Documents for GBCI Submission

Step 5: Submit to GBCI for Construction Review

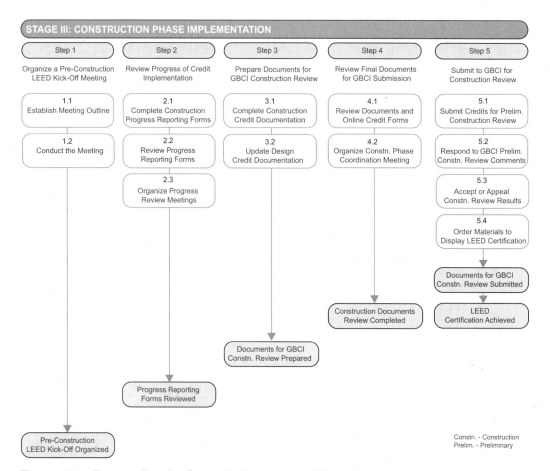

Figure 5.1   Process flow for Stage III: Construction Phase implementation.
Credit: Green Potential LLC (www.green-potential.com).

Step 1: Organize a Pre-Construction LEED Kick-Off Meeting

Objective	The main objective of this step is to inform the construction team of the procedures to follow and the information to record in order to implement the requirements of the LEED construction credits. The meeting will also help address any questions of the construction team and establish protocols for communication between the various groups.
Recommended Activities	1.1 Establish Meeting Outline. 1.2 Facilitate Kick-Off Meeting.
Roles and Responsibilities	LEED PM: Organize pre-construction LEED kick-off meeting. Construction Team: Participate in meeting. Design Team and Owner: Participate in meeting.
Duration	2 weeks
Outcome	Pre-construction LEED kick-off meeting organized.

ACTIVITY 1.1—ESTABLISH MEETING OUTLINE

The first step is to prepare an outline for the pre-construction kick-off meeting to ensure that all participants are aware of the goals and objectives of the meeting. The LEED PM should prepare this outline. This section is written with the assumption that there is a single LEED PM who is managing the process through both design and construction phases. However, in some project scenarios, this may not be the case. The LEED PM who was involved with managing the project during the design phase may not be associated with it during construction. The GC may have been provided with construction documents and asked to build a LEED project, with no direction or training in LEED-related construction procedures. In such circumstances, it is recommended that the GC procure the services of a LEED consultant to conduct this kick-off meeting and assist with documentation during construction. If the GC has in-house LEED specialists, then they may conduct the meeting as well.

The LEED PM should prepare a detailed agenda with input from the GC to ensure that the suggested items address all the necessary topics. The purpose of this meeting is to:

- Introduce the LEED aspects of the project to the construction team.
- Discuss responsibilities of all crew members and designate an on-site LEED champion/construction manager.
- Discuss the construction waste management plan.
- Discuss the construction indoor air quality management plan.

- Provide specific direction to subcontractors on expectations regarding materials used on-site.
- Discuss the communication protocol and the schedule for submittal of LEED-related information.
- Address any issues, concerns, or questions from the construction team concerning implementation.
- Encourage enthusiasm for the project and provide a sense of ownership to the crew for the success of the project.

The LEED PM should discuss with the GC the length of the meeting and the topics to be covered prior to sending out the outline. The outline should include:

- Date, location, and time of the meeting
- Goals of the meeting
- Detailed agenda with time slots for planned activities

The LEED PM should procure from GC a list of all people who should be invited to the meeting and send out invitations. Below is a sample outline for a pre-construction kick-off meeting.

Sample Outline for Pre-Construction LEED Kick-Off Meeting

Project Name: YZ Corp Office
Pre-Construction LEED Kick-Off Meeting Outline

Date: Friday, February 5, 2010   Time: 9:30 AM–12:00 PM Central Time
Location: Offices of LMN Construction
Address: 1010 Congress Avenue, Suite 200, Atlanta, GA

Goals of the Meeting:

1 Understand LEED credit-related responsibilities
2 Discuss construction waste management plan
3 Discuss indoor air quality management plan
4 Discuss subcontractor responsibilities
5 Discuss forms, checklists, submittals
6 Future steps—immediate action items, project schedule

Agenda:

1 Introductions	10 minutes
2 Discuss LEED construction credits	50 minutes
3 General contractor responsibilities	15 minutes
4 Subcontractors responsibilities	15 minutes
5 Discuss submittals	30 minutes
6 Next steps and questions	30 minutes

ACTIVITY 1.2—FACILITATE KICK-OFF MEETING

After sending the outline, the next activity is to facilitate the pre-construction kick-off meeting. Prior to the meeting, the LEED PM must compile all necessary material for discussion at the meeting. Most of the items that are required to conduct the pre-construction kick-off meeting should have been prepared during the DD phase as part of the specifications and the LEED construction reporting forms. (Refer to Chapter 4, Step 6 activities.) The following items should be prepared and discussed at the meeting:

- General responsibilities with LEED credit checklist
- Construction waste management procedures
- Indoor air quality management procedures
- Material and product procurement
- Documentation requirements
- Communication protocol

General Responsibilities

The construction team working on a LEED project is required to follow *green* construction practices, procure products, and provide documentation, in addition to the standard construction procedures. Although the construction team is indirectly responsible for execution of all LEED credits, it is directly responsible for the implementation of up to 15 credits that are categorized as *construction credits* by the LEED rating system.

The first step is to clearly define these additional responsibilities. The LEED PM may prepare a detailed list of responsibilities using the team roles and responsibilities section described in Chapter 2 as a guiding framework. Table 5.1 provides an example of a list of general responsibilities prepared for a project's pre-construction kick-off meeting.

TABLE 5.1   EXAMPLE OF GENERAL LEED RESPONSIBILITIES OF CONSTRUCTION TEAM

NO.	GENERAL RESPONSIBILITIES
1	Procure all materials and products as outlined in drawings and specifications. (No substitutions allowed unless prior approval sought from architect or LEED PM.)
2	Implement the measures outlined in the erosion and sedimentation control plan.
3	Prepare and implement construction waste management plan to recycle demolition and construction waste.
4	Prepare and implement construction indoor air quality management plan.
5	Follow the procedures highlighted in Commissioning Plan with respect to testing and commissioning of systems.
6	Document all procedures implemented by means of photographs, inspection logs, and reports.
7	Provide documents requested for material/product data, manufacturer data sheets, etc. Provide a record of all receipts.

LEED Construction Credit Checklist

In addition, the LEED PM may prepare a checklist of the construction credits that the team is pursuing, highlighting the responsibilities of the construction team. Figure 5.2 illustrates an example of a checklist prepared for a project. This checklist was used as a ready reference by the GC for the duration of the project.

Project Name:		Points	Responsibilities
LEED Construction Credit Responsibilities Checklist			
The following is a list of construction credits applicable to the project			
General Contractor to use this as a quick reference for credits and responsibilities			
Sustainable Sites			
Prereq 1	Construction Activity Pollution Prevention		Implement erosion and sedimentation control measures; document via photographs
Credit 5.1	Site Development—Protect or Restore Habitat	1	Maintain limits of disturbance as per civil dwg; document via photographs
Energy and Atmosphere			
Prereq 1	Fundamental Commissioning of Building Energy Systems		Follow Commissioning Plan
Credit 3	Enhanced Commissioning	2	Follow Commissioning Plan
Materials and Resources			
Credit 2	Construction Waste Management	2	Implement construction waste management; divert as much as waste as possible from landfills; provide hauling receipts
Credit 3	Materials Reuse	2	Procure materials as per specs; track information via forms provided; provide manufacturers data
Credit 4	Recycled Content	2	
Credit 5	Regional Materials	2	
Credit 6	Rapidly Renewable Materials	1	
Credit 7	Certified Wood	1	
Indoor Environmental Quality			
Credit 3.1	Construction IAQ Management Plan—During Construction	1	Implement indoor air quality management measures during construction; document via photographs
Credit 3.2	Construction IAQ Management Plan—Before Occupancy	1	Perform building flush-out
Credit 4.1	Low-Emitting Materials—Adhesives and Sealants	1	Procure materials as per specifications; track information via forms provided; provide manufacturers data
Credit 4.2	Low-Emitting Materials—Paints and Coatings	1	
Credit 4.3	Low-Emitting Materials—Flooring Systems	1	
Credit 4.4	Low-Emitting Materials—Composite Wood and Agrifiber Products	1	

Figure 5.2   Example of LEED construction credit responsibilities checklist.
Credit: Green Potential LLC (www.green-potential.com).

TABLE 5.2 EXAMPLE OF CONSTRUCTION WASTE MANAGEMENT PROCEDURES

NO.	CONSTRUCTION WASTE MANAGEMENT PROCEDURES
1	All crew members to recycle construction waste in designated containers. Construction waste includes, but is not limited to, wood, metal, gypsum board, concrete, packaging waste, and on-site personnel waste; lists of acceptable/unacceptable materials will be posted.
2	All subcontractors will be given copy of the waste management plan to be distributed among their on-site crew members.
3	Clearly labeled (bilingual) recycling containers will be provided and locations for their placement will be identified.
4	Subcontractors to notify on-site supervisor of additional containers needed; recyclable waste should not be put in trash dumpsters for any reason.
5	GC's on-site supervisor will oversee waste management implementation.
6	Copy of construction water management plan will be posted on site.
7	GC to keep record of all hauling receipts and provide LEED PM with monthly waste reduction reports.

Construction Waste Management Procedures

The next step is to prepare a comprehensive list of procedures to be followed on-site by the team to manage construction waste. This list is necessary if the team is pursuing the credit *MR C2, Construction Waste Management*, which requires diverting construction waste from landfills by either recycling, reuse, or donation. Prior to preparing this list, the LEED PM should get the GC's input on their site-specific plan for managing construction waste, including the items that will be recycled, how they will be separated, and any special requirements of the haulers. It is recommended that the GC provide the LEED PM with a final construction waste management plan that will be followed on the job site prior to the kick-off meeting. This is to ensure that the procedures discussed at the meeting are as specific as possible to the job site. The procedures should also be coordinated with the specification section 01 74 19. Table 5.2 provides an example of a list of construction waste management procedures that was prepared for a project.

Indoor Air Quality (IAQ) Management Procedures

The indoor air quality management procedures should be included if the team is pursuing credits *EQ C3.1, Indoor Air Quality Management—During Construction* and *EQ C3.2, Indoor Air Quality Management—Before Occupancy*. The LEED PM should use the procedures described in the specification Section 01 81 19 to prepare this list. The purpose is to present key items that must be followed by all on-site crew members to ensure that the required indoor air quality management measures are implemented. If a draft IAQ management plan was included in the specifications, then the GC may use that to modify and finalize a plan specific to the project and provide it to the LEED PM. According to the LEED Reference Guide, the indoor air quality measures should follow the recommendations of the Sheet Metal and Air Conditioning National Contractors Association (SMACNA) IAQ Guidelines for Occupied Buildings under Construction. Table 5.3 provides an example of a list of indoor air quality management procedures that was prepared for a project.

TABLE 5.3   EXAMPLE OF INDOOR AIR QUALITY MANAGEMENT PROCEDURES DURING CONSTRUCTION (BASED ON SMACNA GUIDELINES)

NO.	INDOOR AIR QUALITY MANAGEMENT PROCEDURES

1   HVAC Protection

- HVAC system to be shut off during heavy construction or demolition.
- If HVAC system is used, then MERV 8 filters should be provided at return and supply openings; filters should be replaced prior to occupancy.
- Supply diffusers and return openings should be sealed with plastic whenever they are not in use to prevent the accumulation of dust and debris in the duct system.
- No construction or waste materials may be stored in mechanical rooms.
- Ducts, diffusers, and window units should be inspected for deposits of particulate matter and cleaned before occupancy.

2   Source Control

- Only low VOC products, as indicated by the specifications, should be used.
- No idling of motor vehicles allowed near building entrances to prevent emissions from being drawn into the building.
- Equipment should be cycled off when not being used or needed.
- Pollution sources should be exhausted to the outside with portable fan systems.
- Containers of wet products should be kept closed as much as possible.

3   Pathway Interruption

- Dust curtains or other partitions or enclosures should be provided to prevent dust from migrating to occupied spaces.
- Pollutant sources to be located as far away as possible from supply ducts and areas occupied by workers, when feasible.

4   Housekeeping

- Suppress and minimize dust with wetting agents or sweeping compounds.
- All surfaces should be kept clean, including inside mechanical equipment and hard-to-reach places such as behind furniture.
- Protect porous materials, such as insulation and ceiling tile, from exposure to moisture.
- Spills or excess applications of solvent-containing products should be removed.

5   Scheduling

- High pollution activities that utilize high VOC level products (including paints, sealers, insulation, adhesives, caulking, and cleaners) should be scheduled to take place prior to installing highly absorbent materials (such as ceiling tiles, gypsum wall board, fabric furnishings, carpet, and insulation).

Materials and Product Procurement

One of the most critical responsibilities of the construction team is the procurement of appropriate materials and products according to the drawings and specifications. A total of eight LEED credits are directly dependent on this aspect, and if any product or material substitutions are made without approval, the credits may not be achieved. The LEED PM should communicate to the construction team the importance of pro–curing the specified materials. Emphasis should be placed on the VOC levels allowed for all interior adhesives, sealants, paints, and coatings; and crew members should be requested to alert the on-site supervisor to any procurement issues. The GC should post the allowable VOC content list on the job site, and a copy should be provided to all subcontractors.

Project Name:		
LEED Construction Submittals Checklist		
Credit	**Description**	**Date**
SS P1	**Construction Activity Pollution Prevention**	**Due by**
Preconstruction	NA	
Progress	NA	
Final	Photographs of erosion and sedimentation control measures	
MR C2	**Construction Waste Management**	**Due by**
Preconstruction	Construction Waste Management Plan	
Progress	Waste Reduction Reports	
Final	CWM Plan, Photographs, hauling receipts, final waste reduction report	
MR C3-MR C7	**Construction Activity Pollution Prevention**	**Due by**
Preconstruction	Preliminary Line-Item Budget with target goals	
Progress	Preliminary budget sheet updated with real cost data from subcontractors	
Final	Final line item costs with total percentages of materials, manufacturers data collected from subcontractors	
EQ C3.1	**Indoor Air Quality Management (during construction)**	**Due by**
Preconstruction	Indoor Air Quality Management Plan- during construction	
Progress	Inspection checklists	
Final	Photographs, compiled inspection checklists, IAQ plan, cut sheets of filtration media used during construction	
EQ C3.2	**Indoor Air Quality Management (before occupancy)**	**Due by**
Preconstruction	Indoor Air Quality Management Plan - before occupancy	
Progress		
Final	Photographs, IAQ plan with flush-out procedure or testing data	
EQ C4.1- C4.4	**Low-Emitting Materials**	**Due by**
Preconstruction	NA	
Progress	Materials data compiled from subcontractors	
Final	Final list of all materials- product name, VOC content, manufacturer's name and data sheets	

Figure 5.3 LEED construction credit submittals checklist.
Credit: Green Potential LLC (www.green-potential.com).

Submittal Schedule

The next step is to communicate the schedule for submittal of all documents and forms required from the GC and the subcontractors. Figure 5.3 provides an example of a submittals schedule prepared for a project.

Communication Protocol

Finally, the communication protocol or roadmap outlining how the information will flow from one group to the other should be discussed. This is important to make sure that information is not lost and all the construction team members are aware of their responsibilities with regard to submittal documentation. The protocol should include a construction manager on the GC's team to be the LEED champion, who will act as the liaison between the construction team and the LEED PM. Figure 5.4 provides an example of a communication protocol for LEED construction information for a project.

The pre-construction LEED kick-off meeting should be viewed as an opportunity not only for the LEED PM to present the various procedures and documentation protocols but also to clarify any questions or concerns of the construction team. The LEED PM should facilitate the process of open dialogue and communication to ensure that the

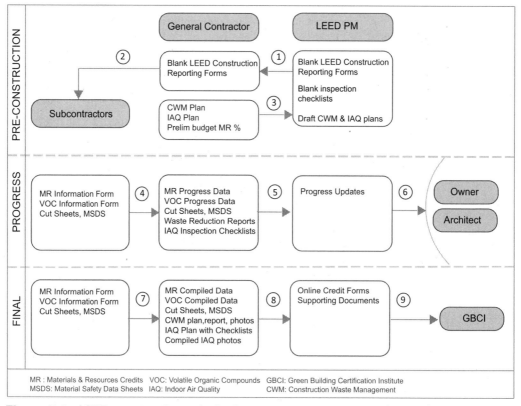

Figure 5.4   LEED construction submittals communication protocol.
Credit: Green Potential LLC (www.green-potential.com).

TABLE 5.4 EXAMPLE OF FAQS RELATED TO CONSTRUCTION CREDITS

Question: What areas are included in "inside the weatherproofing system" with respect to using low VOC adhesives, sealants, paints? Is it from the interior gyp-board and within or from exterior sheathing and within?

Answer: All sealants, adhesives, paints, and coatings used on surfaces from the exterior weatherproofing membrane to the inside of the building have to meet the VOC requirement set forth by LEED credits EQ C4.1–EQ C4.2. The *LEED Reference Guide* further states that "this includes all surfaces in contact with indoor air, such as flooring; walls; ceilings; interior furnishings, suspended ceiling systems and the materials above those suspended ceilings; ventilation system components that contact the ventilation supply or return air; and all materials inside wall cavities, ceiling cavities, floor cavities, or horizontal and vertical chases. It also includes caulking materials for windows and ceiling or wall insulation"[1]. Any material used outside of the weatherproofing membrane like siding does not have to meet the VOC requirement.

Question: Does the VOC requirement apply to products/materials stained off-site?

Answer: No, the VOC requirement is only for adhesives, sealants, paints, and coating used on-site. If any pre-stained wood or other material is obtained, those are not required to meet the VOC requirements set forth in EQ 4.1 and EQ 4.2.

Question: Can the use of crushed concrete/brick/asphalt from previously demolished building as a base for the new project qualify for any credits?

Answer: Yes, reusing demolished asphalt and crushing it for reuse either on-site or on any other site will contribute toward *MR C2, Construction Waste Management*, where demolition waste is diverted from landfill. It will also contribute towards credit *MR C5, Regional Materials*.

Question: Would the reuse of land-clearing debris like soil, vegetation, or rocks qualify for any credit?

Answer: Reuse of land-clearing debris like soil, vegetation, or bedrock will not be applicable for either MR C2 or MR C4, but can be applied to MR C5. The reason for this is that MR C2 is applicable only to demolition and construction waste, not naturally found site features; MR C4 is applicable only to products that have recycled content in them.

project is built according to expectations. Table 5.4 provides some examples of queries by the GC and subcontractors at a pre-construction LEED kick-off meeting.

Commissioning Activities

Note that this pre-construction kick-off meeting should be held in addition to the pre-construction commissioning kick-off meeting that may be organized by the Commissioning Authority (CxA). The CxA should organize the meeting at an appropriate time prior to the procurement of systems to be commissioned. The CxA and LEED PM may coordinate and arrange the two meetings together. Fundamental commissioning is a prerequisite (EA P1), and the contractors must adhere to the commissioning plan and specifications provided by the CxA.

Step 2: Review Progress of Credit Implementation

Objective	The objective of this step is to review the implementation and documentation of LEED credit requirements periodically to ensure the targeted LEED goals will be achieved. The GC is primarily responsible for collecting information from subcontractors and submitting progress reports.
Recommended Activities	2.1 Complete Construction Progress Reporting Forms.
	2.2 Review Progress Reporting Forms.
	2.3 Organize Progress Review Meetings.
Roles and Responsibilities	GC and subcontractors: Provide construction progress reporting forms.
	LEED PM: Review progress reporting forms.
Duration	Lasts the entire construction period.
Outcome	Progress of implementation of LEED requirements during construction monitored.

ACTIVITY 2.1—COMPLETE CONSTRUCTION PROGRESS REPORTING FORMS

During the entire construction period, the GC should collect information from sub-contractors, maintain records, compile information into progress reports, and submit them to the LEED PM. This is to ensure that the progress of the project in meeting the construction credit requirements is monitored. This activity may be completed by the LEED champion designated from the GC's team. The GC should use the forms provided (LEED construction reporting forms described in Step 6 of Chapter 4) to submit information to the LEED PM. At a minimum, the following reports should be provided:

- Construction waste reduction progress report
- Indoor air quality management report
- Materials credits progress report
- VOC data progress report

Construction waste reduction progress report   The GC should provide waste reduction reports showing total amount of waste generated on-site and the amount of waste that was diverted from landfill. The type of waste, the quantity (in tons or cubic yards) of waste, and whether it was diverted (reused, recycled, or donated) or land-filled should be included. Receipts of hauling activities and calculations showing percentage of waste that was diverted from landfill to the total waste generated on-site must also be provided. Each report should include the time period for which the report was prepared. The GC may have to obtain this information from the recycling hauler and then present it in the format requested by the LEED PM. This activity starts from the time demolition begins and lasts until construction is complete. Refer to Figure 4.10 in Chapter 4 for an example of a waste reduction report.

Indoor air quality management report   The GC should periodically inspect the indoor air quality control measures implemented, identify any issues that may have been corrected, and record the observations on a checklist. Include time period and also provide dated and labeled photographs of measures implemented. Refer to Figure 4.11 in Chapter 4 for an example of an indoor air quality inspection report.

Materials credits progress report   The GC should collect information with respect to materials credits (MR C3–MR C7) from subcontractors and compile them into a line-item spreadsheet with CSI division, material name, material cost, as well as the reuse, recycled content, or regional dollar value of the material. The report should include the most current percentage (based on cost) of materials that contribute to the MR C3–MR C7 credits. Refer to Figure 4.12 in Chapter 4 for an example.

VOC data progress report   The final report that the GC must provide should contain information related to the VOC content of adhesives, sealants, paints, and coatings used in the interior of the building. The GC should collect information from subcontractors (Figure 4.15) and compile all items in a common spreadsheet. The product name, manufacturer, specific VOC (g/L) content, and allowable VOC limit should be included. In each progress report the same spreadsheet may be updated with new products that may have been used since the last progress report. Refer to Figure 4.14 in Chapter 4 for an example of a compiled VOC data report.

The contractor may submit all these four reports at one time to the LEED PM. The LEED PM can request the progress reports on a monthly basis or at the time of any payment draw made by the GC.

ACTIVITY 2.2—REVIEW PROGRESS REPORTING FORMS

After receiving the progress documentation from the GC, the LEED PM should review each item to monitor the progress towards achieving the targeted credits. Specific items to review include, but are not limited to, the following.

Credit Percentages

The overall percentages of waste reduction, recycled content, regional materials, etc. should be reviewed to ensure that the project is progressing towards achieving the desired credit thresholds. The PM should determine whether any of the desired percentages have been met already. For instance, after six months of construction he/she may observe that the project has procured 20% (by cost) of materials with recycled content. This is sufficient reason to get the two points available for the credit. Similarly, he/she may realize that the amount of materials recycled is less than the minimum 50% diversion rate required by the MR C2 credit. The LEED PM can then bring this to the attention of the GC, who, in turn, can reevaluate their waste management procedures to increase the diversion rates. The progress reports serve as an important tool/tracking mechanism to ensure that the targeted credit goals can be met.

Incorrect Materials in Recycling Calculations

The LEED PM should also review waste reduction reports to check whether any materials may have been inaccurately included in the calculations. For instance, rock found on site and reused or sold for use to others may not be counted towards this credit. Only waste generated as a result of demolition and construction activities can be included in the recycling calculations.

Materials Reported from Inapplicable CSI Divisions

The MR credits progress report should be reviewed to check whether any materials outside of the applicable Divisions may have been mistakenly included. Only materials that fall under the following CSI Masterformat 2004 Divisions should be included:

> Divisions 3 to 10
>
> Division 31: Section 31.60.00 Foundations
>
> Division 32: Sections 32.10.00 Paving, 32.30.00 Site Improvements, 32.90.00 Planting

Mechanical, electrical, or plumbing equipment should not be included. The purpose of the review is to correct inaccurate information as construction progresses in order to avoid spending time that, perhaps, may have to be spent later at the end of construction.

Materials with VOC Level More Than Allowable Limits

The VOC data progress reports should be reviewed to determine whether any items that do not meet the VOC limits have been included so that the error can be brought to the immediate attention of the GC. For credits EQ C4.1 through EQ C4.4, all products used on-site in the interior of the building weatherproofing system have to meet the VOC level requirements. The LEED PM should make a note of cases where the VOC limit on an adhesive or sealant may have been inadvertently exceeded, and use the VOC budget methodology outlined in the *LEED Reference Guide* to bring the levels into compliance.

ACTIVITY 2.3—ORGANIZE PROGRESS REVIEW MEETINGS

Progress review meetings may be scheduled by the LEED PM to discuss any issues and bring them to the attention of the GC. The meetings will also ensure that there is proper communication of issues, concerns, or questions that the contractor or sub-contractors may have during construction. The meetings may be scheduled either on-site, via telephone, or web and can occur at a frequency that is coordinated with the progress reports submission schedule. If the contractor is submitting progress documentation on a monthly basis, then the coordination meetings may occur monthly as well.

Site Visits

While the majority of the progress meetings can occur remotely (via phone or web), it is recommended that at least a few are scheduled on-site so that the LEED PM may verify proper implementation of measures on-site. Note that the site visit reviews are not required to meet LEED credit requirements or part of the official GBCI certification process. These site visits are only recommended here to serve as an additional quality control measure. This may also be done by the architect or owner's representative as part of their regular construction administration visits. This section is written with the assumption that the LEED PM will be conducting the site reviews. Including the pre-construction kick-off meeting (Activity 1.3 discussed in this chapter), a minimum of four site visits are recommended:

Site Meeting One: Pre-construction LEED kick-off meeting   This is the kick-off meeting describing the contractor and subcontractor responsibilities and has been described in detail in Step 1 of this chapter.

Site Meeting Two: Review implementation of SS P1, SS C5.1, MR C2   The second site visit should be scheduled after the construction activities have begun to review the stormwater pollution prevention measures, the site disturbance boundary limits, and the waste-recycling measures. The LEED PM should review all measures to ensure they have been initiated appropriately according to the requirements of the credits.

Site Meeting Three: Review IAQ implementation measures during construction The third site visit should be scheduled after the building has been closed in and interior work has begun, to inspect measures implemented to prevent pollutant contamination and preserve indoor air quality. The LEED PM should look for specific measures, like covered ductwork, proper storage of absorptive materials, appropriate filtration media, etc.

Site Meeting Four: Review IAQ implementation measures before occupancy This final visit may be scheduled at the end of construction and before occupancy when the contractor is performing a building flush-out, or a third party testing agency is testing the indoor air quality of the spaces.

   Note that these site visits do not include site visits that the CxA will conduct for the purposes of commissioning. Commissioning-related activities should be completed by the construction team according to the Cx plan provided by the CxA.

Step 3: Prepare Documents for GBCI Construction Review

Objective	The objective of this step is to complete, prepare, and compile all documentation after construction is complete and submit to GBCI for Construction Review. Project team members are responsible for uploading documentation related to construction credits and updating any significant changes to the design credits.
Recommended Activities	3.1 Complete Construction Credit Documentation. 3.2 Update Design Credit Documentation.
Roles and Responsibilities	Construction Team: Provide final construction credit documentation. Design Team: Update/complete any design credit and/or construction credit documentation. LEED PM: Coordinate document preparation; compile information as necessary.
Duration	4 weeks
Outcome	Final documents for submission to GBCI prepared.

ACTIVITY 3.1—COMPLETE CONSTRUCTION CREDIT DOCUMENTATION

After the project has reached *substantial completion*, team members should complete the documentation required to demonstrate compliance with the construction credits assigned to them. The LEED PM should coordinate this and ensure that all credit documentation is uploaded to LEED Online. The GC is responsible for a majority of the construction credits, but there are some credits related to *site* and *energy* that require the involvement of other team members. It is recommended that team members complete the online documentation during the CD phase itself and supplement it with additional documentation or changes after the construction is complete. Table 5.5 lists all the credits and prerequisites that fall under the *construction* credit category. The team members responsible for each of these credits should complete the required drawings, documents, and online credit forms.

TABLE 5.5   TYPICAL LEED CONSTRUCTION CREDITS

CREDIT CATEGORY		POINTS
Sustainable Sites		
Prereq 1	Construction Activity Pollution Prevention	
Credit 5.1	Site Development—Protect or Restore Habitat	1
Credit 7.1	Heat Island Effect—Non-roof	1
Energy and Atmosphere		
Prereq 1	Fundamental Commissioning of Building Energy Systems	
Credit 3	Enhanced Commissioning	2
Credit 5	Measurement and Verification	3
Credit 6	Green Power	2
Materials and Resources		
Credit 1.1	Building Reuse—Maintain Existing Walls, Floors, and Roof	1 to 3
Credit 1.2	Building Reuse—Maintain 50% of Interior Non-Structural Elements	1
Credit 2	Construction Waste Management	1 to 2
Credit 3	Materials Reuse	1 to 2
Credit 4	Recycled Content	1 to 2
Credit 5	Regional Materials	1 to 2
Credit 6	Rapidly Renewable Materials	1
Credit 7	Certified Wood	1
Indoor Environmental Quality		
Credit 3.1	Construction IAQ Management Plan—During Construction	1
Credit 3.2	Construction IAQ Management Plan—Before Occupancy	1
Credit 4.1	Low-Emitting Materials—Adhesives and Sealants	1
Credit 4.2	Low-Emitting Materials—Paints and Coatings	1
Credit 4.3	Low-Emitting Materials—Flooring Systems	1
Credit 4.4	Low-Emitting Materials—Composite Wood and Agrifiber Products	1
Innovation in Design*		
Credit 1.1	Innovation in Design: Specific Title	1
Credit 1.2	Innovation in Design: Specific Title	1
Credit 1.3	Innovation in Design: Specific Title	1
Credit 1.4	Innovation in Design: Specific Title	1
Credit 1.5	Innovation in Design: Specific Title	1
Credit 2	LEED Accredited Professional	1
Total		**35**

*Innovation in Design Credits C1.1 to C1.5 could be either design or construction credits depending on the project.

The following are key *construction credit* documents that may need to be prepared by team members.

Site and Landscape

Prerequisite SS P1 requires the implementation of erosion and sedimentation control measures on-site. Typically, the civil engineer is responsible for providing construction documents detailing the measures to be implemented. The GC should implement the measures identified in the drawings and document them by means of photographs and/or inspection reports. The civil engineer may complete the online credit form by uploading drawings, inspections reports, and photographs obtained from the GC.

Credit SS C5.1 requires the implementation of specific measures that protect and/or restore natural habitat. Greenfield sites may achieve this by limiting site disturbance; previously developed sites may achieve this by restoring a minimum of 50% of the project site with native and adapted vegetation. If the project is on a greenfield site, then a drawing showing the limits of disturbance should be provided by the civil engineer in the CD set. The GC should implement and enforce the limits during construction and document it by means of photographs and/or inspection reports. After construction is complete, all documents may be uploaded by the civil engineer. The GC or LEED PM may also collect the information and upload it online.

If the project is built on a previously developed site, then a landscape plan identifying areas that will be protected or restored with native/adapted vegetation should be provided by the landscape architect in the CD set. After construction, the landscape architect may need to update drawings and calculations based on any changes that may have occurred since design.

Credit SS C7.1 requires the implementation of specific measures that contribute toward mitigating the urban heat island effect. This may be achieved by providing either shaded and/or high reflectance paving or covered parking. Usually, the documentation for this credit is completed by either the landscape architect or the architect. Drawings and initial calculations may have been completed during the CD phase. However, after construction these should be updated based on actual construction data.

Commissioning

One of the most important activities during construction is the commissioning of the building systems by the CxA. This is required for the prerequisite *EA P1, Fundamental Commissioning* and also for credit *EA C3, Enhanced Commissioning*, if the team is pursuing that credit. By the end of construction, the CxA should complete all commissioning-related activities and prepare a final commissioning summary report verifying the proper installation and operation of the building systems.

Measurement and Verification

If the team chooses to attempt *Credit EA C5, Measurement and Verification*, then the responsible team member should provide a measurement and verification plan along with a plan for corrective measures. This credit may be completed by the MEP engineer or by a third-party provider that may have been engaged by the owner.

Materials and Indoor Environmental Quality

Finally, the majority of the *construction credit* documentation is related to the credits: MR C1.1 to MR C7 and EQ C3 to EQ C4.4. The GC should implement and document these credits as explained in Steps 1 and 2, and use the progress reports to compile all information.

ACTIVITY 3.2—UPDATE DESIGN CREDIT DOCUMENTATION

After completing all the construction credit documentation, the next step is to conduct a review of the *design credit* documents that were previously submitted to GBCI. The LEED PM should perform this review in order to determine whether there have been any significant changes between what was designed and what was constructed. If there are no significant changes, then the LEED PM may continue with the submission and the previously *accepted* design credits will be *awarded* at this stage. Note that teams may make changes to design credits that were previously marked as *anticipated* and may also attempt additional design credits if desired. The LEED PM should update any project information on LEED Online that may have changed since the previous submission, including project area, building footprint areas, etc.

Step 4: Review Final Documents for GBCI Submission

Objective	The main objective of this step is to review both design and construction credit documents and ensure all information clearly demonstrates compliance with credit requirements. The LEED PM should conduct this final review before submission to GBCI.
Recommended Activities	4.1 Review Documents and Online Credit Forms.
	4.2 Organize Construction Phase Coordination Meeting.
Roles and Responsibilities	Review and compile final documents.
Duration	3 weeks
Outcome	Documents for GBCI submission reviewed.

ACTIVITY 4.1—REVIEW DOCUMENTS AND ONLINE CREDIT FORMS

Similar to the review done before submitting the *design* credits to GBCI, the LEED PM must perform a final review of all the credit forms, documents, drawings, and narratives that have been uploaded to LEED Online by the team. This review is the final check before all the documents are submitted to GBCI and the final opportunity for any changes/corrections to be made by the team. Refer to Chapter 4, Step 10 for information on what to look for while reviewing credit documentation. In general, the LEED PM should ensure that

- There is consistency in information across all credits.
- All *required* documents have been uploaded.
- Documents and narratives demonstrate credit requirements.
- Information in drawings is consistent with information in credits.
- Changes have been made to previously attempted design credits.
- All targeted LEED credits have been completed.

ACTIVITY 4.2—ORGANIZE CONSTRUCTION PHASE COORDINATION MEETING

After the review of the construction drawings, specifications, and online documentation, the LEED PM should compile a list of all comments and conduct a final coordination meeting to discuss any last-minute changes required. Most likely there will not be many changes in the design credits because they were already submitted for review to GBCI. The continuous progress monitoring and review of the construction credit documentation will help reduce the number of changes required of the construction team as well. This meeting may be scheduled via phone, web, or on-site, whichever is deemed most effective by the LEED PM.

Step 5: Submit to GBCI for Construction Review

Objective	The objective of this step is to submit documentation to GBCI for Construction Review. This is the final step for the construction phase as well as the LEED project management process.
Recommended Activities	5.1 Submit Credits for Preliminary Construction Review. 5.2 Respond to GBCI's Preliminary Construction Review Comments. 5.3 Accept or Appeal Construction Review Results. 5.4 Order Materials to Display LEED Certification.
Roles and Responsibilities	LEED PM: Submit all documents to GBCI.
Duration	4 weeks
Outcome	Documents submitted to GBCI for Construction Review.

ACTIVITY 5.1—SUBMIT CREDITS FOR PRELIMINARY CONSTRUCTION REVIEW

The submittal to GBCI for the Construction Review is composed of two reviews: one is a preliminary Construction Review and the second one is a final Construction Review. At the end of the preliminary Construction Review, each reviewed prerequisite and credit will be designated by GBCI as *anticipated, pending,* or *denied,* and accompanied with technical advice. All project information forms will be designated as *approved* or *not approved* and accompanied with technical advice as well. The preliminary Construction Review gives the project design team a chance to provide clarifications to any questions that the GBCI review team may have.

The first step before submitting the credits is to verify that all credits have been satisfactorily completed by the team members. Then, the LEED PM should mark all the credits as *complete*. Once all the construction credits that were attempted are marked as *complete*, the online instructions screen will display a description of the next steps. It will indicate whether or not all the attempted construction credits are complete and ready to be submitted for preliminary Construction Review. Even if there is one attempted credit that is still *in progress*, the option to continue the submission will not appear. If there is a credit that is incomplete or was attempted in the beginning of the project but is no longer applicable, then that credit must be *unattempted* to allow the successful submission of the rest of the credits. Payment options for the certification fee will be provided. The LEED PM should choose the appropriate payment option and submit all the credits for the preliminary Construction Review to GBCI.

ACTIVITY 5.2—RESPOND TO GBCI'S PRELIMINARY CONSTRUCTION REVIEW COMMENTS

After the submittal for preliminary Construction Review, GBCI will provide feedback through LEED Online. Comments will be provided on each attempted credit, and it will be indicated whether that credit is *anticipated, denied,* or *pending* and needs clar-fication. All project teams have the option to accept the results of the preliminary Construction Review as final. However, at the discretion of the project team, a response to the preliminary Construction Review may be submitted via LEED Online. The team is allowed 25 business days to respond with clarifications. The LEED PM should coordinate the project team's response to GBCI by contacting the team members responsible for the credits that require clarifications. Once all clarifications have been prepared, the information on those credits should be resubmitted to GBCI. The response of the design team must include revised documentation for any prerequisite or project information form marked as *pending* or *denied* during the preliminary Construction Review. Alternatively, project teams have the option of withdrawing such credits from the application.

ACTIVITY 5.3—ACCEPT OR APPEAL CONSTRUCTION REVIEW RESULTS

If the project team chooses to accept the results of the preliminary Construction Review as final, then this activity will not be required. However, if the team elects to respond to the preliminary Construction Review, then, GBCI will review the clarifications provided by the project team and issue final Construction Review comments designating each prerequisite and credit as *awarded* or *denied* along with technical advice from the review team. The project team has the option of either accepting the final Construction Review, or if they are not satisfied with GBCI's decision, then they can appeal the decision by paying a fee of $500 per credit within 25 business days of GBCI's posting of the final review decision.

If no appeals are made, then the LEED PM should accept the Construction Review results and announce the achievement of the appropriate LEED Certification.

ACTIVITY 5.4—ORDER MATERIALS TO DISPLAY LEED CERTIFICATION

Following the acceptance of the final Construction Review results, GBCI will send information to the LEED PM on how to order plaques, certificates, and other market-ing material available for LEED-certified projects. The LEED PM should communicate the information to the owners and the project team so that the required items can be pro-cured and installed at the building.

Summary

The last stage of the LEED project management process presents all the activities that need to be completed by the contractor, subcontractors, design team, and LEED PM during construction and before occupancy. Although there are certain LEED credits that require post-occupancy implementation measures (like *EA C5, Measurement and Verification* and *EQ C7.2, Thermal Comfort Survey*), the activities described here are performed only until the final LEED certification is awarded. The certification is not dependant on the implementation of the post-occupancy measures. However, it is recommended that the owners and building management implement the post-occupancy measures to demonstrate compliance with the intents stated during submission to GBCI and also to derive the benefits of implementing these measures.

The duration of activities listed in this stage will vary from project to project, depending on their respective construction schedules. Figure 5.5 describes a checklist that project managers may use in the implementation of activities in this final stage. Figure 5.6 presents a representative schedule that project managers may adapt to suit their respective construction schedules.

Note

[1] USGBC, LEED Reference Guide for Green Building Design and Construction (USGBC, 2009), 475.

Stage III Construction Phase Implementation		Due Date	Not Started	In Progress	Complete	Remarks
Step 1	Organize a Pre-Construction LEED Kick-Off Meeting	4/5/2011				
1.1	Establish Meeting Outline	4/5/2011		x		
1.2	Facilitate Kick-Off Meeting	4/5/2011		x		
Step 2	Review Progress of Credit Implementation	Monthly		x		
2.1	Complete Construction Progress Reporting Forms			x		
2.2	Review Progress Reporting Forms		x			
2.3	Organize Progress Review Meetings		x			
Step 3	Prepare Documents for GBCI Construction Review					
3.1	Complete Construction Credit Documentation					
3.2	Update Design Credit Documentation					
Step 4	Review Final Documents for GBCI Submission					
4.1	Review Documents and Online Credit Forms					
4.2	Organize Construction Phase Coordination Meeting					
Step 5	Submit to GBCI for Construction Review					
5.1	Submit Credits for Preliminary Construction Review					
5.2	Respond to GBCI Preliminary Construction Review Comments					
5.3	Accept or Appeal Construction Review Results					
5.4	Order Materials to Display LEED Certification					

Figure 5.5   Process checklist for Stage III.
Credit: Green Potential LLC (www.green-potential.com).

		Constn. Start						Substantial Completion	Constn. Complete			
		← Duration of Construction →										
	Duration*	Q1			Q2			Q3		Q4		
Stage III Construction Phase Implementation		MO1	MO2	MO3	MO4	MO5	MO6	MO7	MO8	MO9	M10	M11	M12
Step 1 **Organize a Pre-Construction LEED Kick-Off Meeting**	**2 weeks**	■											
1.1 Establish Meeting Outline	1 week	▨											
1.2 Facilitate Kick-Off Meeting	2 weeks	▨											
Step 2 **Review Progress of Credit Implementation**	**10 months**												
2.1 Complete Construction Progress Reporting Forms	10 months												
2.2 Review Progress Reporting Forms	10 months												
2.3 Organize Progress Review Meetings	10 months												
Step 3 **Prepare Documents for GBCI Construction Review**	**4 weeks**										■		
3.1 Complete Construction Credit Documentation	4 weeks										▨		
3.2 Update Design Credit Documentation	4 weeks										▨		
Step 4 **Review Final Documents for GBCI Submission**	**8 weeks**											■	
4.1 Review Documents and Online Credit Forms	4 weeks											▨	
4.2 Organize Construction Phase Coordination Meeting	4 weeks											▨	
Step 5 **Submit to GBCI for Construction Review**	**4 weeks**												◆
5.1 Submit Credits for Preliminary Construction Review	4 weeks												
5.2 Respond to GBCI Preliminary Construction Review Comment	Varies**												
5.3 Accept or Appeal Construction Review Results	Varies**												
5.4 Order Materials to Display LEED Certification	Varies**												◆

* Duration is approximate. Actual duration may vary based on project schedules
** Dependent on when comments are received from GBCI

■ Process Step Duration
▨ Activity Duration
◆ Milestone

Figure 5.6 Process schedule for Stage III.

Credit: Green Potential LLC (www.green-potential.com).

LEED PROCESS: ADAPTABILITY, APPLICABILITY, AND BEST PRACTICES

This chapter highlights the benefits of a streamlined project management process and its applicability to rating systems other than LEED-NC. It also provides a brief synopsis of the flexibility that LEED-NC offers in terms of allowing a range of building types to achieve certification. Finally, it describes how the commissioning process may be integrated and presents best practices for LEED projects.

Other LEED Rating Systems

The project management process described in this book essentially caters to the LEED-NC rating system. However, it can easily be adapted and used by a number of other rating systems. Figure 6.1 provides a snapshot of the key differences and similarities between some of the LEED rating systems.

LEED CORE & SHELL

The LEED for Core & Shell (LEED-CS) program is applicable to projects in which only the design and construction of the building's core and shell are controlled by the developer. The developer does not control the interior finish-out of spaces. Examples of projects that may use this program are commercial office buildings, retail centers, warehouses, etc.

The LEED-CS rating system includes additional number of credits and requirements compared to LEED-NC. Some credits are very similar to LEED-NC while others are not. Because of the speculative nature of LEED-CS projects, where the future design and occupancy conditions are uncertain, the rating system allows for assumptions in

Applicability	Rating System	Certification Process
LEED-NC New construction or major renovation of commercial buildings Examples- offices, institutional buildings, hotels, residential buildings (4story or higher)	Prereqs Credits Points SS 1 14 26 WE 1 3 10 EA 3 6 35 MR 1 8 14 EQ 2 15 15 Use credit specific implementation activities explained in Part II	Registration Application (Split D&C or combined) GBCI Review (Split D&C or combined) GBCI Award Use LEED project management process explained in Part I
LEED-CS Commercial buildings where owner has control over only the core and shell and not interior tenant finish-out Examples- office, retail center, warehouse, labs etc	Prereqs Credits Points SS 1 15 28 WE 1 3 10 EA 3 7 37 MR 1 6 14 EQ 2 12 12 May use some credit specific implementation activities explained in Part II	Registration Precertification Application (Split D&C or combined) GBCI Review (Split D&C or combined) GBCI Award May use LEED project management process explained in Part I; only additional optional step is precertification
LEED-Schools New construction or major renovation of school buildings	Prereqs Credits Points SS 2 16 24 WE 1 5 11 EA 3 6 33 MR 1 8 12 EQ 3 19 23 May use some credit specific implementation activities explained in Part II	Registration Application (Split D&C or combined) GBCI Review (Split D&C or combined) GBCI Award May use LEED project management process explained in Part I
Others LEED-EBOM for existing buildings operations & maintenance LEED-Homes for single family homes LEED- ND for neighborhood development	Significant difference in credit categories, credit prerequisites, requirements, points and thresholds compared to LEED-NC	Significant difference in processes to manage and submit to GBCI compared to LEED-NC

Figure 6.1   Comparison of different LEED rating systems.

various credit aspects such as occupancy counts and energy modeling. Guidance for making these assumptions is provided in the appendix section of the *LEED Reference Guide.*

The key difference in the certification process between LEED-CS and LEED-NC projects is that GBCI offers what is known as *precertification* for the former. Precertification is an optional step that project teams may choose to pursue and is not a mandatory requirement. It is a formal recognition by GBCI of the developer's intent to pursue LEED-CS certification. The developers can use this precertification document as a marketing tool to attract potential tenants. If the team elects to pursue precertification, then they need to submit documentation to GBCI based on declared goals and intent to use green strategies. The results, however, do not guarantee confirmation or indicate a commitment from the GBCI to award LEED-CS certification. Project teams must continue to complete all design application materials and submit them for review to the GBCI in order to achieve certification.[1] The process is similar to LEED-NC except for the precertification step. Therefore, LEED-CS projects may be managed using the same process described in this book, with the minor addition of the *precertification* activity during early design phases.

LEED FOR SCHOOLS

The LEED for Schools program provides a framework to design and build both new school buildings and major renovations of existing school buildings. The LEED-NC and LEED-Schools programs share a common reference guide, *The LEED Reference Guide for Green Building Design and Construction.*

The requirements of credits and prerequisites for LEED for Schools vary from LEED-NC. But the overall process of obtaining certification is similar to LEED-NC. Therefore, the three stages of the LEED project management process described in this book may be adapted for use by LEED for Schools projects as well.

LEED PORTFOLIO PROGRAM (PILOT)

The LEED Portfolio Program, currently under development, was established by USGBC with the primary goal of providing a cost-effective path to achieve LEED certification on a volume scale. This would enable investors, developers, owners, and managers to systematically *green* their building portfolios. The Portfolio Program pilot was launched in November 2006, and according to the USGBC website, "it included 40 participating companies and institutions and covered 1700 buildings and approximately 135 million square feet of building space."[2]

The Portfolio program is a powerful tool that, when released, will help in transforming the green building market further. The project management process described in this book can be adapted to facilitate the Portfolio Program as well. The application will be split between design and construction, just as for single building certification. However, the key difference would be that the design application will be for a prototype design that can be precertified, while the site and construction credits will be submitted after construction has been completed at multiple locations. The PNC Bank's Green Branch® project described in Chapter 8 demonstrates the Portfolio Program process. The Program allowed the bank to achieve LEED certification for 72 of its buildings in an efficient and cost-effective manner.

Other rating systems, such as LEED for Existing Buildings: Operations & Maintenance (LEED-EBOM), LEED for Neighborhood Development (LEED-ND), and LEED for Homes, have credits, requirements, and processes that are significantly different from the LEED-NC rating system. There are some similarities, however, and cross-pollination might be possible between the rating systems.

LEED-NC for Different Building Types

The LEED-NC rating system provides a broad and flexible framework that allows a number of different building types, such as commercial, institutional, and residential, to achieve certification. While some credits are easy to achieve across all building types, there are others that are not applicable or easy to achieve for certain building types. For instance, while the use of Minimum Efficiency Reporting Value (MERV) 13 filtration media for air handling units to achieve EQ C5 may be feasible in office buildings, it may not be practically feasible in mid-rise residential projects that use smaller HVAC systems.

A credit-by-credit comparison of three hypothetical building types was done in order to understand how different building types might use the LEED-NC rating system. The building types analyzed were: office and institutional buildings (university, public, or recreational), high-rise multifamily buildings (above seven stories) and mid-rise multifamily buildings (between four to seven stories). Each building type is governed by a unique set of constraints and parameters that define the credits that are easy or difficult to achieve. In order to perform the comparison, the following assumptions were made with respect to the context of all the building types:

- All building types meet the minimum program requirements and all prerequisites.
- All buildings are located in an urban setting with access to community services and public transport and are on previously developed land.
- All buildings will be able to achieve the Innovation in Design (ID) credits and Regional Priority (RP) credits.
- All buildings are representative of typical design and construction scenarios for their respective project types. Special circumstances, innovative designs, and unique approaches to design are not taken into account.

The requirements of each of the credits were reviewed and compared based on the following three criteria:

1 Easy (E): The term *easy* is used loosely here to indicate that these credits may be achieved by all building types if they satisfy the credit's requirements. It does not mean that the project will be able to achieve it with no effort or cost; it only implies that the nature of the building type will not affect their achievability. All building types should be able to achieve these credits.

2 Moderate (M): This category of credits may be impacted by the building type. There is a 50% chance that a project may be able to achieve a particular credit depending on the specific project scenario.

3 Difficult (D): This category refers to credits that are difficult to achieve based on typical design and construction. These are credits that may not be directly applicable or require innovative approaches to meet the credit's intent.

Figures 6.2 and 6.3 show a comparison of various credits that may be achieved by each building type using the LEED-NC rating system. This is a broad illustration to show that not all credits may be feasible for all project types. Some may not be applicable; others may need significant investment or may be difficult to implement. The comparison shows that almost all projects may achieve Platinum level certification, depending on the green features they implement.

LEED-NC 2009 Credit Checklist	Points possible	Office & Institutional (University, Public)			Multifamily (Highrise - over 7 stories)			Multifamily (Midrise - 4-7 stories)		
	110	E	M	D	E	M	D	E	M	D
Sustainable Sites										
Prereq 1 Construction Activity Pollution prevention	P	Y			Y			Y		
Credit 1 Site Selection	1	1			1			1		
Credit 2 Development Density and Community Connectivity	5	5			5			5		
Credit 3 Brownfield Redevelopment	1	1			1			1		
Credit 4.1 Alternative Transportation—Public Transportation Access	6	6			6			6		
Credit 4.2 Alternative Transportation—Bicycle Storage and Changing Rooms	1	1			1			1		
Credit 4.3 Alternative Transportation—Low-Emitting and Fuel-Efficient Vehicles	3	3			3			3		
Credit 4.4 Alternative Transportation—Parking Capacity	2	2				2			2		★
Credit 5.1 Site Development—Protect or Restore Habitat	1	1			1			1		
Credit 5.2 Site Development—Maximize Open Space	1	1			1			1		
Credit 6.1 Stormwater Design—Quantity Control	1	1			1			1		
Credit 6.2 Stormwater Design—Quality Control	1	1			1			1		
Credit 7.1 Heat Island Effect—Non-roof	1	1			1			1		
Credit 7.2 Heat Island Effect—Roof	1	1			1			1		
Credit 8 Light Pollution Reduction	1	1				1				1	★
Water Efficiency										
Prereq 1 Water Use Reduction—20% Reduction		Y			Y			Y		
Credit 1 Water Efficient Landscaping	2 to 4	2	2		2	2		2	2		★
Credit 2 Innovative Wastewater Technologies	2		2				2			2
Credit 3 Water Use Reduction	2 to 4	4			4			4		
Energy and Atmosphere										
Prereq 1 Fundamental Commissioning of Building Energy Systems	Y	Y			Y			Y		
Prereq 2 Minimum Energy Performance	Y	Y			Y			Y		
Prereq 3 Fundamental Refrigerant Management	Y	Y			Y			Y		
Credit 1 Optimize Energy Performance	1 to 19	15	4		10	9		10	9		★
Credit 2 On-Site Renewable Energy	1 to 7		7			7			7	
Credit 3 Enhanced Commissioning	2	2			2			2		
Credit 4 Enhanced Refrigerant Management	2	2				2				2	★
Credit 5 Measurement and Verification	3	3				3			3		★
Credit 6 Green Power	2	2			2			2		

E= Easy   M=Moderate   D= Difficult

★ SS C4.4 Total parking on-site may be more readily reduced for office and institutional buildings compared to multifamilies. Carpool/carshare services may find more acceptance in office buildings compared to multifamily buildings.

★ SS C8 Exterior Lighting requirements can be easily met by all building types. Interior lighting requirements are relatively tough to achieve in multifamily settings because of the requirement of automatic switch-off of lighting and/or shading devices. High rise residential buildings may have larger budgets to introduce automatic shading devices but mid-rise multifamilies may not be able to provide these. Further providing automatic switch off of residential lighting may be tough.

★ WE C1 It is equally difficult or easy to incorporate these credit requirements into design. Most projects are able to achieve 50% reduction in water use for irrigation using efficient irrigation and landscape practices; the challenge is for 100% reduction which requires reuse of rainwater, or treated greywater. Implementation of these is dependent on the project goals and budget.

★ WE C2 The challenge of achieving this credit lies in the high first-costs associated with on-site sewage treatment and/or greywater treatment and reuse. Most teams refrain from pursuing this. Because office buildings and institutional buildings may have larger capital budgets and would benefit from the operational savings, it is moderately easy to implement in those projects. These strategies may not be readily implemented in multifamily buildings because of initial capital expense and the operational saving incentive split with the tenants.

★ WE C3 It is equally difficult or easy for projects to achieve high levels of water use reduction; depends on strategies that may be used - high efficiency flush fixtures, low flow fixtures and waterless urinals, etc.

★ EA C1 It has typically been seen that greater levels of energy efficiency may be achieved in office and institutional buildings compared to multifamily buildings because of the nature of their building HVAC systems and design.

★ EA C4 The HVAC systems typically specified for office, institutional or high-rise multifamily buildings allow the use of refrigerants that have low Global Warming Potential and Ozone Depleting Potential. The HVAC systems specified for mid-rsie multifamilies typically do not allow the use of such refrigerants.

★ EA C5 Although not impossible to achieve, there is a moderate level of difficulty associated with implementing measurement & verification measures in residential settings compared to other building types because of metering each residential unit.

Figure 6.2 LEED-NC comparison matrix for different building types—part 1.

Credit: Green Potential LLC (www.green-potential.com).

LEED-NC 2009 Credit Checklist	Points possible	Office & Institutional (University, Public)			Multifamily (Highrise - over 7 stories)			Multifamily (Midrise - 4-7 stories)			
	110	E	M	D	E	M	D	E	M	D	
Materials and Resources											
Prereq 1 Storage and Collection of Recyclables	P	Y			Y			Y			
Credit 1.1 Building Reuse—Maintain Existing Walls, Floors, and Roof	1 to 3		3			3			3		
Credit 1.2 Building Reuse—Maintain 50% of Interior Non-Structural Elements	1		1			1			1		
Credit 2 Construction Waste Management	1 to 2	2			2			2			
Credit 3 Materials Reuse	1 to 2		2			2			2		
Credit 4 Recycled Content	1 to 2	2			2			2			
Credit 5 Regional Materials	1 to 2	2			2			2			
Credit 6 Rapidly Renewable Materials	1		1			1			1		
Credit 7 Certified Wood	1	1				1				1	★
Indoor Environmental Quality											
Prereq 1 Minimum Indoor Air Quality Performance	P	Y			Y			Y			
Prereq 2 Environmental Tobacco Smoke (ETS) Control	P	Y			Y			Y			
Credit 1 Outdoor Air Delivery Monitoring	1	1				1				1	★
Credit 2 Increased Ventilation	1	1				1				1	★
Credit 3.1 Construction IAQ Management Plan—During Construction	1	1			1			1			
Credit 3.2 Construction IAQ Management Plan—Before Occupancy	1	1			1			1			
Credit 4.1 Low-Emitting Materials—Adhesives and Sealants	1	1			1			1			
Credit 4.2 Low-Emitting Materials—Paints and Coatings	1	1			1			1			
Credit 4.3 Low-Emitting Materials—Flooring Systems	1	1			1			1			
Credit 4.4 Low-Emitting Materials—Composite Wood and Agrifiber Products	1		1			1			1		
Credit 5 Indoor Chemical and Pollutant Source Control	1	1				1				1	★
Credit 6.1 Controllability of Systems—Lighting	1	1			1			1			
Credit 6.2 Controllability of Systems—Thermal Comfort	1		1		1			1			
Credit 7.1 Thermal Comfort—Design	1	1			1			1			
Credit 7.2 Thermal Comfort—Verification	1	1					1			1	★
Credit 8.1 Daylight and Views—Daylight	1		1			1			1		
Credit 8.2 Daylight and Views—Views	1		1		1			1			★
Innovation and Design Process	6	6			6			6			
Regional Priority Credits	4	4			4			4			
Total	110	84	26	0	68	39	3	68	32	10	

E= Easy   M=Moderate   D= Difficult

★ MR C7 More cost-effective to implement in office, institutional and high rise buildings that are typically steel structures with minimal wood products. Difficult or cost intensive in mid rise multifamilies that are typically wood-framed.

★ EQ C1 Relatively high first-costs for multifamiliy buildings as CO_2 monitors would be required for each apartment unit.

★ EQ C2 Easier to implement in mechanically ventilated buildings. Most residential units in mid-rise multifamily buildings are naturally ventilated and typical construction (window sizes etc.) must be carefully planned to achieve increased ventilation rates.

★ EQ C5 Requires use of MERV 13 filters across air handling units. May not be feasible in mid-rise multifamily buildings which use HVAC systems where the use of this filtration media might cause pressure drop issues.

★ EQ C7.2 As per rating system, this credit is not applicable in residential projects

★ EQ C8.2 Has been easier to achieve in residential settings, because almost all regulary occupied areas are typically designed to have windows and views to the outside; office and institutional buildings need to be designed such that views are maximized; tough to achieve in scenarios where multiple regularly occupied areas are without access to views because of interior partitions and walls.

Figure 6.3 LEED-NC comparison matrix for different building types—part 2.

Credit: Green Potential LLC (www.green-potential.com).

Building Commissioning

Building Commissioning is defined as the "systematic process of ensuring that a building's complex array of systems is designed, installed, and tested to perform according to the design intent and the building owner's operational needs."[3] The commissioning process starts from the beginning of design until the building is occupied. The process involves establishing expectations for the performance of the building systems and also procedures to determine whether those expectations have been met. The commissioning process can be applied to nearly any building system—from the envelope to the lighting and HVAC systems. The general commissioning procedures include drawing reviews, inspection during construction, training of operations and maintenance personnel, etc.

Commissioned buildings are documented to be more energy efficient, more comfortable, easier to maintain, and less expensive to own and operate in the long term. It may appear that the commissioning process requires too much time and money. However, reviewing, checking, and correcting work throughout the design and construction process significantly reduce the amount of rework, the number of change orders, or the number of post-occupancy issues. Energy Design Resources' *Design Briefs: Building Commissioning* illustrates the financial benefits of commissioning through the example of two representative projects of the Division of Facilities and Services of Montgomery County. It was estimated that commissioning saved the Division $1.57 per square foot due to reduced mechanical/electrical change orders and claims, plus another $0.48 per square foot in energy costs during the first year of occupancy. The typical commissioning costs (ranging from $0.30 to $0.90 per square foot) were more than offset by the savings.[4]

Commissioning is a prerequisite for LEED projects, and the process needs to be documented to demonstrate compliance. *EA P1, Fundamental Commissioning* is the specific prerequisite that provides guidance on the various activities that need to be completed. In addition, teams have the option to implement additional commissioning procedures to derive maximum benefit from the process. The additional commissioning procedures are described in credit *EA C3, Enhanced Commissioning.* Regardless of whether or not a project team elects to pursue EA C3, fundamental commissioning, as described by EA P1, is still essential. For the purposes of LEED, commissioning must be completed for the following systems at a minimum:

- HVAC and refrigeration systems and associated controls
- Lighting and daylighting controls
- Domestic hot water systems
- Renewable energy systems

Although the project management process described in Chapters 2 through 5 identifies some activities that are part of the commissioning process, it does not cover all the steps. The commissioning (Cx) process is very detailed and is usually managed by a designated Commissioning Authority (CxA), and it is expected that the CxA will identify specific commissioning steps and activities to integrate with the overall project process.

LEED COMMISSIONING PROCESS

The following describes how the commissioning activities align with the various design phases and the LEED process. These steps have been sourced from the *LEED Reference Guide*[5] and the intent in presenting them here is to show the relationship between the commissioning process and the LEED project management process explained in this book. Figure 6.4 illustrates the process for commissioning and how it aligns with the various design phases.

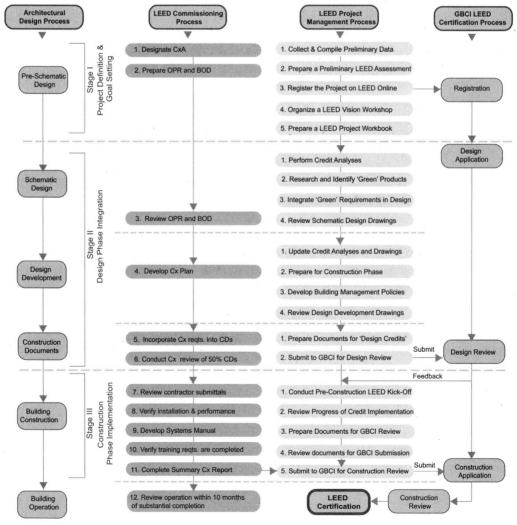

Figure 6.4 LEED commissioning process.
Credit: Green Potential LLC (www.green-potential.com).

Stage I: Project Definition and Goal Setting
Step 1: Designate Commissioning Authority

During pre-schematic design phase, the owner should designate a Commissioning Authority to lead, review, and oversee the commissioning process. Refer to Chapter 2 for who can be the CxA and his/her responsibilities. This step is common to both EA P1 and EA C3.

Step 2: Prepare Owner's Project Requirements (OPR) and Develop Basis of Design (BOD)

The next step is to prepare the OPR and BOD for the project. The OPR should be prepared by the owner, along with assistance from the CxA and/or the design team. The LEED PM may also assist with preparing the OPR. (Refer to Chapter 3, Activity 1.1 on items to be included in the OPR.)

The BOD is a document that the design team should prepare, identifying the basis of the design for the systems to be commissioned. This document should be updated as the design progresses and should include, at a minimum, the primary design assumptions, applicable codes, standards, and narrative descriptions of performance criteria for systems to be commissioned. This step is common to both EA P1 and EA C3.

Stage II: Design Phase Integration—SD Phase
Step 3: Review OPR and BOD

During the SD phase, the CxA should review the OPR and BOD to ensure that the BOD reflects the OPR. These should be reviewed again for completeness prior to the approval of contractor submittals for any commissioned systems. This step is common to both EA P1 and EA C3.

Stage II: Design Phase Integration—DD Phase
Step 4: Develop and Implement Cx Plan

During the DD phase, the CxA should prepare a commissioning plan outlining the overall process, schedule, organization, responsibilities, and documentation requirements of the commissioning process. The commissioning plan should be regularly updated by the CxA to reflect any changes as the design progresses. This step is common to both EA P1 and EA C3.

Stage II: Design Phase Integration—CD Phase
Step 5: Incorporate Commissioning Requirements into Construction Documents

During the CD phase, the commissioning requirements should be incorporated into the construction documents. The requirements are typically included in the specifications manual under general conditions. Refer to Chapter 4, Activity 6.2 for a description of the specification section. The CxA may assist the design team in preparing this section and it should include: contractor's responsibilities, submittal review procedures, operation and maintenance documents required, number of meetings to be scheduled, construction verification procedures, functional performance testing requirements, training requirements, etc. This step is common to both EA P1 and EA C3.

Step 6: Conduct Cx Design Review prior to Mid-Construction Documents

The CxA should conduct a review of the OPR, BOD, and design drawings during the mid-CD phase to ensure that the owner's project requirements are addressed adequately in the BOD and design documents. This step is required for EA C3 only. If the team elects not to pursue this credit, then this step is not essential to meet the requirements of EA P1.

Stage III: Construction Phase Implementation: Construction Phase
Step 7: Review Contractor Submittals

During construction, the CxA should review contractor submittals of the systems that are to be commissioned to ensure that the systems conform to the OPR and BOD as well as fulfill operation and maintenance requirements. This step is required for EA C3 only. If the team elects not to pursue this credit, then this step is not essential to meet the requirements of EA P1.

Step 8: Verify Installation and Performance of Commissioned Systems

This step occurs throughout construction as the systems to be commissioned are installed. This is the heart of the commissioning process where the installation and performance of systems are verified. The activities may be completed by the CxA and/or the installing contractor depending on the project type. It typically involves three types of activities:

1 Pre-functional inspection: This occurs at the start-up of individual units and equipment and involves using *pre-functional checklists* to ensure that the systems have been installed properly.
2 Functional performance testing: This occurs after all system components are installed and ready for operation to verify that systems are operating as intended.
3 Evaluation of results: This involves verifying whether the installed systems meet the criteria set forth in the OPR and BOD. The CxA typically performs this activity.

Step 9: Develop Systems Manual for Commissioned Systems

The next step is to develop a systems manual that explains the functioning of the systems so that the future staff may operate them optimally. The manual may be prepared by the CxA and/or the contractor. The manual should provide information for each commissioned system and include at a minimum: the final version of BOD, single-line diagrams, the sequence of operations and original setpoints, the recommended schedule of maintenance requirements, and the recommended schedule for retesting and calibrating sensors. This step is required for EA C3 only. If the team elects not to pursue this credit, then this step is not essential to meet the requirements of EA P1.

Step 10: Verify that Requirements for Training are Completed

The next step is for the CxA to verify and document that training for the operating personnel has been completed. The training requirements should be established by the CxA in consultation with the owner. This step is required for EA C3 only. If the team elects not to pursue this credit, then this step is not essential to meet the requirements of EA P1.

Step 11: Complete a Summary Commissioning Report

The final step of the construction phase is to complete a summary commissioning report that documents all the commissioning activities with results from the testing and verification procedures. The report should also include any deficiencies or issues and how they were addressed. This step is required for both EA P1 and EA C3.

Stage III: Construction Phase Implementation: Occupancy Phase
Step 12: Review Building Operation

The final step of the commissioning process is to review the building operation within 10 months after substantial completion and verify whether the systems are operating as intended. The CxA should complete this review and also prepare a plan to resolve any outstanding issues. This step is required for EA C3 only. It is not essential to meet the requirements of EA P1.

Benefits of the LEED Project Management Process

The main purpose of defining a process to manage LEED projects in a simplistic, streamlined manner is to facilitate the widespread implementation of green building design, in general, and the LEED rating system, in particular. The tools, steps, and framework outlined in this book will, hopefully, help project teams integrate green building principles into the design, construction, and operation of buildings more easily. Ultimately, it is hoped that the LEED process will become the new *normal*. The project management process presented in this book offers the following four distinct advantages.

1. FACILITATES COMMUNICATION AND COLLABORATION

The key to the success of a LEED project is communication and collaboration between team members. This is essential to fully explore synergies between various design strategies and to create a building that is optimized for performance. Often team members work in isolation and are consulted on an as-needed basis.

The process outlined in this book ensures that there is constant communication and interaction between team members by means of various workshops and coordination meetings that occur throughout the duration of the project. Team members are brought together more than once during the process, to explore synergies, define responsibilities, set goals, and brainstorm to find optimal design solutions. The designation of a LEED project manager further helps by ensuring there is a *green* champion who is managing the process proactively and facilitating communication between team members.

2. ENSURES INTEGRATION WITH DESIGN

The iterative process of performing analyses, developing the design, and reviewing drawings ensures that green building principles are truly integrated with the design.

The analyses are performed in early design phases to provide team members with parameters to develop the design. Documents, drawings, and analyses are reviewed by the designated LEED PM periodically—at completion of design phases (SD, DD, CD), during construction, and finally, at project completion. The reviews not only ensure that green principles are integrated into the design and construction but also help verify that LEED requirements are suitably demonstrated. This ensures that clear, concise, and accurate information is submitted to the GBCI, thereby reducing potential rework and clarifications.

3. FACILITATES INTEGRATION WITH SCHEDULE

One of the key determinants of the success of any project is whether it was delivered within the targeted schedule without any major delays. The process outlined in this book provides specific steps and identifies key milestones that can be readily integrated into any design and construction schedule. This will enable project managers to allocate resources and to plan their projects with sufficient time built into the schedule, thereby avoiding additional service hours.

4. IMPROVES EFFICIENCY OF PROJECT DELIVERY

The process offers all the necessary tools and procedures required to implement a LEED project with practical tables, worksheets, and checklists that teams may use as they work through a LEED project. This will help reduce the time that project managers may have to spend in understanding the process and developing their own tools. Therefore, the process in general, helps improve the overall efficiency with which LEED projects are managed and delivered.

Credit Interpretations and Troubleshooting

It is well recognized that the LEED rating system is flexible and provides a broad framework for implementation by a number of building types (also illustrated in the previous sections). However, this flexibility sometimes leads to questions on interpreting credit requirements according to individual project circumstances. Teams have the opportunity to ask technical and administrative questions specific to their project by sending a *credit interpretation request* via LEED Online. The GBCI team reviews the request and offers comments in the form of *Credit Interpretation Rulings* (CIRs) on whether the project's interpretation meets the credit intent or not. Prior to 2009, the CIRs were considered precedent setting. However, now the CIRs are no longer considered precedent setting, and each ruling is applicable to only that project. As of 2010, a fee of $220 is charged by GBCI for each CIR.

The CIRs are the *official* route that project teams can use to request interpretations. However, before pursuing this route, team members and the LEED PM could potentially troubleshoot minor credit interpretation questions internally using any of the following approaches.

SCHEDULE A PROJECT TEAM BRAINSTORMING SESSION

The LEED PM may organize a brainstorming session with all team members to discuss the credit issue at hand. The expertise and experience of the team members is a valuable resource that the team could tap into. Team members should discern how their interpretation meets the intent of the credit. Usually, it has been observed (from study of previous CIRs) that the LEED rating system is flexible enough to accommodate special project circumstances and unique approaches, as long as the intent of the credit is met without significantly compromising on the requirements.

REFER TO ONLINE RESOURCES

With the evolution of the LEED rating system, a number of resources have populated the market, such as forums, magazines, blogs, and newsletters that talk about green building and LEED specifically. Team members may tap into the expertise offered by these resources to potentially find a solution to their credit issues. Some examples of good online resources that provide a wealth of information are: www.buildinggreen.com, www.leeduser.com, www.wbdg.com, and http://green.harvard.edu/theresource.

UTILIZE THE GBCI DESIGN REVIEW

The Split Review option offered by GBCI, where project teams can submit design credits and construction credits separately is a great feedback tool for project teams. It allows team members to present their design approach and receive comments and clarifications from the GBCI. The project teams may then take the appropriate course of action or make changes as needed.

If all of the above resources are not sufficient to resolve the issue or if getting a credit interpretation is imperative for the progress of the design, then the LEED PM may request an interpretation through the CIR channel.

Best Practices for LEED Projects

Green building design requires a paradigm shift in the way buildings are designed, managed, built, and operated. The shift is needed in terms of both processes and mindsets. The following are some best practices recommended for teams working on green/ LEED projects.

DESIGNATE A GREEN CHAMPION/LEED PROJECT MANAGER

LEED project teams should designate an individual as the green champion or project manager. The responsibility of this individual is to spearhead the LEED process and provide proactive direction, facilitation and support to the rest of the project team in integrating LEED requirements into the design. Refer to Chapter 2 for detailed responsibilities of the LEED PM.

LOAD THE FRONT END OF THE DESIGN

The second best practice is to load the front end of the design. This means that all the major decisions related to design and even construction should be evaluated and made as early as possible during the design process. This can happen if all the team members are involved right from the onset of the project to provide inputs. It will also help save time that perhaps may be spent later in making costly changes. Figure 6.5 shows a comparison between the front-end loaded design process for a LEED project and a linear design process for a traditional project.

OPTIMIZE BUILDING DESIGN

The most significant best practice is for architects to design buildings that are inherently optimized for energy performance. It is well-known and documented that the simple

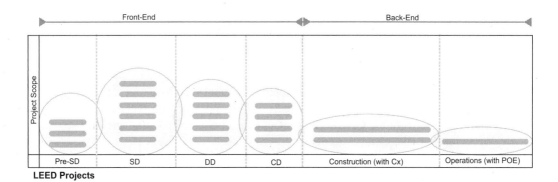

Figure 6.5   **Front-end loaded design approach to LEED projects.**
Credit: Green Potential LLC (www.green-potential.com).

fundamentals of architectural design such as orienting the building appropriately, designing the envelope according to the local climate, maximizing heat gain during winter, utilizing daylighting, and other passive design strategies, significantly help in reducing the building energy use. However, because of various reasons, such as aesthetics, site constraints, or owner needs, these important strategies are often sacrificed. This results in the utilization of massive active energy systems that increase the energy consumption and capital costs of the building. Project teams must consider passive strategies during the early design stages to create the conceptual design with *energy efficiency* in mind. As the design progresses, project teams must also be cognizant of how every design decision might affect energy use or other *green* aspects of the project.

OPTIMIZE ENERGY SYSTEMS

HVAC systems and lighting account for most of the energy consumption by a building. Passive design optimization in combination with building system optimization can significantly reduce the energy needs of a building. The use of on-site renewable energy systems to offset the energy needs can further help in reducing the total energy consumption. Figure 6.6 illustrates the design approach to reduce energy consumption in buildings.

UTILIZE DECISION SUPPORT TOOLS

Several tools are available in the market to assist design teams in making decisions based on the predicted energy performance of the building. Some software tools help in making early design decisions, and there are others that assist with complex energy

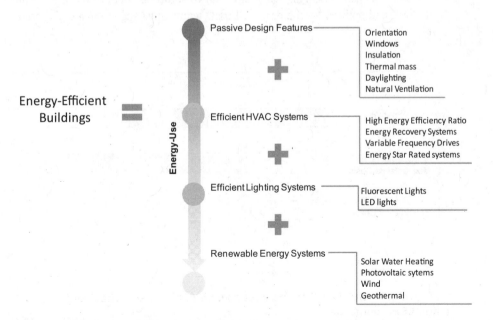

Figure 6.6 **Design approach for energy-efficient buildings.**
Credit: Green Potential LLC (www.green-potential.com).

system detailing. The widespread adoption of these tools by designers is still in its infancy stages. However, it is essential that they become part of the building design process. Building energy simulation tools, such as eQuest, Visual DOE, and IES VE, as well as daylighting simulation tools, such as DaySim, Radiance, etc., are some popular analysis software. Further, building information modeling software, also known as BIM, can allow architects, engineers, and building owners to evaluate various design options and their effects on energy consumption, daylighting, thermal comfort, and more. BIM modeling can take place at the early stages of design and continually during more detailed phases. Popular BIM software products include Bentley Architecture, Graphisoft ArchiCAD, VectorWorks ARCHITECT, as well as Autodesk's Revit and Architectural Desktop.

CONSIDER LONG-TERM OPERATIONAL BENEFITS

Financiers, developers, and owners must consider the operational costs along with the first costs when allocating funds for their projects. The reason for this is that the additional first costs associated with a LEED project may be recouped with savings from the operational costs over the building life. Project teams must evaluate each green strategy based on its long-term benefit and choose solutions that reduce the total cost of ownership. Figure 6.7 shows a representative comparison of the costs associated with a LEED and a non-LEED building.

ENCOURAGE CROSS DISCIPLINE SYNERGIES

Good green building design requires cross-disciplinary effort and collaboration between various team members. This will help achieve solutions that are more innovative, cost effective, and optimal for the project. LEED teams must engage in a collaborative effort

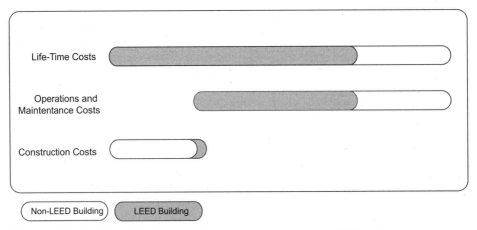

Figure 6.7   Building life-time cost comparison between LEED and non-LEED buildings.

and encourage the exploration of synergies between various disciplines through regularly scheduled charrettes, workshops, and meetings.

OPT FOR THE SPLIT REVIEW OPTION OFFERED BY GBCI

GBCI offers the option to split the review process into two parts: one at the end of the design phase and one at the end of the construction phase. It is recommended that teams opt for the Split Review option in order to take advantage of the feedback offered by GBCI and make changes as needed. On the other hand, if teams choose the Combined Review option, then opportunities for making changes are limited.

INTEGRATE LEED MILESTONES INTO THE SCHEDULE

Another best practice for teams is to identify important LEED-related milestones, such as charrettes, meetings, drawing reviews, submissions, etc., and integrate them with the design and construction schedule. The schedule will help the team work toward the milestones within the overall project timeline.

REVIEW DESIGN DRAWINGS AND SPECIFICATIONS

Finally, the last practice that is essential for a LEED project is a periodic review of the drawings and specifications to ensure that the LEED requirements have been integrated into the design. The LEED project manager must spearhead this effort and review design and construction drawings as the design progresses.

Notes

[1] USGBC, "LEED for Core & Shell," http://www.usgbc.org/DisplayPage.aspx?CMSPageID=295.
[2] USGBC, "Portfolio Program," http://www.usgbc.org/DisplayPage.aspx?CMSPageID=1729.
[3] Energy Design Resources, *Design Briefs: Building Commissioning* (Energy Design Resources, 2002), http://www.energydesignresources.com/Resources/Publications/DesignBriefs/tabid/74/articleType/ArticleView/articleId/109/Design-Briefs-Building-Commissioning.aspx (accessed April 2010).
[4] Ibid.
[5] USGBC, LEED *Reference Guide for Green Building Design and Construction* (USGBC, 2009).

FUTURE OF GREEN:

2010 AND BEYOND

The first decade of this century saw the resurgence of the green building movement and the emergence of rating systems, such as LEED, BREEAM, GreenStar, etc., across the world. Many national, state, and local governments have embraced green buildings and are providing several incentives for their widespread development. As a result, green buildings have gained acceptance among public agencies, private sector developers, and institutional owners. The LEED rating system can be regarded as a *game changer* that has catapulted *green* to the spotlight. Over the last decade, LEED buildings have had a tremendous impact on reducing CO_2 emissions, water consumption, and material use. In 2009 alone, annual CO_2 savings from LEED buildings was estimated to be approximately 2.9 million tons from energy efficiency and renewables. Total water savings from LEED was estimated at 15 billion gallons, composed of 0.5% of annual non-residential water use. Over 60% of construction and demolition waste was diverted from LEED projects, totaling 25 million tons to date and is expected to reach almost 800 million cumulative tons by 2030. The productivity benefits from LEED buildings to-date are estimated between $230 million to $450 million.[1]

The adoption of the LEED rating system may be regarded as an important first step in reducing the impact of buildings on global warming and climate change. However, there is still a long way to go before a momentous impact can be made. This chapter presents various ideas, technologies, solutions, principles, and policies that are currently in the early stages of implementation and have a massive potential to take green building development to the next level. These ideas have been presented in the following broad categories:

- Building design and construction
- Building technologies
- Building operations
- Building finance
- Building policies/regulation

Building Design and Construction

PERFORMANCE-BASED DESIGN

In the coming years, the building design process will likely see a dramatic shift, wherein design decisions will be based on the anticipated performance of the building in terms of energy use, electricity consumption, water use, and CO_2 emissions. Several sophisticated software tools such as eQuest, VisualDOE, Revit, etc., are already available in the market to help designers make these performance-based decisions. The next decade is going to see an even greater rise in the use of such tools. Modeling the building's energy performance right from the beginning of design will become standard, whether a building is LEED or not. The modeled performance of the building will dictate its design, and operational efficiency will become an important factor during early design discussions.

INTEGRATED DESIGN APPROACH

The integrated design approach, currently in its nascent stages, is expected to soon become mainstream. Although the number of projects following this approach is still miniscule, it will hopefully continue to grow and find wider acceptance by owners, developers, and design professionals alike. The evolution of this process will most likely occur in an organic fashion. Even though a conscious decision to follow the integrated design approach may not be made, design teams will most likely end up following this approach because *green* building design almost demands it.

GREEN CODES AND STANDARDS

The development of the International Green Construction Code (IGCC) is a significant step in the direction of making green building design a code requirement. Therefore, developing green and energy-efficient buildings may no longer be a voluntary decision but a mandatory requirement. Several green building standards, such as ASHRAE 189.1 and California's CalGreen, are taking shape to spearhead the process of widespread implementation of green building practices within the design and construction community. Most of these standards are currently voluntary, but soon cities and states will be embracing and eventually mandating them as standards that must be met to obtain a permit for construction.

NET ZERO ENERGY BUILDINGS

The ultimate goal that the building design community should strive for is *zero energy* buildings. Net Zero Energy buildings are defined as buildings that have zero net energy consumption annually. There are several examples of buildings all over the world that have already achieved net zero energy usage. However, the technologies and systems required to achieve this outcome require significant upfront investment. This is a significant barrier to entry that is impeding the extensive implementation of net zero energy concepts. A number of initiatives, such as Department of Energy's (DOE) Zero Energy Commercial Building Initiative, the Consortium of Zero Energy Buildings and Architecture 2030, are conducting significant research to provide guidance in this regard. The next twenty years will likely see a tremendous growth in net zero buildings. Green codes, regulations, and voluntary programs such as LEED will constantly evolve themselves to push the building community toward achieving the coveted net zero status.

BUILDINGS ANCHORED IN SUSTAINABLE COMMUNITIES

Generally speaking, buildings cannot be green or energy efficient in isolation. For buildings to truly reduce their environmental impact, they need to be located in communities that are planned with sustainability in mind. During the early 70s, cities and communities were planned in a decentralized manner where people worked in the city and lived in the suburbs. Over the years, this has led to increased commute times, congestion from increased vehicle use, and increased strain on the infrastructure. The next two decades will see an increased emphasis on designing communities that are self-sustaining, transit oriented, pedestrian-friendly, and mixed-use developments.

Building Technologies

ADVANCED ENERGY-EFFICIENT SYSTEMS

Since LEED's inception in 1998, the last ten years have seen a remarkable growth in several innovative technologies, products, and materials to cater to rapidly changing building needs. Widespread adoption of new technologies or products is often impeded by the high first-costs. It is important for the building community to find innovative solutions that are not only efficient but are also cost-effective. The next twenty years is anticipated to be an era of *green innovation* especially in the areas of HVAC, lighting, and building envelope technologies. The demand for energy-efficient buildings will fuel the development of cost-effective, energy-efficient products and systems.

RENEWABLE ENERGY SYSTEMS

The development of cost-effective and efficient renewable energy technologies will be the single most important factor that will help in achieving the goal of net zero buildings. Although there has been significant advancement in solar, wind, and geothermal technologies during the last ten years, there is still a huge gap that needs to be bridged in the areas of system efficiency and cost effectiveness. Apart from the development of these technologies, there is also a need for training a workforce that is adept at installing, operating, and maintaining these systems. The next decade will likely usher in tremendous growth of cost-effective and efficient renewable energy technologies that will help reduce the environmental footprint of buildings.

SMART METERS

The building community's effort to design and build energy-efficient buildings needs to be complemented with energy conservation efforts by the occupants. Typically, it has been observed that occupants are more actively involved in energy conservation efforts if they are aware of their energy use patterns. Smart meters that allow building occupants to track their real-time energy use will help in encouraging energy conservation. Although practicing energy conservation measures is tied to behavioral and cultural patterns of occupants, incorporating smart-meters will allow tenants to manage their energy footprint actively.

Building Operations

POST-OCCUPANCY EVALUATION/PERFORMANCE MONITORING

From an operational perspective, buildings are rarely evaluated after occupancy to see whether they are (a) functioning according to the owner's requirements, (b) performing as anticipated, and (c) satisfying the needs of the occupants. Building owners and managers are usually weary of dissatisfied occupants and potential expensive changes that they may have to make. However, it is important to realize that evaluating the performance of the building after it has been occupied is an important tool in making future decisions. This is especially important for managers of a multiple portfolio of buildings, as the evaluation of one building may provide them with valuable feedback that can be implemented at other properties. Although this type of monitoring is slowly picking up, post-occupancy evaluations are expected to become part and parcel of our building management process.

GREEN LEASES

One of the biggest barriers in the implementation of green building measures by developers is the issue of split incentives where the benefit of operational efficiency is realized by the tenants and not necessarily by the developers. This dilemma often prevents developers from exploring all the green building opportunities, as they might not be able to recoup the high initial costs. Several new leasing models referred to as *green* leases are currently under development to address this issue. Green leasing is defined by the California Sustainability Alliance as "the integration of energy and water efficiency, emissions reduction, waste minimization, and other sustainability objectives throughout the entire commercial leasing process. Green leasing dictates that building performance become transparent to all parties involved in the lease transaction. This performance includes the efficiency in which the main engineering plant operates (HVAC, plumbing, lighting, etc.), the environmental standards of building materials, as well as the effectiveness of the building operations and management programs."[2]

The Building Owners and Managers Association (BOMA) International has released a lease guide, *Guide to Writing a Commercial Real Estate Lease, Including Green Lease*

Language, featuring step-by-step instructions to help building owners and managers write green operations and management practices into their lease agreements. BOMA's new lease guide was developed to address commonly cited barriers to implementing green building practices. The guide serves as both a legal-language tool to help building owners and managers maintain a green building through best operations and management practices, and also as an education tool for working with brokers and tenants to outline what is expected of tenants in a high-performance green building and the responsibilities of all parties involved in the ongoing efforts to keep it green and encourage continuous improvement.[3]

The next decade is expected to see further evolution of the green lease concept, ensuring that both developers and tenants are able to maximize their benefits from a green building.

Building Finance

LIFE-CYCLE COSTING

Typically the budgets for buildings are allocated based on first time costs of design and construction. The cost associated with operations of a building over its typical 20- to 30-year life is rarely taken into consideration when allocating the capital budget of a project. The awareness of operational savings achieved by energy-efficient buildings is making building owners (especially those who manage their own buildings) rethink their capital budgets. Owners are realizing the long-term value associated with an initial high-cost item and making decisions based on long-term, life-cycle costing methodologies. However, the percentage of projects following this approach is still small, and there is a need to develop better life-cycle costing tools that can assist owners in making decisions based on long-term benefits. It is hoped that the momentum that has been built in this area will continue to pick up, and sooner rather than later, life-cycle costing will become part of how building budgets are finalized.

POWER PURCHASE AGREEMENTS

An important factor impeding the rapid installation of renewable energy systems such as photovoltaics (PV) is the high initial costs associated with them. Even after maximizing available government incentives and rebates as well as taking long-term benefits into account, solar PV systems are still high-cost items for owners. A new financing structure, known as a *power purchase agreement,* has emerged over the last few years. Through this agreement owners can purchase power at a fixed price (for a fixed term) from a third-party company that would own, install, maintain, and operate the PV systems on the owner's site or building. This will allow the owners to offset their energy needs by means of a renewable source without having to pay high amounts for installation or maintenance. They would pay a fixed price for that electricity regardless of market fluctuations. Although still in its early inception stages, this concept will certainly be the frontrunner in the extensive implementation of solar PV systems. Just as there is need for evolution of renewable energy technologies, there is also a need for evolution of the financing structures for these systems. Such financing structures are necessary to make renewable technologies affordable and accessible.

Building Policies/Regulation

MANDATORY GREEN BUILDING

From a policy and regulation perspective, there is little doubt that the push to build green by various local, state, and federal governments is very strong. At present, all federal buildings are required to achieve LEED certification. Similarly, several cities are offering incentives to developers to build green. While all these are steps in the right direction, the bigger goal would be to mandate a certain minimum level of green building requirements through green codes, standards, or other such measures. It is anticipated that minimum-level green requirements will become mandatory in the next few decades.

PERFORMANCE DATA OR GREEN LABEL FOR BUILDINGS

One important decision that the USGBC made in 2009 was to require building owners to disclose the performance of their LEED buildings as a precondition to achieve certification. The intent was to understand whether LEED buildings that are designed to perform better than other similar buildings, are in fact performing as anticipated. Similar steps are required at a policy/regulation level as well, where local, state, or federal governments should stipulate that building owners display their energy performance in comparison to other buildings of similar type. Potential buyers or tenants of a property would then be able to base their decisions on the energy performance rating of the building. This would be akin to the Energy Star label typically seen on appliances, which informs users of their expected performance. Similarly, an energy performance label (updated on an annual basis) should be provided for all buildings, so that potential buyers/renters can make informed decisions. The Energy Star rating program available for buildings at present provides a similar label, but the program is voluntary. Buildings in the United Kingdom (UK) are already required to display their energy performance rating. It is likely that there will be a sharp rise in this requirement of displaying building energy performance labels in the coming decade.

NET METERING/SMART GRID

Many European countries such as Germany have been able to push the widespread use of photovoltaic solar systems by owners of commercial buildings and individual homes through the development of smart-grids. Building owners are allowed to sell excess power generated on their site back to the grid. Meters track electricity use both ways—from grid to building and building to grid. Some states in the United States are currently investing in developing smart grid technologies. Although in a nascent stage, smart grids and net metering are expected to become the norm in the next twenty to thirty years.

CO$_2$ EMISSIONS TAX

The fundamental need for designing green buildings is to reduce their impact on the environment. The increasing use of energy by buildings has led to a depletion of fossil-fuel-based resources as well as increased generation of CO_2 emissions. There is scientific evidence to prove that these emissions have resulted in global warming that is causing the earth's temperatures to rise at a dramatic pace. There are proposals for industries and businesses to pay taxes on their CO_2 emissions when they exceed a certain predetermined amount. On the flip side, when buildings are able to achieve significant reductions in emissions, they would be able to trade their emission reductions with other high emission-producing industries. The levying of a CO_2 tax will become more likely in the next decade as it is essential to bring about a meaningful change from a global warming perspective.

Notes

[1] Watson, Rob. *Green Building Impact Report*. Greener World Media, Inc., 2009.
[2] California Sustainability Alliance, "What Is Green Leasing," http://sustainca.org/green_leases_ toolkit/what_green_leasing (accessed June 28, 2010).
[3] Building Owners and Managers Association International, "New BOMA Green Lease Guide offers solutions for writing sustainability into leasing agreements," http://www.boma.org/news/pressroom/ Pages/press062208-3.aspx (accessed June 28, 2010).

SUCCESS STORIES

Introduction

This chapter presents select case studies of projects that have achieved LEED certification in the last two years. The various stages and steps that have been described until now (in the previous chapters) have been condensed from several projects and simplified to present typical scenarios of project management to readers. To get a deeper understanding about the LEED process, however, it is useful to get a real-world perspective on some of the processes followed, strategies implemented, and challenges faced by teams in the field. The journey of these projects has been presented from start to finish in order to give readers a flavor of how these projects were able to meet their green building goals. It is hoped that readers will be able to interlink these processes with their own project and utilize the framework from these case studies, as they pursue the path toward green/LEED building design. The following case studies have been chosen:

- Great River Energy (GRE) Headquarters, Maple Grove, MN[1]
- Rockefeller Hall Renovation, Harvard Divinity School, Cambridge, MA[2]
- PNC Green Branch® Bank Branches, U.S.[3]

These projects were chosen because they each present a unique approach to and perspective on green building design. Each project attained LEED certification in unique ways, and each can be considered a pioneering effort in its own right.

The GRE headquarters in Minnesota was the first building in the state to achieve LEED Platinum certification; it incorporates many innovative green building strategies, such as daylighting, geothermal, wind, and solar technologies. In fact, it was winner of the prestigious American Institute of Architects' Committee on the Environment (AIA/COTE) Top Ten Green Projects in 2009. The project demonstrates the integrated design approach at its best, where the owners, design team, occupants, and facility managers worked exhaustively and collaboratively to achieve their green goals. This case study—an office building that was built ground-up on a greenfield site—is a perfect showcase for building owners, developers, and project teams that are involved with the design and construction of new office and other commercial projects.

The second case study, Rockefeller Hall on the Harvard Divinity School (HDS) campus in Cambridge, received LEED Gold certification in 2009 and is part of a campus-wide effort by Harvard University to *green* all their buildings. In fact, Harvard has stipulated that all their projects with budgets over five million dollars be LEED-certified. The HDS project is a major renovation project, where several energy efficiency features were added to the 1970s-built structure to improve the building's energy performance. Since it was a major renovation, this project used the LEED-NC rating system instead of the LEED for Existing Buildings: Operations & Maintenance (LEED-EB:O&M). This case study is a classic example of how existing buildings may be renovated to achieve green buildings that not only are better performing but also offer better value in terms of operational and intangible savings.

Finally, the last case study, PNC's Green Branch® Bank branches, is the world's very first volume-build LEED project that led to the development of the now LEED–Portfolio Program pilot. It illustrates how green building design may be applied across a portfolio of buildings in a simplified, streamlined manner to save time and money for all stakeholders. The design team created a LEED-ratable Green Branch® prototype that served as the standard prototype to be constructed throughout the supply chain. Through this program, PNC has been able to achieve LEED certification for 72 of its banks at various locations and has over 23 in the pipeline. This case study is a great showcase for volume builders, such as retailers, restaurants, and banks that utilize an expansion-based business model relying heavily on physical locations to engage with customers. Table 8.1 provides a snapshot of the three projects and their key features.

TABLE 8.1   SNAPSHOT OF CASE STUDY PROJECTS

KEY FEATURES	GREAT RIVER ENERGY HEADQUARTERS	ROCKEFELLER HALL RENOVATION— HARVARD DIVINITY SCHOOL	PNC BANK—GREEN BRANCH® BANK BRANCHES
Location	Maple Grove, MN	Cambridge, MA	15 states
Building Type(s)	Commercial Office	Office/School	Banks
Project Type	New Construction	Major Renovation	New Construction
Area	166,000 square feet	26,715 square feet	Portfolio of buildings
Scope	4-story building	4-story building	Varies
Setting	Suburban	Campus	Varies
Year Completed	2008	2008	Ongoing
LEED Rating	LEED Platinum	LEED Gold	LEED Silver (minimum)

Case Study One: Great River Energy Headquarters, Maple Grove, MN

PROJECT OVERVIEW

Great River Energy (GRE) is an electric generation and transmission cooperative that serves two-thirds of the geographic area of Minnesota. In 2007–2008, the organization built a brand new headquarters building to meet the company's growing needs in Maple Grove, MN. In 2008, the building achieved the highest certification level of LEED Platinum from USGBC, making it the first building to achieve such recognition in Minnesota.

GRE's mission is "to provide its members with reliable energy at competitive rates in harmony with a sustainable environment." Over the last few years, the escalating costs of building large-scale generation facilities and the growing pressure to reduce CO_2 emissions, had made it challenging for the company to create new generation facilities for its customers. As a result, energy conservation and efficiency became GRE's primary "fuels." The company realized that the more efficiently buildings, homes, and businesses were built, the less energy they would use, which would help them delay building expensive new power plants. GRE decided that there would be no better place to exemplify the benefits of energy efficiency and conservation than in their own new headquarters building in Minnesota.

The 166,000-square-foot building was designed to use approximately 50 percent less energy than a comparable facility built to state code requirements and approximately 90 percent less water than a similarly sized corporate campus. It features an in-lake geothermal HVAC system, in-floor displacement ventilation, daylight harvesting, 72 kilowatts of on-site solar panels, and a 200-kilowatt wind turbine. The GRE team was able to accomplish all this for an incremental cost of less than 10 percent more than a conventional corporate headquarters of the same size. Figures 8.1 and 8.2

Figure 8.1 GRE headquarters—a 200-kW wind turbine provides renewable energy to the building.
Credit: Lucie Marusin / Perkins + Will.

Figure 8.2 GRE headquarters—reflection of the wind turbine in the glazing of the building.
Credit: Don F. Wong.

show the GRE headquarters building with the wind turbine that provides renewable energy to it.

The primary objectives that GRE wanted to achieve for their headquarters building were as follows:

- To build a state-of-the-art building in energy efficiency, sustainability, and conservation that included at least one innovative design feature that had never been done before
- To contribute to the advancement of sustainable green building and be an educational tool for others to see how easily and relatively inexpensively these technologies could be applied
- To display their commitment to renewable energy

- To dispel the notion that energy efficiency and conservation meant sacrificing beauty, functionality, or practicality
- To create a place that people would want to visit and a building that employees would love to come to every day

PROJECT MANAGEMENT

Project Team

One of the first steps that GRE took was to assemble an internal team that understood the vision of the project and how it directly affected the company's bottom line. "We started by assembling a core team that understood what it takes to successfully complete an ambitious project." GRE's Director of Business Operations Mike Finley said, "You need to engage all the disciplines in a way that creates synergies in the project." In addition to this internal core team, a subcommittee of the GRE Board of Directors was also assembled to evaluate decisions and determine the best options for the company and its 28 member cooperatives. Once the internal team was assembled, the members set out to find an owner's representative, an architect, and a general contractor that understood GRE's goals.

GRE felt it was important to hire an owner's representative, when building for LEED certification, because of the latter's understanding of the approvals and the partners needed to introduce new construction concepts. GRE interviewed three owner's representative companies and selected the Minneapolis-based Tegra Group due to the organization's experience in building corporate campuses in the Twin Cities area. Tegra Group also appealed to GRE leaders because they laid out a plan to keep the ambitious project on budget. Sticking to a budget was much more than just a financial goal for Great River Energy—it would achieve the project's goal to stand as an example of the practicality of building for LEED certification. Tegra further helped establish criteria for assessing partners to ensure that the architect, the general contractor, and everyone else who worked on the building were committed to the vision and had the experience to make it work.

Perkins + Will, a global firm with a Minneapolis office, was named the architect on the project because of the company's experience with high-performance buildings and willingness to design a cutting-edge structure. The firm was also experienced with integrated front-end loaded design that encouraged feedback and input from employees throughout the design and construction phases. This process closely matched the way GRE conducted business by valuing everyone's input. The alignment of cultures between Perkins + Will and Great River Energy sparked a synergy that drove the project. Perkins + Will also managed the LEED process for the project.

McGough Construction of St. Paul, MN, was selected as the general contractor due to its commitment to sustainable building practices, its roster of LEED accredited professionals, and its track record with energy efficient buildings. Table 8.2 identifies the key team members.

Project Process

Perkins + Will conducted a Vision Workshop where high-level goals for the building were outlined by the GRE leadership. It was at this meeting that GRE's CEO, David

TABLE 8.2 GRE HEADQUARTERS—KEY TEAM MEMBERS	
Owner	Great River Energy
Owner's Representative	Tegra Group
Architect	Perkins + Will
General Contractor	McGough Construction
Civil Engineer	RLK
Landscape Architect	Close Landscape Architects
MEP Engineer	The Weidt Group
Structural Engineer	BKBM
Commissioning Authority	KFI

Saggau, outlined his vision to build the most energy-efficient building possible. That single goal from the vision meeting influenced every decision related to the building. He also challenged designers to do something with energy efficiency that had never been done in the world to contribute to the evolution of green design. It was also at the vision workshop that Perkins + Will explained the levels of LEED certification. While explaining each point and outlining the differences between Silver, Gold, and Platinum LEED certification, the architects made it clear that Platinum LEED certification was attainable for the project and site, but it would be challenging to achieve. The site was immediately ineligible for certain points due to the fact that there was no existing structure, nor was it a brownfield site. The team would need to make the most of the available site resources.

After the vision workshop, McGough and Perkins + Will set up tours of office buildings in the area so employees could identify features they liked and disliked. Comments from the tours helped drive the design process. After the tours were completed, Perkins + Will interviewed every Great River Energy manager and conducted employee focus groups to identify additional design drivers. Focus groups revealed several design elements that employees valued, which guided the design. For instance, daylight was repeatedly named as a feature employees valued in their workspace. This reinforced the commitment to daylighting in the design; it would meet employee preference in addition to promoting energy efficiency. Employees also identified privacy, collaboration, access to colleagues, and outside views as features they valued in a workspace.

During conceptual design, Perkins + Will and the firm's engineering partner, Dunham Associates, identified the best opportunities for LEED points on the site. They identified two opportunities that would drive the building's efficiency: daylight harvesting and lake-source geothermal displacement heating and cooling. With these defined, the building began to take shape and a lot of decisions were influenced by these design concepts. Figure 8.3 is a typical floor plan showing a new pattern for two-

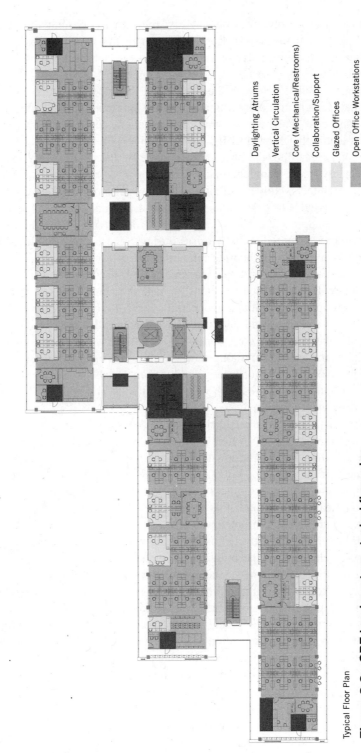

Typical Floor Plan

Figure 8.3 **GRE headquarters—typical floor plan.**
Credit: Perkins + Will.

Daylighting Atriums

Vertical Circulation

Core (Mechanical/Restrooms)

Collaboration/Support

Glazed Offices

Open Office Workstations

to four-story office buildings obtained by relocating *core functions,* such as storage, utility rooms, and file rooms, from the center of the floor plate to the ends, thereby allowing the creation of daylight atriums and the optimization of the footprint and envelope area.

Perkins + Will along with the mechanical engineers at Dunham Associates collected data on a matrix that outlined every design option and its effect on energy efficiency and cost. This allowed the team to take a cost-benefit approach and select features that both achieved LEED points and increased efficiency.

Decisions regarding the project were made by the design team as a whole. Great River Energy's core team lent their expertise on a variety of topics and led discussions with the architects, builders, engineers, and owner's representative. Ideas were thoroughly discussed before decisions were made as a team. The team took a proactive and collaborative approach that encouraged early problem solving during design, before the issues become potentially costly crises during construction. Employee feedback was continually sought throughout design and construction.

The materials for the building were selected and specified with LEED credits in mind. Perkins + Will provided specifications for materials, and McGough Construction priced out several options. McGough and Perkins + Will created conceptual budgets based on specified materials and determined what materials were available that had low levels of volatile organic compounds (VOCs). They weighed cost parameters against material characteristics and long-term durability to establish whether it made sense to use recycled materials or locally produced materials for each building feature. The findings of all materials research were submitted to and approved by Perkins + Will before they were purchased.

A formal process was established to check materials during the construction. A McGough employee was available to answer subcontractors' questions about green products. With a single representative charged with routinely monitoring materials and ensuring that the correct products were being used, crews could get their material-related questions answered quickly. Also, during every biweekly safety meeting, construction means and methods that were related to compliance with indoor air quality (IAQ) measures were reinforced.

GRE had to also effectively communicate with the city officials about the energy-conserving elements of the building. Though city officials were largely in favor of Great River Energy moving in, there were some who initially resisted the design concept—specifically the wind turbine and lake-source geothermal heating and cooling system. LEED was also new to many city officials. Because the design elements were largely unfamiliar, the team made an effort to connect the building's features to their community benefits and educational potential. Great River Energy and Perkins + Will presented a design concept during a hearing attended by an assortment of city officials ranging from the city planner to the fire chief. The advantages of a high-performance Platinum LEED building in Maple Grove were outlined and well received by those in attendance. Work with city planners quickly uncovered a solution to accommodate the proposed lake-source geothermal heating and cooling system. After explaining the system's minimal effects on wildlife and aesthetics, the city offered an easement agreement that would allow the necessary 36 miles of piping to be coiled below the surface of Arbor Lake. Another hurdle was the approval of the proposed wind turbines. Great River Energy presented three options for the building's wind energy resources to the Maple Grove City Council. There was an option with three turbines, another with two, and a third with a single turbine to be used on an interim basis. The city council agreed to the third option, and the turbine was granted a five-year permit to provide energy for the building.

LEED SCORECARD

The building achieved LEED Platinum Certification using the LEED-NC v2.2 rating system. The project maximized all the available points in the *Energy* and *Water Efficiency* categories. It achieved a total of 57 out of the 69 possible points. Figure 8.4 shows a snapshot of the credit points achieved in the different categories.

Sustainable Sites

The location of the GRE building afforded access to multiple community amenities and public transportation. The building was constructed on a previously developed site within a half mile of a residential zone with an average density of 10 units per acre and within a half mile of 10 basic services. The Maple Grove Transit Center, located across the street from Great River Energy's headquarters building, offered access to multiple bus routes serving the greater Twin Cities area. Additionally, Great River Energy worked with a private transportation service to provide a shuttle bus that travels between its Maple Grove headquarters and its Elk River facility to accommodate employees who live outside the metropolitan area.

Bike racks as well as shower and changing facilities were provided to encourage employees to use alternate modes of transportation. Bikes were also provided for employees to check out during their lunch hour to run errands nearby without using a vehicle. In addition, preferred parking spaces were provided for low-emitting and fuel-efficient vehicles. The parking provided on-site was carefully sized not to exceed what was absolutely necessary for the building.

More than half of the vegetation (approximately 6.5 acres) on the GRE grounds was native or ecologically adapted to Minnesota. By reducing the size of the parking lot, Great River Energy was able to increase the amount of green space, exceeding the city of Maple Grove's open space requirement by more than 25 percent. Stormwater management strategies were implemented to control both the quantity and quality of stormwater from the site.

A thermal polyolefin (TPO) white roof with a minimum solar reflectivity index of 78 was specified to reduce the building's contribution to the urban heat island effect. The parking lot and site lighting were woven into the landscape design. Low-profile, cut-off fixtures focused light onto the property, not into the sky. The approach taken toward lighting was not to reduce the lighting levels at grade as much as to effectively focus and manage the light through design. A total of 12 out of the 14 possible points were achieved in this category by the GRE headquarters.

LEED Scorecard

Great River Energy Headquarters

Location	Maple Grove, MN
Rating System	LEED-NC, v 2.2
Certification achieved	Platinum
Total Points achieved	57 out of 69 points possible

	SS Sustainable Sites	12/14
Prereq 1	Construction Activity Pollution Prevention	Y
Credit 1	Site Selection	1
Credit 2	Development Density & Community Connectivity	1
Credit 3	Brownfield Redevelopment	0
Credit 4.1	Alternative Transportation, Public Transportation Access	1
Credit 4.2	Alternative Transportation, Bicycle Storage & Changing Rooms	1
Credit 4.3	Alternative Transportation, Low-Emitting & Fuel-Efficient Vehicles	1
Credit 4.4	Alternative Transportation, Parking Capacity	1
Credit 5.1	Site Development, Protect or Restore Habitat	1
Credit 5.2	Site Development, Maximize Open Space	1
Credit 6.1	Stormwater Design, Quantity Control	1
Credit 6.2	Stormwater Design, Quality Control	1
Credit 7.1	Heat Island Effect, Non-Roof	0
Credit 7.2	Heat Island Effect, Roof	1
Credit 8	Light Pollution Reduction	1

	WE Water Efficiency	5/5
Credit 1.1	Water Efficient Landscaping, Reduce by 50%	1
Credit 1.2	Water Efficient Landscaping, No Potable Use or No Irrigation	1
Credit 2	Innovative Wastewater Technologies	1
Credit 3.1	Water Use Reduction, 20% Reduction	1
Credit 3.2	Water Use Reduction, 30% Reduction	1

	EA Energy & Atmosphere	16/17
Prereq 1	Fundamental Commissioning of the Building Energy Systems	Y
Prereq 2	Minimum Energy Performance	Y
Prereq 3	Fundamental Refrigerant Management	Y
Credit 1	Optimize Energy Performance	10
Credit 2	On-Site Renewable Energy	3
Credit 3	Enhanced Commissioning	1
Credit 4	Enhanced Refrigerant Management	1
Credit 5	Measurement & Verification	0
Credit 6	Green Power	1

	MR Materials & Resources	6/13
Prereq 1	Storage & Collection of Recyclables	Y
Credit 1.1	Building Reuse, Maintain 75% of Existing Walls, Floors & Roof	0
Credit 1.2	Building Reuse, Maintain 100% of Existing Walls, Floors & Roof	0
Credit 1.3	Building Reuse, Maintain 50% of Interior Non-Structural Elements	0
Credit 2.1	Construction Waste Management, Divert 50% from Disposal	1
Credit 2.2	Construction Waste Management, Divert 75% from Disposal	1
Credit 3.1	Materials Reuse, 5%	0
Credit 3.2	Materials Reuse, 10%	0
Credit 4.1	Recycled Content, 10% (post-consumer + ½ pre-consumer)	1
Credit 4.2	Recycled Content, 20% (post-consumer + ½ pre-consumer)	0
Credit 5.1	Regional Materials, 10% Extracted, Processed & Manufactured Regionally	1
Credit 5.2	Regional Materials, 20% Extracted, Processed & Manufactured Regionally	1
Credit 6	Rapidly Renewable Materials	0
Credit 7	Certified Wood	1

	EQ Indoor Environmental Quality	14/15
Prereq 1	Minimum IAQ Performance	Y
Prereq 2	Environmental Tobacco Smoke (ETS) Control	Y
Credit 1	Outdoor Air Delivery Monitoring	1
Credit 2	Increased Ventilation	1
Credit 3.1	Construction IAQ Management Plan, During Construction	1
Credit 3.2	Construction IAQ Management Plan, Before Occupancy	1
Credit 4.1	Low-Emitting Materials, Adhesives & Sealants	1
Credit 4.2	Low-Emitting Materials, Paints & Coatings	1
Credit 4.3	Low-Emitting Materials, Carpet Systems	1
Credit 4.4	Low-Emitting Materials, Composite Wood & Agrifiber Products	1
Credit 5	Indoor Chemical & Pollutant Source Control	0
Credit 6.1	Controllability of Systems, Lighting	1
Credit 6.2	Controllability of Systems, Thermal Comfort	1
Credit 7.1	Thermal Comfort, Design	1
Credit 7.2	Thermal Comfort, Verification	1
Credit 8.1	Daylight & Views, Daylight 75% of Spaces	1
Credit 8.2	Daylight & Views, Views for 90% of Spaces	1

	ID Innovation & Design Process	4/5
Credit 1.1	Innovation in Design, Green Building Education	1
Credit 1.2	Innovation in Design, Exemplary Performance for EA C6	1
Credit 1.3	Innovation in Design, Concrete Use Reduction	1
Credit 1.4	Innovation in Design	0
Credit 2	LEED® Accredited Professional	1

Credits achieved   Credits not achieved

Figure 8.4 Great River Energy Headquarters—LEED scorecard.

Water Efficiency

Native and adapted vegetation were incorporated into the landscape design, and a high-efficiency permanent irrigation system was installed to help the vegetation get established and to provide occasional supplementary watering. The design also incorporated reuse of stormwater for irrigation of turf grass and the establishment of new plantings, dramatically reducing the use of potable water for irrigation. Great River Energy made an agreement with the city of Maple Grove to use captured rainwater for irrigation, as necessary. Site stormwater runoff equal to the amount of water needed for irrigation is directed through a filtration pond and into the city's stormwater pond before being used for irrigation.

A 20,000-gallon cistern was designed to collect rainwater. The water is filtered and circulated through an eco-friendly water treatment system using a small amount of hydrogen peroxide to sanitize the water. The water is then used in the building's toilets and urinals, limiting the potable water that ends up in the city's sewer system. Because of this system, the building is designed to use 80 percent less potable water for sewage conveyance.

The engineers specified low-flow 0.5-gallon-per-minute aerators and motion sensors on bathroom faucets as well as dual flush meters on the toilets. All of these measures are predicted to reduce the amount of water used in the building by more than 35 percent. The building was able to achieve all the 5 points available in this category.

Energy

GRE was able to achieve 16 out of 17 points available in this category. The two significant design features that helped the building achieve a more than 47.5 percent reduction in energy cost were: (a) daylighting and (b) a high efficiency HVAC system that combined lake-source geothermal energy and under-floor displacement ventilation. Energy Modeling was started early in the design process to inform design decisions and system selection.

The building is predicted to receive approximately 14% of its energy from on-site renewable resources (approximately 10% from wind and about 3.5% from the photovoltaic array). The wind turbine produces 200 kilowatts (kW) at full output, and the photovoltaic solar array produces 72 kW of energy at full capacity. By implementing renewable energy into its headquarters, Great River Energy wanted to make a visible statement about the future of power generation. The turbine, manufactured in Denmark, was a NEG Micon M700. It was first installed and put in service in the Netherlands before being purchased and shipped to Canada. Later, the turbine was purchased by Energy Maintenance Services (EMS), who refurbished and sold it to Great River Energy. The gears in the gearbox were remanufactured, and the generator was rewound to change the machine from a two-speed to a one-speed wind turbine to increase efficiency. Space on the roof was covered with a photovoltaic solar array. Solar *trees* were also installed on the building grounds for demonstration purposes.

During the early design phase of the project, a third party Commissioning Authority was hired to make sure that all of the green features within the building functioned properly. The engineers created a measurement and verification plan to evaluate the building's energy performance. However, the project did not earn the EA C5 credit; the project's alternative approach to M&V did not meet the specific criteria required by the USGBC and the team decided not to appeal the point.

Materials and Resources

Recycling receptacles and a recycling room were provided to encourage recycling and reduce the waste that is hauled to landfills. In addition to the required receptacles for paper, corrugated cardboard, glass, plastics, and metal, the building included compostable waste bins near the cafeteria and in pantries. Organic waste is taken to a composting facility where it is transformed into compost that is sold to residential and commercial customers.

More than 95 percent of construction waste was recycled, due largely to a successful recycling program on the construction site. The program was successfully implemented by placing recycling facilities in the right places, educating workers, and monitoring progress. The construction team set rigorous standards on the site. To ensure that as much waste was recycled as possible, on-site recycling bins were provided so all workers could correctly dispose of waste without spending time collecting and transporting it to an off-site recycling center. Once the facilities were in place, the general contractor educated crews by showing them the correct place to dispose of waste, specific to the materials they planned to use. As new subcontractors began work on the site, they were given a copy of the waste management plan and a tour of the recycling facilities. Recycling was also a recurring topic at every site safety meeting.

As an added measure, the general contractor monitored materials and tailored recycling facilities according to subcontractor needs. For example, when crews began installing geothermal piping in Arbor Lake, they alerted the general contractor that they would be creating a lot of plastic waste and recommended that additional plastic recycling receptacles be obtained.

The building contains a variety of materials made from recycled content, including rebar framing made from recycled steel and countertops made from recycled glass and concrete. In all, approximately 18 percent of the total value of materials used in the building had postconsumer or pre-consumer recycled content. Approximately 22 percent of the value of materials used in the building were extracted, processed, and manufactured within 500 miles of the building. Approximately 87 percent of the wood used throughout the building was Forest Stewardship Council (FSC) certified. In total, the project was able to achieve 6 out of the total 13 points possible.

Indoor Environmental Quality

Permanent monitors were installed to ensure that the air circulated throughout the building was of optimal quality. Fresh-air terminals feeding the core-building's underfloor air systems measure the fresh air to each portion of the building and react to maintain a healthy indoor air quality by measuring the carbon dioxide level in occupied spaces. Carbon dioxide sensors were strategically located throughout the occupied spaces. Figure 8.5 shows a typical office section with raised-access floor systems that offer flexibility and individual control of supply air by occupants.

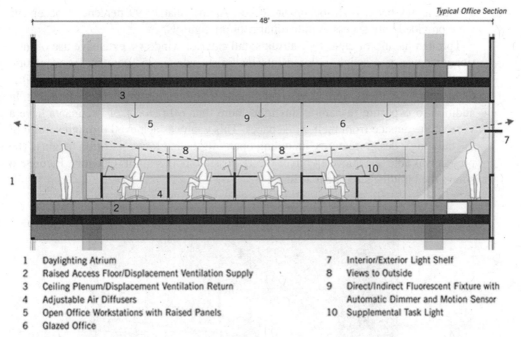

Typical Office Section

1	Daylighting Atrium	7	Interior/Exterior Light Shelf
2	Raised Access Floor/Displacement Ventilation Supply	8	Views to Outside
3	Ceiling Plenum/Displacement Ventilation Return	9	Direct/Indirect Fluorescent Fixture with
4	Adjustable Air Diffusers		Automatic Dimmer and Motion Sensor
5	Open Office Workstations with Raised Panels	10	Supplemental Task Light
6	Glazed Office		

Figure 8.5 GRE headquarters—typical office section.
Credit: Perkins + Will.

Compared to the minimum rates required by ASHRAE Standard 62.1-2004, the building was designed to provide 10% more outdoor air and 30% more fresh air to the breathing zone. In order to maintain high air quality during the construction process, builders and contractors created an indoor air quality (IAQ) management plan. All absorptive materials, such as insulation, carpeting, and ceiling tile, were stored properly to avoid moisture damage. Enclosed, well-ventilated areas were designated for cutting materials to control dust and other harmful particles. Additional measures that helped control air quality included capping ductwork and the use of temporary ventilation until the permanent system was completed and operable under normal conditions.

All interior paints, coatings, adhesives, and sealants applied within the building were low in volatile organic compounds (VOCs). VOC limits were clearly stated in each section of the materials specifications. All low-emitting materials were checked and tracked in the field to ensure that the specified materials were used. All carpet and cushion met the testing and product requirements of the Carpet and Rug Institute's Green Label Plus program. Composite wood and agrifiber products, such as particleboard, medium density fiberboard (MDF), and plywood used on the project, contained no added urea-formaldehyde resins. All laminating adhesives used were also free of added urea-formaldehyde resins.

Task lighting controllable by the occupants was provided to meet individual task needs. Manually operated in-floor diffusers were located in every individual workspace as well as in shared multi-occupant spaces. Approximately 99 percent of occupants were provided with access to individual comfort controls.

Through the use of clerestory atriums, tall exterior windows, extensive use of interior windows, and a light-colored interior finish palette, 25 foot-candles of natural daylight reach more than 84 percent of all regularly occupied spaces in the building. Computer simulations were used to predict daylight levels in much of the building. In addition to exposure to natural daylight, more than 90 percent of occupants have a view of the outside from their workspaces. This goal was achieved through several of the building's design features, including open offices and lower partition heights. The project was able to achieve 14 out of the 15 possible points in this category. Figure 8.6 shows the building's daylight atriums.

Figure 8.6 Interior photo showing the building's daylight atriums.
Credit: Don F. Wong.

Innovation in Design

The GRE building achieved four of the five points available in this category. A *green* education program was created in which GRE offered employees and guests the opportunity to interact with building information through kiosks located in the first floor lobby. The interactive screens show real-time data on energy savings throughout the building and production from the wind turbine and photovoltaic panels. Other applications include system animations that illustrate how the building's features work and recommendations for environmental savings in a person's own home. Educational signage was also placed inside the building and along walking trails on the grounds to educate the public about the building's energy and water conservation features. Additionally, the building is open to public tours on a scheduled basis.

FINDINGS AND CONCLUSION

Understand the Occupants

The future occupants of a LEED-certified building are an invaluable resource for establishing design drivers and finding efficiencies. The sooner a design team can get access to a building's inhabitants, the more effectively they can design with them in mind. The availability and involvement of GRE employees helped make the new head-quarters a success.

Invest Time in Schematic Design

The more time an owner and design team can spend on schematic design the better. With an emphasis on schematic design, the interior systems can be more thoroughly analyzed. This also allows for more accurate cost parameters on a project because it will reduce the need to change plans during construction. The GRE team was heavily invested in ensuring that the bulk of *green* design decisions were made as early as possible and established buy-in from all stakeholders from the onset.

Do Not Lose Sight of Non-Sustainable Elements

The continued pursuit of green building depends on its success right now. Owners are hesitant about building for LEED certification because of its reliance on new building systems and construction methods. In order for green building to gain momentum and make a significant impact, the structures must be shown to be as successful as traditional construction. Buildings must function properly with no negative consequences from the inclusion of sustainable elements to ensure the proliferation of green building.

Challenge Everything

During the course of the project, the GRE team realized that high-efficiency buildings are, by definition, non-traditional. They are new and different. In order to accomplish the goals of building an efficient structure they constantly asked: "Is there a better way?" There may be traditional building methods that are better options for a project, but the evolution of green building depends on every project seeking better, more efficient construction methods.

Case Study Two: Rockefeller Hall Renovation—Harvard Divinity School, Cambridge, MA

PROJECT OVERVIEW

Rockefeller Hall, on the Harvard Divinity School (HDS) campus, was originally constructed in 1970, designed by distinguished architect Edward Larrabee Barnes. In 2007–2008, the building underwent a complete renovation to meet new programming needs of Harvard Divinity School. The school has seven buildings on campus, and Rockefeller Hall is one of the seven, with a total area of 26,715 square feet. The four-story building originally housed a graduate dormitory on the upper three floors, with a dining room, kitchen, and seminar rooms on the first floor. The purpose of the renovation was to create more up-to-date, welcoming spaces on the first level, convert the dormitory space for administrative office use, and improve accessibility throughout. The project team was able to creatively redesign the building to meet modern needs, while keeping a large portion of the building mass.

HDS intended to create a better performing building that would reduce operational costs, improve the comfort and health of occupants, and also minimize environmental impact. In keeping with this commitment to sustainability, the Rockefeller project implemented numerous energy conservation measures and pursued LEED certification to achieve a LEED Gold rating in November 2009. Figures 8.7 and 8.8 show the building exterior after the renovation.

Figure 8.7   Rockefeller Hall—view from the HDS campus green.
Credit: Venturi, Scott Brown and Associates, Inc.

Figure 8.8   Rockefeller Hall—east façade view.
Credit: Venturi, Scott Brown and Associates, Inc.

According to the school's Building Committee, the main goals of the Rockefeller renovation were to:

- Make the building fully accessible
- Convert floors two–four from graduate dormitory to administrative use
- Reconfigure the first-floor dining area along with kitchen and food service facilities upgrades
- Upgrade life safety and building energy systems
- Improve the building envelope
- Create a new campus green

PROJECT MANAGEMENT

Project Team

Harvard Divinity School formed a core *Building Committee,* which consisted of all those who were responsible for making the decisions on the project. This committee included both HDS's internal facilities and operations personnel as well as individuals from the central Harvard University campus services group. HDS specifically engaged the services of a project manager who was responsible for managing the entire project from start to finish and ensuring that it was delivered on time and within budget.

This core committee then, in turn, assessed the project's needs and arrived at specific goals and objectives for the project. One of the key goals identified in the early meetings was to design and construct a sustainable renovation with a focus on optimized energy performance and improved occupant comfort. In writing the request for proposal (RFP) for the architect selection, the Building Committee stated that the project would be a sustainable project and that the design team should have experience in designing sustainable or LEED-certified buildings. The architects were identified, and subsequently, other members of the project team—the MEP engineer, the structural engineer, the civil engineer, the landscape architect, and the construction manager— were engaged.

Beginning in the Design Development phase, the Building Committee also engaged the services of a sustainability consultant from Harvard University's Office for Sustainability, Green Building Services. The Office for Sustainability (OFS) Green Building Services supports Harvard University in designing, building, and operating their buildings in an energy-efficient manner. The OFS facilitated the LEED process by assisting the HDS team with research and decision-making support. The OFS also managed the documentation and handled the certification process with the USGBC.

Energy simulations were performed by the MEP engineers, and an independent party was engaged to provide commissioning services for the project. Figure 8.9 presents a graphic description of the project team structure, and Table 8.3 lists the key team members.

Process

Although the project was designed as a *green* building, with the LEED credit framework in mind, the HDS Building Committee's initial thoughts were not to pursue the official documentation and certification route with the USGBC (who was then administering certification). At the time, in 2005, the LEED certification process was still

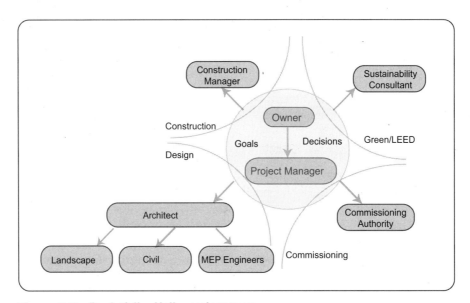

Figure 8.9 Rockefeller Hall—project team.

TABLE 8.3 ROCKEFELLER HALL—KEY TEAM MEMBERS	
Owner	Harvard Divinity School
Project Manager	Harvard Real Estate Services
Architect	Venturi, Scott Brown and Associates, Inc.
Civil Engineer	Stephen Stimson Associates
Landscape Architect	Green International Affiliates, Inc.
MEP Engineer	SEI Companies (now WSP Flack & Kurtz)
Structural Engineer	Foley Buhl Roberts & Associates, Inc.
Lighting Consultant	Grenald Waldron Associates
Food Service Consultant	Ricca Newmark Design
Cost Consultant	Faithful & Gould
Code Consultant	Harold Cutler, PE
IT/AV/Acoustics Consultant	Accentech Inc.
Sustainability Consultant	Harvard Office of Sustainability, Green Building Services
Commissioning Authority	Facility Dynamics Engineering
Construction Manager	Shawmut Construction

evolving and the committee's perception was that the documentation process was too tedious. However, they did set operational efficiency and occupant comfort as their primary goals for the renovation. All design decisions were made with long-term benefit in mind, and the team performed cost-benefit analyses to make optimal decisions. Since operational efficiency was a key goal, energy simulations were performed to determine the most energy-efficient measures. The design team used the LEED credit framework to make decisions, and HDS was actively involved throughout the process. As the design progressed, HDS made the decision to seek official certification with USGBC, and the sustainability consultants then facilitated the documentation submission process. The sustainability consultants also provided the necessary tools to the design team and the contractor in the implementation of construction-related LEED activities. Overall, the success of the certification depended on the collaborative effort of all the various team members and the active involvement of the owners in the project.

The project was in programming and design phases from May 2005 to May 2007, and construction lasted for a year, from 2007 to 2008. The LEED certification was awarded in November 2009. Figure 8.10 shows the first-floor plan of the building.

LEED SCORECARD

The building achieved LEED Gold certification using the LEED-NC v2.2 rating system. The project maximized its points in the *Energy* and *Indoor Environmental Quality* categories, as it aligned with the owners' goal of creating a better performing

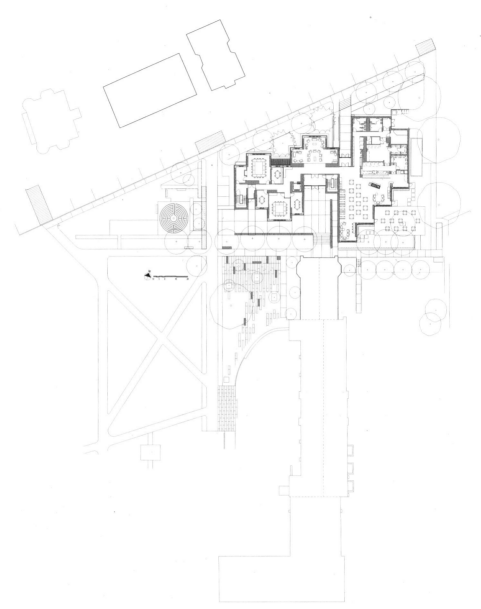

Figure 8.10 Rockefeller Hall—first-floor plan.

Credit: Venturi, Scott Brown and Associates, Inc.

building that reduced operational costs and improved the quality of the environment for the occupants. The building obtained a total of 49 out of the 69 possible points. Figure 8.11 shows a snapshot of credit points in the different categories.

Sustainable Sites

The location of Rockefeller Hall contributed towards the achievement of a number of Sustainable Sites credits. The building has access to multiple amenities and ample

LEED Scorecard

Rockefeller Hall Renovation, Harvard Divinity School

Location	Cambridge, MA
Rating System	LEED-NC, v 2.2
Certification achieved	Gold
Total Points achieved	49 out of 69 points possible

SS	Sustainable Sites	9/14
Prereq 1	Construction Activity Pollution Prevention	Y
Credit 1	Site Selection	1
Credit 2	Development Density & Community Connectivity	1
Credit 3	Brownfield Redevelopment	0
Credit 4.1	Alternative Transportation, Public Transportation Access	1
Credit 4.2	Alternative Transportation, Bicycle Storage & Changing Rooms	1
Credit 4.3	Alternative Transportation, Low-Emitting & Fuel-Efficient Vehicles	1
Credit 4.4	Alternative Transportation, Parking Capacity	1
Credit 5.1	Site Development, Protect or Restore Habitat	0
Credit 5.2	Site Development, Maximize Open Space	1
Credit 6.1	Stormwater Design, Quantity Control	1
Credit 6.2	Stormwater Design, Quality Control	0
Credit 7.1	Heat Island Effect, Non-Roof	0
Credit 7.2	Heat Island Effect, Roof	1
Credit 8	Light Pollution Reduction	0

WE	Water Efficiency	2/5
Credit 1.1	Water Efficient Landscaping, Reduce by 50%	0
Credit 1.2	Water Efficient Landscaping, No Potable Use or No Irrigation	0
Credit 2	Innovative Wastewater Technologies	0
Credit 3.1	Water Use Reduction, 20% Reduction	1
Credit 3.2	Water Use Reduction, 30% Reduction	1

EA	Energy & Atmosphere	13/17
Prereq 1	Fundamental Commissioning of the Building Energy Systems	Y
Prereq 2	Minimum Energy Performance	Y
Prereq 3	Fundamental Refrigerant Management	Y
Credit 1	Optimize Energy Performance	10
Credit 2	On-Site Renewable Energy	0
Credit 3	Enhanced Commissioning	1
Credit 4	Enhanced Refrigerant Management	1
Credit 5	Measurement & Verification	0
Credit 6	Green Power	1

MR	Materials & Resources	7/13
Prereq 1	Storage & Collection of Recyclables	Y
Credit 1.1	Building Reuse, Maintain 75% of Existing Walls, Floors & Roof	1
Credit 1.2	Building Reuse, Maintain 100% of Existing Walls, Floors & Roof	1
Credit 1.3	Building Reuse, Maintain 50% of Interior Non-Structural Elements	0
Credit 2.1	Construction Waste Management, Divert 50% from Disposal	1
Credit 2.2	Construction Waste Management, Divert 75% from Disposal	1
Credit 3.1	Materials Reuse, 5%	0
Credit 3.2	Materials Reuse, 10%	0
Credit 4.1	Recycled Content, 10% (post-consumer + ½ pre-consumer)	1
Credit 4.2	Recycled Content, 20% (post-consumer + ½ pre-consumer)	0
Credit 5.1	Regional Materials, 10% Extracted, Processed & Manufactured Regionally	1
Credit 5.2	Regional Materials, 20% Extracted, Processed & Manufactured Regionally	1
Credit 6	Rapidly Renewable Materials	0
Credit 7	Certified Wood	1

EQ	Indoor Environmental Quality	13/15
Prereq 1	Minimum IAQ Performance	Y
Prereq 2	Environmental Tobacco Smoke (ETS) Control	Y
Credit 1	Outdoor Air Delivery Monitoring	0
Credit 2	Increased Ventilation	1
Credit 3.1	Construction IAQ Management Plan, During Construction	1
Credit 3.2	Construction IAQ Management Plan, Before Occupancy	1
Credit 4.1	Low-Emitting Materials, Adhesives & Sealants	1
Credit 4.2	Low-Emitting Materials, Paints & Coatings	1
Credit 4.3	Low-Emitting Materials, Carpet Systems	1
Credit 4.4	Low-Emitting Materials, Composite Wood & Agrifiber Products	1
Credit 5	Indoor Chemical & Pollutant Source Control	0
Credit 6.1	Controllability of Systems, Lighting	1
Credit 6.2	Controllability of Systems, Thermal Comfort	1
Credit 7.1	Thermal Comfort, Design	1
Credit 7.2	Thermal Comfort, Verification	1
Credit 8.1	Daylight & Views, Daylight 75% of Spaces	1
Credit 8.2	Daylight & Views, Views for 90% of Spaces	1

ID	Innovation & Design Process	5/5
Credit 1.1	Innovation in Design: Commuter Choice Program	1
Credit 1.2	Innovation in Design: Green Power (100%)	1
Credit 1.3	Innovation in Design: MR C5, Regional Materials, Exemplary Performance	1
Credit 1.4	Innovation in Design: WE C3, Water Use Reduction, Exemplary Performance	1
Credit 2	LEED® Accredited Professional	1

Credits achieved   Credits not achieved

Figure 8.11 Rockefeller Hall—LEED scorecard.

Figure 8.12   Rockefeller Hall—old parking lot was removed to create a green space.
Credit: Venturi, Scott Brown and Associates, Inc.

public transportation. A Harvard shuttle provides access to all parts of the Harvard University campus, and the nearby Massachusetts Bay Transportation Authority (MBTA) subway and bus routes provide access across the Boston metro area.

All occupants of Rockefeller Hall have access to Harvard's comprehensive Commuter Choice Program, which provides incentives and discounts for all modes of alternative transportation as well as for carpooling and use of fuel-efficient vehicles.

A large asphalt parking lot was removed, recycled, and replaced with a green space, supporting recreation and community-building activities that helped to provide a campus identity. Figure 8.12 is a view of the new green space. The stormwater system was designed to collect stormwater by trench drains, area drains, and deep sump catch basins that were piped to an underground recharge system. In summary, 9 out of the 14 possible points were achieved in this category.

Water Efficiency

The Rockefeller Hall project incorporated dual flush toilets, low-flow plumbing fixtures, and waterless urinals to achieve a reduction of domestic water consumption by approximately 44% over standard Energy Policy Act (EPAct) 1992 fixtures. This is the equivalent of saving over 167,000 gallons per year. Figure 8.13 describes the flow and flush rates of fixtures used in Rockefeller Hall in comparison with the EPAct standard referenced by LEED. The project achieved two out of the possible five

Fixture Type	Flush & Flow Rates	
	Rockefeller Hall	EPAct 1992 Standard
Water Closet [GPF]	Dual Flush- 1.6 & 1.1	1.6
Urinal [GPF]	0	1
Bathroom Sink [GPM]	0.5	2.5
Kitchen Sink [GPM]	1.5	2.5

GPF: Gallons per flush

GPM: Gallons per minute

Total potable water consumption reduction 43.7%

Figure 8.13   Rockefeller Hall—fixture flow rates.
Credit: Harvard University, Office of Sustainability.

points in this category. Note that an earlier version of the rating system, LEED-NC v2.2, was used in which there was no prerequisite under this category. Under the current version, LEED-NC v3, all projects are required to reduce their potable water consumption by 20% over the EPAct standard.

Energy and Atmosphere

The team incorporated numerous energy conservation measures to reduce operating costs and greenhouse gas emissions. According to the OFS, the Rockefeller Hall was modeled to use 42.2% less energy than an ASHRAE 90.1-2004 code-compliant building, saving roughly $106,000 and 355 metric tons of carbon dioxide equivalent greenhouse gasses compared to an ASHRAE baseline. When designing the building, Harvard used energy modeling and life-cycle costing to evaluate various design options and quantify their 20-year net benefit. The team eventually chose to include ten energy conservation measures that had an increased initial cost but saved energy and utility costs over time. These ten measures cost an additional $115,000 but were calculated to have a net present value of more than $700,000 to the Divinity School, when the operational impacts were included. The project achieved 13 out of the possible 17 points in this category.

The following were the ten energy conservation measures that were integrated within the building design.

Variable frequency drives (VFDs)   The project utilizes variable frequency drives (VFDs) on the chilled water and hot water pumps and fans, which slow down when demand is low and allow different zones to operate separately. For example, the upper office floors can be isolated while the first-floor public space is energized for an extended schedule of use.

Enhanced chilled water controls   Pressure independent characterized control valves (PICC) were installed to send the right amount of chilled water through each cooling coil, maximizing efficiency and improving dehumidification.

Energy recovery wheel   An enthalpy wheel installed on the central ventilation air handling unit recovers heat, cool air, and humidity, so less energy is required to condition fresh outdoor air.

Fan coil unit cycling   The fans heating and cooling the space have been programmed to automatically turn off whenever the temperature set point is satisfied, rather than run continuously when the system is on.

Kitchen hood controls   A *Melink* system was installed on the kitchen exhaust fan, varying the speed according to the presence of heat and smoke.

Demand control ventilation   Carbon dioxide sensors in the cafeteria were incorporated to vary the quantity of fresh air according to the number of people in the space.

Increased heating water Delta T   By installing fan coil units with larger heating coils, the heating hot water gives up more heat each time through the coils (30 degrees F instead of 20) and reduces pump energy.

Lighting optimization   Lighting power density was reduced by 20% compared to ASHRAE 90.1 code requirements by installing highly efficient linear fluorescent and compact fluorescent lights.

Lighting controls   Occupancy sensors were installed to control lighting fixtures in all occupied spaces.

Improved building envelope   Insulation of walls and roofs was increased, as was the performance of windows when compared with both ASHRAE 90.1 code requirements and the existing conditions.

The project also included a number of design features that improved energy efficiency but were not considered an added cost, such as utilizing the efficient central plant chilled water system with water side economizers, a white Firestone TPO roof membrane that reduces heat gain, new Otis Gen2 machine-room-less elevators with regenerative braking, and Xlerator hand dryers that save significant cost and energy compared to other hand dryers or paper towels. To help promote renewable energy and offset some of the environmental impact from the Divinity School campus, renewable energy credits are purchased annually equal to 100% of the campus electricity load.

Materials and Resources

HDS minimized the quantities of materials and energy required to construct a new building by renovating an existing building. Over 95% of the existing walls, floors, and roof were retained. The project was able to achieve 7 out of the 13 possible points in this category.

The construction manager was able to divert from landfills almost 85% of the total waste generated during demolition and construction, with most of the waste being reused or recycled. In total, 418 tons of waste was removed, which included asphalt from the parking lot, brick, metals, and old furniture. Further, HDS made a decision to donate as many building elements as possible before the demolition work of the interiors began. The Building Committee engaged Institutional Recycling Network (IRN) to plan the donation. After the last meal was served at Rockefeller Hall, IRN began to dismantle the functional kitchen equipment and also took interior wood doors and all the sinks and toilets. According to metrics provided by IRN, "a total of 148 pieces of kitchen equipment and interior doors were removed from Rockefeller, and a total weight of 17,169 lbs, the equivalent of 8.58 tons of surplus material, was donated." HDS later found out from IRN that most of the donated kitchen equipment is now being used by a university in Costa Rica. Figures 8.14 to 8.17 show the dining hall before and after the renovation.

The fundamental criteria used by the design team in specifying materials and products for Rockefeller were to select materials that had recycled content, were locally manufactured, and/or were also low emitting. About 19% of the total material value in the project consisted of postconsumer and/or pre-consumer recycled content materials, and 42% of the total material value consisted of materials manufactured within

Figure 8.14 The dining hall before renovation.
Credit: Venturi, Scott Brown and Associates, Inc.

Figure 8.15   The dining hall after renovation.
Credit: Venturi, Scott Brown and Associates, Inc.

Figure 8.16   The kitchen after renovation.
Credit: Venturi, Scott Brown and Associates, Inc.

Figure 8.17   View of the dining hall.
Credit: Venturi, Scott Brown and Associates, Inc.

500 miles of the project site. Figure 8.18 lists some of the products and materials used in Rockefeller Hall.

Indoor Environmental Quality

One of the original goals of HDS was to provide a better indoor environment for the building's occupants. The design team incorporated strategies that improved the overall air quality of the interior spaces and thermal comfort for the occupants. Further, the construction team implemented measures, both during construction and prior to occupancy, to minimize pollutants and provide a healthy environment.

An air quality management plan was implemented by the contractors to keep pollutant levels down during the renovation and to ensure that the building was fully ventilated prior to occupancy. MERV 13 filters were installed on the building's air handling units to improve air filtration.

	Recycled Content		Regional Distance
	Postconsumer	Preconsumer	
Acoustical Ceiling Tiles (Armstrong)		68%	
Arreis MDF (Sierra Pine)		100%	
Greenguard certified Batt Insulation (Roxul)		40%	
Carpet Tile (Lees) Recycled Content		35%	
Metal Framing Stud Dietrich) Recycled Content	17%	37%	
Metal Framing Stud Drywall (USG)	95%	4%	248 Miles

Figure 8.18   Products used in Rockefeller Hall.
Credit: Harvard University, Office of Sustainability.

All products used in the interior of the building—adhesives, paints, carpets, etc.—were products with low or no volatile organic compounds (VOCs). Doors and other composite wood products were specified to be free of urea-formaldehyde. Figure 8.19 lists some of the low-emitting products used in Rockefeller Hall. The project was able to achieve 13 out of the 15 possible points in this category.

FINDINGS AND CONCLUSION

Active involvement of owners   From the very beginning, HDS, the owner, was actively involved in and committed to a *green* renovation. It was one of their project goals, and all design decisions were made based on this guiding principle. The design team had clear direction on how to proceed and what the overall objectives for the project were.

***Green* integrated with design**   The design team used the LEED credit framework to guide the design for the project. They did not take a prescriptive approach to LEED but instead placed emphasis on making *whole building* decisions based on what was *good* for the project, while taking long-term benefits into consideration. Benefits were viewed from both the financial perspective (savings in operational costs) as well as the social and environmental perspectives (occupant and community improvement).

Collaborative team effort   The team members attribute the success of the project to the collaborative effort and collective participation of everyone involved to achieve the targeted goals.

***Green* decision support**   Both the design team and HDS benefited from the constant research support and technical advice offered by the OFS, the sustainability

Product Manufacturer	VOC Content (g/L)	VOC Limit (g/L)	Standard
Paints & Coatings			
Eco Spec Flat --219, Benjamin Moore	0	50	Green Seal GS-11
Eco Spec Eggshell --223, Benjamin Moore	0	150	Green Seal GS-11
M29 D.T.M. Acrylic Semi-gloss rust inhibitor	207	250	Green Seal GS-11
Adhesives & Sealants			
3M Fire Barrier CP 25WB + Caulk	0	250	SCAQMD Rule #1168
Heavy Duty Construction Adhesive, Greenchoice	6.6	70	SCAQMD Rule #1168

Figure 8.19   Low-emitting materials used in Rockefeller Hall.
Credit: Harvard University, Office of Sustainability.

consultants. The OFS provided feedback by performing cost-benefit analyses that helped the client make the most cost-effective decisions. The sustainability consultants established protocols, provided documentation tools to the contractor, and facilitated the process of achieving certification from the USGBC.

Case Study Three: PNC Green Branch® Bank Branches

PROJECT OVERVIEW

PNC Financial Services Group is a long-established leader in integrating energy-efficient, green building principles into its real estate management and construction planning. PNC's first foray into sustainable corporate-wide practices started with a single 650,000-square-foot financial processing center that broke ground on July 9, 1998, the same year the United States Green Building Council (USGBC) began certifying commercial buildings using LEED (Leadership in Energy and Environmental Design) criteria. By November 2000, PNC's Firstside Center was complete and earned a LEED Silver rating as well as the distinction as the first financial facility to be certified and the largest LEED-certified green building in the world. Figure 8.20 is a photograph of PNC's Firstside Center.

Figure 8.20 **PNC Bank's Firstside Center earned LEED Silver in 2000.**
Credit: PNC Bank.

Figure 8.21 LEED Gold headquarters of PNC Global Investment Servicing.
Credit: PNC Bank.

PNC's mutual fund servicing company, PNC Global Investment Servicing (sold in June 2010), then decided to build its 112,000-square-foot global headquarters to LEED Gold standard—the first in the state of Delaware. Figure 8.21 is a photograph of the headquarters.

Not long after the financial processing center opened in 2000, employee retention and operations costs noticeably improved compared to the facility it replaced. The benefit to PNC's bottom line was accompanied by a significant increase in media and industry interest in PNC's green building efforts, and a business case for sustainability became clear. Encouraged by these successes, PNC Realty Services, led by Gary Saulson, made a pioneering commitment to the environment in 2003 by vowing to seek LEED certification for all newly constructed stand-alone branches.

PNC initiated the branch bank greening process by adding green features to its existing *legacy branch* prototype for six buildings under construction. The company found that branches with added green features were pushed through the entitlement and permitting process faster and received a positive response from the media and the community at openings. Several branches offered communities the opportunity to experience a green building for the first time. Figure 8.22 is a photograph of PNC's *legacy branch*.

Spurred by these positive responses and with more than 100 new branches in the pipeline, PNC decided to take the bold step of completely redesigning its branch banks to create a LEED ratable Green Branch® bank branch prototype. A significant departure from the legacy branch style, the new design would integrate green features from

Figure 8.22 PNC's legacy branch.
Credit: PNC Bank.

the ground up and serve as the standard prototype to be constructed throughout the supply chain.

The decision to seek LEED certification across a portfolio of more than 100 buildings raised several challenges. First, despite strong acceptance of LEED by building owners and developers, the LEED certification process was primarily created to be applied to *one-off* design and construction projects delivered by a single project team working sequentially. Replicating LEED certification across PNC's portfolio of 100 new Green Branch® bank locations would require preparing and submitting each LEED application individually, regardless of whether the design had previously been rated through an earlier construction process, and in a manner somewhat disconnected from PNC's actual building delivery process. This requirement was too onerous and expensive for PNC and other volume builders, such as retailers, restaurants, and banks that utilize an expansion-based business model relying heavily on physical locations to engage with customers. Saulson regularly received calls and correspondence from retailers asking how PNC dealt with the paperwork required in one-off certifications when they were replicating a building many times. He concluded a path to least resistance needed to be created to encourage LEED certification from volume builders.

Additionally, success in the highly competitive arena of retail banking requires a clear focus on speed and efficiency, as competitors who enter a market first are likely to gain greater market share. The process of certifying every location individually can delay the design and construction process just long enough to allow a competitor to set up shop. Also, further schedule delays could result, as each location must be customized to the site and the building is often constructed in a short period of time by local construction teams who were not part of the design process and who may have never worked on a LEED building.

In order for PNC's plan to certify 100 bank branches to succeed, the LEED certification process would need to embrace, rather than confront, the nature of volume design and construction processes used by PNC and other companies like it. It became clear that a new LEED management process was needed that could guarantee consistent implementation of LEED design strategies across multiple sites, but that also allowed for enough flexibility to respond to local site conditions in a way that would not compromise the LEED rating or owner delivery schedule.

PROJECT MANAGEMENT

The Project Team

With the goal established, PNC's Gary Saulson assembled a dream team to help create a volume LEED certification solution. With a reputation for high-quality design and implementation for a full range of retail design services, Gensler was hired to design the prototype and envision a brand for the Green Branch® banks. Paladino and Company, who worked with PNC on Firstside Center and the six green legacy bank branches, was invited to help ensure the world's first volume LEED project was a success. The firm was a significant contributor to the development of the LEED® Green Building Rating System and had the in-depth knowledge and expertise required to design a new LEED management process and to collaborate with the USGBC to gain support. Local green engineering firm CJL joined the team to provide mechanical, electrical, and plumbing design, and help perform the technical analysis required to validate potential green design strategies. Table 8.4 illustrates the key team members of the prototype project.

Project Process—Embracing the Volume Build Model

The key to resolving the conflict between the LEED certification process, the volume design and construction supply chain requirements, and speed was to develop a LEED

TABLE 8.4 PROTOTYPE PROJECT TEAM	
Owner	PNC Financial Services Group
Architect	Gensler
MEP	CJL Engineering
Structural	Gilsanz, Murray, Steficek, LLP
Contractor	Clemens Construction Company
Commissioning Agent	PNC Realty Services
Sustainability Consultant	Paladino and Company

certification process that would disconnect the certification of the standard, or proto-type, design and all related LEED design phase credits from the certification of the construction process and associated LEED construction phase credits. To that end, Paladino and Company worked collaboratively with the USGBC on PNC's behalf to build support for a two phase LEED certification process in which an initial prototype design would be developed and submitted for *pre-certification* prior to any actual loca-tions being constructed. Upon completion of construction, the project teams would then submit reduced construction submittals to validate construction credits for each branch to the USGBC for final rating of the branch based on specific site variables, along with the pre-rated prototype credits. The sum total of points represents corpo-rate-wide policies, standardized building components, and local site conditions to cre-ate individualized final scores for each building to max out at a LEED Gold level, while assuring that there are always enough standardized points to reach a LEED-certified level.

The decision to separate the certification process into two phases that may not occur in close sequence implied that an optimization of the prototype design to respond to both multiple site variables and to align with PNC's sustainability and design goals would be required. In addition, it implied the need for a quality control process that would ensure that each contractor could adequately construct the prototype without previous LEED project experience. Figure 8.23 illustrates the difference between con-ventional single-building certification process and volume certification.

Design Optimization

Under the proposed certification process, the LEED Green Branch® bank branch prototype design would need to meet certain environmental performance goals to achieve the desired LEED rating while at the same time incorporating a quality con-trol component that would ensure that each branch could achieve the desired rating regardless of site characteristics. Therefore, the PNC team recognized that the initial design goals for the new prototype should provide general performance guidelines against which all proposed strategies would be tested, and included:

- LEED Silver certification or above
- Minimized climate impact of each location
- Strategies validated based on life-cycle cost analysis
- Predictability of performance regardless of location

With the initial goal to ensure that any single site would perform at least as well as any other built under the same prototype design, the PNC design team, led by Gensler, began by trying to anticipate and accommodate a broad range of site variables that might be encountered, and that might impact the effective construction of the proto-type. Through this process, the team identified 4,608 different combinations of site factors, covering four possible climate zones, twelve site orientations, and hundreds of possible microclimates. Immediately, it became apparent that designing a prototype to respond to all anticipated site factors would be daunting and likely detrimental to the

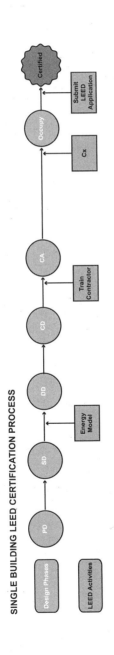

SINGLE BUILDING LEED CERTIFICATION PROCESS

VOLUME BUILDING LEED CERTIFICATION PROCESS

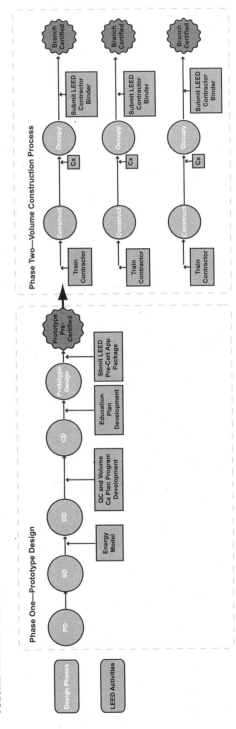

Figure 8.23 Conventional LEED certification vs. volume certification process.

Credit: Paladino and Company.

overall environmental performance of the design and speed at which delivery of a building was needed.

To address this challenge, the PNC design team incorporated three key green design concepts, or big moves: orientation neutral/high performance envelope, controllability, and optional floor-plan configurations.

Orientation Neutral/High Performance Envelope

An important component of the prototype's success was the orientation neutral envelope concept. With an equal quantity of glazing on all sides of a building, solar gain can be predicted regardless of site orientation, and HVAC systems can be sized accordingly. This concept helps reduce the factor of safety that is typically used to compensate for worst-case condition orientation of the building in the design of HVAC systems.

A high-performance envelope was designed to minimize heat loss during the winter, including a four-element glazing system that permits daylight to flood the branch but retains three times as much heat as standard double-glazed systems. The glazing system is prefabricated and integrates with a prefabricated and highly insulated wall panel system. The combination of the two prefabricated shell elements creates a tight seal with insulation levels that are consistent in appearance, performance, and delivery schedule—three critical elements of any successful prototype design. Prefabrication also cuts on-site construction waste and construction time, thereby increasing overall

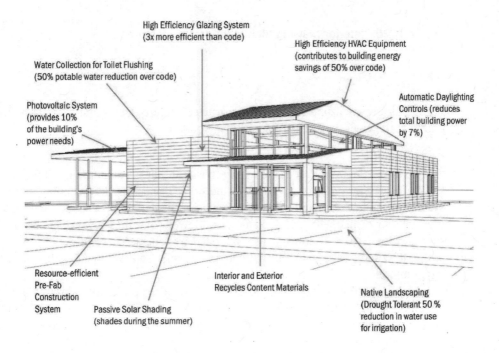

Exterior LEED Attributes

Figure 8.24 **Exterior drawing highlighting green features.**
Credit: Gensler.

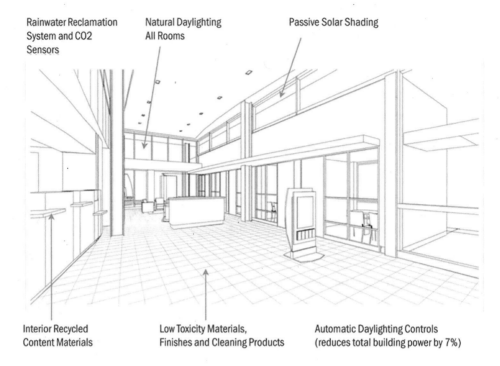

Rainwater Reclamation System and CO2 Sensors

Natural Daylighting All Rooms

Passive Solar Shading

Interior Recycled Content Materials

Low Toxicity Materials, Finishes and Cleaning Products

Automatic Daylighting Controls (reduces total building power by 7%)

Interior LEED Attributes

Figure 8.25 **Interior drawing highlighting green features.**
Credit: Gensler.

profitability as each branch can be constructed more quickly. Figures 8.24 and 8.25 highlight the green features of the prototype building.

Controllability

To maximize flexibility on a site-by-site basis, the design team handed the responsibility of system optimization over to the branch operators by installing adjustable systems that allow occupants to take control after the building is opened. Gensler specified interior components and systems that allowed for a high level of controllability, such as motorized roll-down shades operated by tellers. These allow occupants the ability to control solar gain and mitigate glare while enjoying the benefits of a well-daylit interior.

Lighting controls were also included in the standard prototype. Through a combination of daylight photo sensors, occupancy sensors, and occupant override switches, energy performance can be gleaned without sacrificing the user experience.

An additional controllability feature was installed to supply exactly the right amount of fresh air into the building. By installing carbon dioxide sensors, the HVAC units increase outside air only when there are enough occupants in the space to warrant more air. When the branch is unoccupied, fresh air can be reduced, creating substantial energy savings.

Optional Floor Plan Configurations

To respond to site variability demands, the PNC design team developed ten typical Green Branch® bank branch floor plans, in addition to fifteen special layouts, which allowed design flexibility while remaining true to the high-level performance goals initially determined. The ten floor plans lessened the impact of potential site conditions

Prototype 1.1

SIZE: 3,487 GROSS SQUARE FEET, PLUS 904 S.F. DRIVE THROUGH ROOF

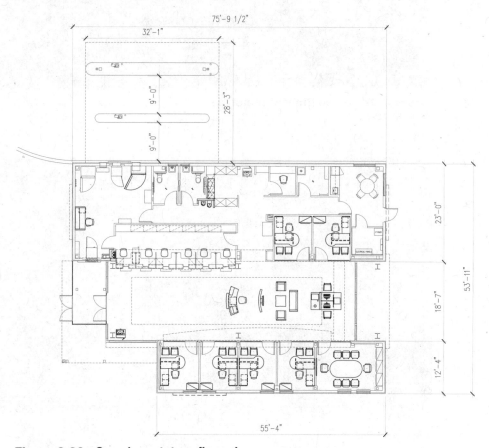

Figure 8.26   Sample prototype floor plan.

Credit: Gensler.

Figure 8.27 Green Branch® prototype.
Credit: PNC Bank.

Figure 8.28 Green Branch® prototype constructed.
Credit: PNC Bank.

through slight architectural revisions that would respond to site constraints, while meeting LEED performance goals. By focusing on climate neutral orientations, each of the ten floor plan configurations perform nearly the same, offering the flexibility to select the floor plan that responded best to branding, accessibility, and parking lot configurations that could be assessed each time. Figures 8.26, 8.27, and 8.28 show a typical floor plan, and photographs of the Green Branch® prototype building.

LEED SCORECARD AND STRATEGY

LEED Prototype Scorecard

While the design of the prototype was driven by high-level performance and design goals rather than the desire to achieve individual LEED credits, the big move strategies used to meet the overall goals for the prototype aligned with specific LEED credits that enabled the prototype to achieve a Silver baseline with optional strategies for Gold—photovoltaic panels and a rainwater harvesting system. Figure 8.29 shows a snapshot of the LEED credits achieved.

Figure 8.29 LEED scorecard.

Credit: Paladino and Company.

Sustainable strategies embedded into the prototype design include:

- **Water**—Water efficient landscaping; 30% water use reduction achieved through low-flow lavatory and kitchen fixtures.
- **Energy**—Optimized energy performance through the high-performance building envelope and glazing, daylighting controls, lighting controls, and highly efficient cooling.
- **Materials**—81.32% of project construction waste diverted; recycled and locally manufactured materials specified.
- **Indoor environment**—Low-emitting adhesives, sealants, carpets, composites, and paints; temperature and humidity monitoring systems and controls for individual zones in the building; extensive daylighting and outdoor views.
- **Innovation in Design**—A green building education program that makes the buildings actively instructional via kiosks and fliers; an outreach program that includes seminar and conference presentations; a green housekeeping program with tracking procedures and evaluation.

Designing a LEED Volume Certification Process

As mentioned before, the solution proposed by PNC to successfully certify buildings constructed through a volume building process was to separate the LEED certification of the design prototype from the submittal and rating of the construction as well as the site-related credits. To achieve buy-in from the USGBC, Paladino worked closely with them to develop and implement the program so that it maintained the rigor of the LEED rating while making the process more efficient for volume building.

Paladino and the USGBC developed a procedure by which strict review of the prototype design was conducted. Additional parametric studies were required to document energy, water, and other savings throughout PNC's banking region and helped demonstrate a minimum prototype LEED score for any branch configuration or site. The ten different branch configurations were modeled by CJL Engineering to establish minimum energy performance levels. The result showed 30–40% savings when compared to ASHRAE 90.1 energy standards, earning the design six to eight LEED energy points. Therefore, the prototype score was set at six energy points, and PNC was allowed to add the additional one to two points to the scorecard, if the branch was located within a climate zone that showed increased performance.

Paladino and the USGBC agreed that the prototype design would receive a *pre-certification* status and that branches built in the future based on any of the ten configurations would then make construction submittals to confirm that the site and construction related credit requirements had been met. With a pre-certified prototype, owners were required to make a separate construction submittal only for each site that was randomly selected by the USGBC for spot-check review. This random audit process allowed PNC to dramatically reduce certification paperwork, while allowing the USGBC enough oversight to assure performance targets were maintained.

Rather than require that audited branches compile a full application, PNC developed a program-wide contractor binder containing all relevant and standardized LEED construction submittals, including documentation of construction waste, indoor air quality, photo documentation, as well as standardized commissioning checklists and reports. The binder then provides the primary documentation for audit branch review by the USGBC. Because the information does not require processing according to standard LEED formats, the cost of reformatting on a per branch basis and further coordination with PNC is kept to a minimum, helping PNC control certification costs across the entire program.

This approach dramatically reduces branch LEED documentation costs, while still maintaining the credibility of the rating system through the randomized USGBC audit process. The audit process helps PNC monitor performance of the program, as without third-party verification, there would be no pressure placed on the design and construction teams to maintain the level of performance that PNC mandated from the beginning. Thus, the USGBC's auditing process provides a win-win situation for both organizations by increasing the credibility of their programs and allowing volume builders to utilize a streamlined certification process.

The LEED Volume Construction Process—Quality Control through Education

Once a LEED ratable prototype Green Branch® bank branch design was developed and a certification process established, the secondary challenge then became one of ensuring the consistent application of the design throughout the construction process for each of the 100 planned branches.

Because the success of many green design strategies and associated LEED credits, such as material procurement, construction waste recycling, and indoor environmental quality, rely heavily on the contractor, Paladino and PNC recommended and developed a quality control process that was centered around a contractor education program. Paladino's proposal to the USGBC asserted that a LEED-rated design could be consistently constructed in multiplicity, if contractors were provided with the tools and expertise to make appropriate site adaptations and to maintain and deliver the documentation required for LEED construction phase credits upon which the prototype rating relied.

When new contractors are enlisted to build their first branch, Paladino leads a training session to educate each contractor about LEED by explaining the green construction policies and procedures that are contained within the specifications and providing LEED documentation tracking tools customized to the PNC prototype. Sample indoor air quality plans, construction waste management plans, and other submittal forms are discussed and shared with the new contractor. No matter their level of experience in delivering a green building, the training process means success is much more easily achievable because the contractor has the necessary tools to guide them. For example, when specifications call for low volatility organic compound (VOC) adhesives and sealants, the objective must be to apply the method consistently across the project to optimize a healthy indoor environment. Recognizing this opportunity, PNC created a list of all low VOC products used in other branches and distributes it to new contractors. Thus, contractors and their subcontractors are spared the search to find compliant products and are, therefore, more likely to use them. Figures 8.30 to 8.33 illustrate some sample documents provided to the contractors.

Figure 8.30   Sample contractor tools.   *Credit: PNC Bank.*

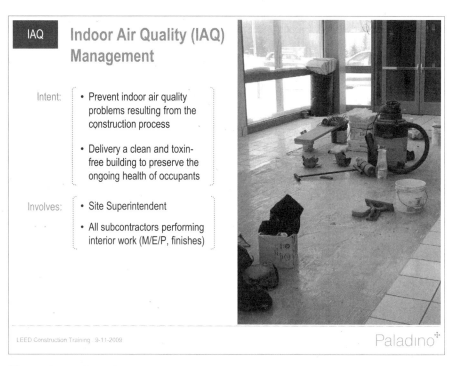

Figure 8.31   Sample contractor tools.   *Credit: PNC Bank.*

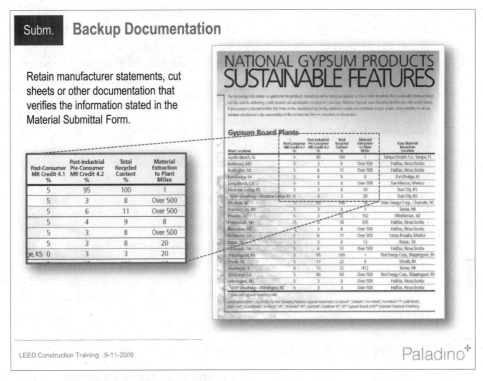

Figure 8.32   Sample contractor tools. *Credit: PNC Bank.*

Figure 8.33   Sample contractor tools. *Credit: PNC Bank.*

The partnership continues until the contractor has completed his/her first two branches with the aim that tools delivered to them are maximally adopted. At that point, the contractor assumes the primary responsibility of maintaining green building practices and LEED credit achievement for subsequent branches. To assure performance is maintained, PNC requires the contractor to prepare the LEED construction binder that documents all relevant credits should the branch ever be fully audited by the USGBC.

Procedural changes in construction practices can seem initially overwhelming to small contractors, but the end result is that the contractor learns to deliver a better product to building owners with little to no added cost. For example, most contractors are surprised to learn that they can increase their revenue through waste recycling by avoiding hefty tipping fees paid to landfills. While it may take more time to identify local recyclers initially, all branches constructed by a given contractor benefit through the process, as it is further refined with each iteration.

The delivery teams have said repeatedly that, for them, the most valuable outcome of the Green Branch® program has been the quality control plan and the contractor education plan. To date, more than 72 branches have been constructed without a single branch failing to achieve certification due to the quality control program created by Paladino.

FINDINGS AND CONCLUSION

Greening a Supply Chain Is a Change Management Effort

PNC's Green Branch® bank branch program revealed that creating an effective green/ LEED volume building effort is less about the buildings themselves and more about affecting organizational change focused on *greening* the people who manage the building supply chain. A top-down mandate and a full commitment from all those involved in the building delivery pipeline, from CEOs to project managers, was critical for the success of the program. PNC's desire to align sustainable business strategies with their corporate mission made the institutional system easier for the design and construction teams to navigate. The contractor training program made it possible to align outside construction teams to the same goals, regardless of their previous green building knowledge. The PNC Green Branch® bank branch program demonstrates that once procedural change is instituted and the people managing the supply chain are focused on green outcomes, it becomes easy to apply green principles to future construction projects and reap the benefits of volume green building.

USGBC and LEED Are Flexible

The fact that LEED did not make a certification program feasible on a volume level did not deter PNC from pursuing their goal to achieve LEED Silver certification for their branch programs. Instead, they reached out to the USGBC to develop a process that would achieve both organizations' objectives. The experience reveals that the USGBC is reasonable, flexible, and willing to work with building owners whose goals align with the organization's mission to transform the built environment and promote market adoption of LEED and green building. With ingenuity, a well-thought-out proposal, and a willingness to collaborate, the USGBC will respond to new ideas that help develop the LEED program further and make it more accessible to other markets, resulting in a win for both building owners and LEED.

Green Rewards Multiply

The PNC Green Branch® bank branch program and LEED Volume development process rewarded the company with many anticipated and unexpected benefits, including but not limited to the following.

Capitalizing on Volume for Cost Efficiency

A volume-build construction program enables bulk-purchase of construction materials; and in a green volume-build program, premium sustainable materials that may be too costly for a one-off project become financially feasible in bulk. In the PNC project, Clemens Construction Company was charged with finding a sustainable supply of plywood through the purchase of Forest Stewardship Council (FSC) certified wood for use in the first branch located in East Bradford, Pennsylvania. Through negotiations with a mill that had FSC wood logs, they were notified that the order volume was not sufficient to warrant milling the lumber. So PNC, in conjunction with Clemens, ordered ten branches worth of plywood and stored it until needed for other locations. This not only resulted in a consistent supply of FSC wood, but helped lock in pricing of plywood during times in which the cost was exponentially increasing. Using this model, Clemens also helped identify local producers of the pre-cast wall panels, which reduced transportation costs as well as transportation emissions. Both the manufacturers of FSC wood and the wall panel system have now become preferred suppliers in a delivery process that will continue to pay financial and environmental dividends well beyond the first branch, as setup costs are only paid once but are applied across all future branch sites.

Transforming a Market: LEED Volume Building and Green Advocates

PNC's visionary program had a transformational effect that reached far beyond the company's portfolio. The LEED Volume pilot program, based on the Green Branch® bank branch process established by PNC and its team, was launched in 2006 with 40 participating organizations and a waiting list of dozens, including retailers, hoteliers, other financial institutions, and restaurants, looking for an efficient way to implement their own green volume certification program across hundreds of buildings. The clamor to join the program proved that many other portfolio owners wanted to participate in LEED but did not know how to do so efficiently.

Through the Paladino-developed training program, several contractors became advocates of the green building movement within their organizations and their regions and are applying what they have learned to other projects. This pay-it-forward mentality drives application of green building practices well beyond PNC's initial training program and pays dividends not only to PNC, who experience consistent delivery of green branches, but also to the green building movement overall, as new advocates are fostered.

GREEN BRANCH® MEDIAN RESPONSE

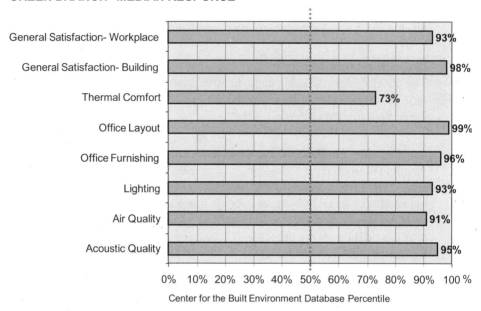

Figure 8.34 **Post-occupancy evaluation data.**
Credit: PNC Bank.

LEGACY BRANCH MEDIAN RESPONSE

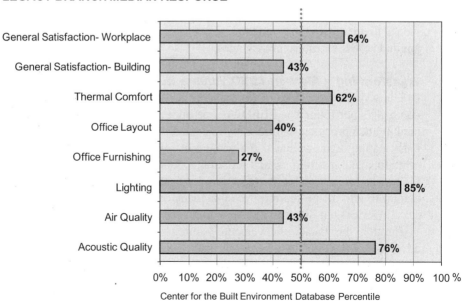

Figure 8.35 **Post-occupancy evaluation data.**
Credit: PNC Bank.

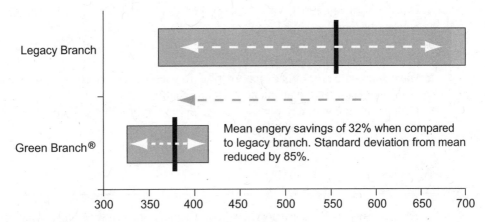

Figure 8.36 **Post-occupancy evaluation data.**
Credit: PNC Bank.

Multiply the impact of the PNC program not only across the 100 Green Branch® bank branch locations already constructed or in the pipeline, but also across the hundreds of buildings in the LEED Portfolio Program pilot; and it is readily apparent that PNC initiated a wholesale shift in the building market, within their business region and beyond.

Assessing the Performance Payoff for Building Green

The PNC Green Branch® program also rewarded PNC by delivering the environmental, financial, and human resource gains they sought when they initiated the program. In 2007, Paladino and Company conducted a post-occupancy performance study with the University of California, Berkeley Center for the Built Environment to assess the impact of the program on employee satisfaction, portfolio performance, and predictability, and found that Green Branch® bank branches outperform conventionally designed banks in areas of worker satisfaction and comfort, leading to greater employee retention. Figures 8.34 to 8.36 illustrate the data obtained from the post-occupancy evaluations. Compared to PNC's legacy prototype:

- General workplace satisfaction jumped from 64% to 93%.
- General building satisfaction leapt from 43% to 98%.
- A mean energy savings of 32% was achieved.
- 44% water savings was achieved.

With 72 PNC Green Branch® bank branches certified to date, the environmental impact continues to multiply as new branches are completed. Combined, the Green

Branch® bank branches are estimated to:

- Conserve 43,744,320 kWh of energy per year, enough to power 128 American households per year
- Conserve 12,069,120 gallons of water per year, enough to provide drinking water to 2,204 Americans each day
- Have sourced 38,704 board feet of FSC forested wood, or enough to preserve an area of forest habitat equivalent to 156 football fields
- Have reduced construction waste equivalent to the lifetime waste produced by 133 Americans, or 19,340,000 lbs

Notes

[1] Great River Energy Headquarters Case Study. Courtesy of Great River Energy and Perkins + Will.

[2] Rockefeller Hall Renovation, Harvard Divinity School Case Study. Courtesy of Harvard Divinity School, Harvard Office of Sustainability, and Venturi, Scott Brown and Associates, Inc.

[3] PNC Bank Green Branch® Bank Case Study. Courtesy of PNC Bank, Paladino and Company, and Gensler.

PART 2

IMPLEMENTATION

The purpose of this part of the book is to be a quick-reference resource on *what needs to be done to implement and document a credit*. It presents activities that may be followed by team members to implement the requirements of credits in the various categories of the LEED-NC rating system. It also describes team member responsibilities, the documents needed, as well as tips and strategies to achieve a credit. All of this has been presented in a user-friendly format as implementation worksheets for each credit.

Most project teams are intimidated by the volume of documentation that must be submitted for a LEED project. In order to demystify the documentation requirements, sample document submittals from the case study projects (described in Chapter 8) have been included as a reference. These sample submittals illustrate that project teams must submit documentation that is accurate, relevant, and required to demonstrate their approach in achieving a credit. However, it must be noted that there are various ways of communicating the same information, in different formats, and at different levels of detail. There is no *one-size-fits-all* documentation methodology that can be applied across all projects. Project teams must provide documents that best represent their approach and compliance with a credit or prerequisite. This part is divided into six sections as outlined below.

Section 1: Credit Implementation—Sustainable Sites
This section describes activities that may be followed by project team members to implement the requirements of credits in the *Sustainable Sites* category of the LEED-NC v3 rating system.

Section 2: Credit Implementation—Water Efficiency
This section describes activities that may be followed by project team members to implement the requirements of credits in the *Water Efficiency* category of the LEED-NC v3 rating system.

Section 3: Credit Implementation—Energy and Atmosphere

This section describes activities that be may followed by project team members to implement the requirements of credits in the *Energy and Atmosphere* category of the LEED-NC v3 rating system.

Section 4: Credit Implementation—Materials and Resources

This section describes activities that may be followed by project team members to implement the requirements of credits in the *Materials and Resources* category of the LEED-NC v3 rating system.

Section 5: Credit Implementation—Indoor Environmental Quality

This section describes activities that may be followed by project team members to implement the requirements of credits in the *Indoor Environmental Quality* category of the LEED-NC v3 rating system.

Section 6: Credit Implementation—Innovation in Design

This section presents sample documents that were submitted by the case study project teams in the *Innovation in Design* category of the LEED-NC v3 rating system.

How to Read the Implementation Worksheets

Each credit has been described in two pages. The first page identifies activities that may be completed during the various phases of the project. It references the same *Stages (Stages I, II, and III)* of the LEED project management process that were described earlier in this book, along with the GBCI certification steps and architectural design phases.

The second page lists the team members primarily responsible for the implementation of a credit and also other members whose support may be required. It also includes a documentation checklist and offers tips, notes, and strategies for implementation of the credit. The page further identifies whether the documents are part of the design or construction submittal. Figures 1 and 2 describe the various elements of both these pages.

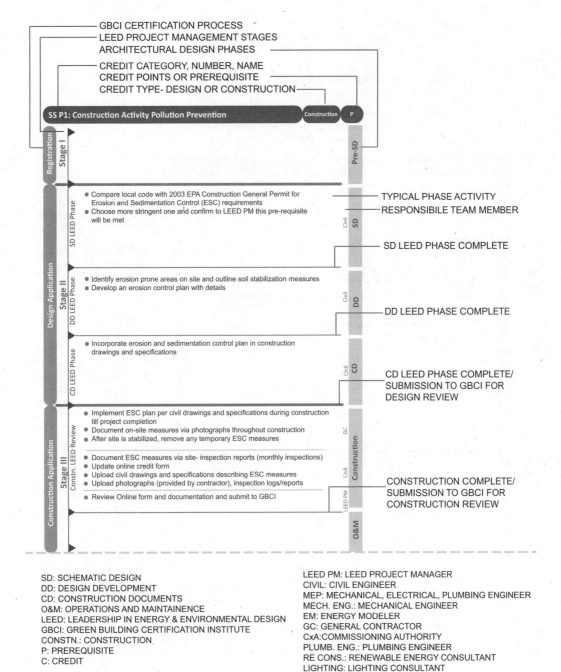

GBCI CERTIFICATION PROCESS
LEED PROJECT MANAGEMENT STAGES
ARCHITECTURAL DESIGN PHASES

CREDIT CATEGORY, NUMBER, NAME
CREDIT POINTS OR PREREQUISITE
CREDIT TYPE- DESIGN OR CONSTRUCTION

SS P1: Construction Activity Pollution Prevention   Construction   P

Registration — Stage I — Pre-SD

Design Application — Stage II

SD LEED Phase — Civil — SD
- Compare local code with 2003 EPA Construction General Permit for Erosion and Sedimentation Control (ESC) requirements
- Choose more stringent one and confirm to LEED PM this pre-requisite will be met

— TYPICAL PHASE ACTIVITY
— RESPONSIBILE TEAM MEMBER

— SD LEED PHASE COMPLETE

DD LEED Phase — Civil — DD
- Identify erosion prone areas on site and outline soil stabilization measures
- Develop an erosion control plan with details

— DD LEED PHASE COMPLETE

CD LEED Phase — Civil — CD
- Incorporate erosion and sedimentation control plan in construction drawings and specifications

CD LEED PHASE COMPLETE/ SUBMISSION TO GBCI FOR DESIGN REVIEW

Construction Application — Stage III

Constn. LEED Review — GC / Civil / LEED PM — Construction
- Implement ESC plan per civil drawings and specifications during construction till project completion
- Document on-site measures via photographs throughout construction
- After site is stabilized, remove any temporary ESC measures

- Document ESC measures via site- inspection reports (monthly inspections)
- Update online credit form
- Upload civil drawings and specifications describing ESC measures
- Upload photographs (provided by contractor), inspection logs/reports

- Review Online form and documentation and submit to GBCI

CONSTRUCTION COMPLETE/ SUBMISSION TO GBCI FOR CONSTRUCTION REVIEW

O&M

SD: SCHEMATIC DESIGN
DD: DESIGN DEVELOPMENT
CD: CONSTRUCTION DOCUMENTS
O&M: OPERATIONS AND MAINTAINENCE
LEED: LEADERSHIP IN ENERGY & ENVIRONMENTAL DESIGN
GBCI: GREEN BUILDING CERTIFICATION INSTITUTE
CONSTN.: CONSTRUCTION
P: PREREQUISITE
C: CREDIT

LEED PM: LEED PROJECT MANAGER
CIVIL: CIVIL ENGINEER
MEP: MECHANICAL, ELECTRICAL, PLUMBING ENGINEER
MECH. ENG.: MECHANICAL ENGINEER
EM: ENERGY MODELER
GC: GENERAL CONTRACTOR
CxA:COMMISSIONING AUTHORITY
PLUMB. ENG.: PLUMBING ENGINEER
RE CONS.: RENEWABLE ENERGY CONSULTANT
LIGHTING: LIGHTING CONSULTANT
ELECTRICAL: ELECTRICAL ENGINEER

Figure 1   Description of page 1 of credit implementation worksheet.

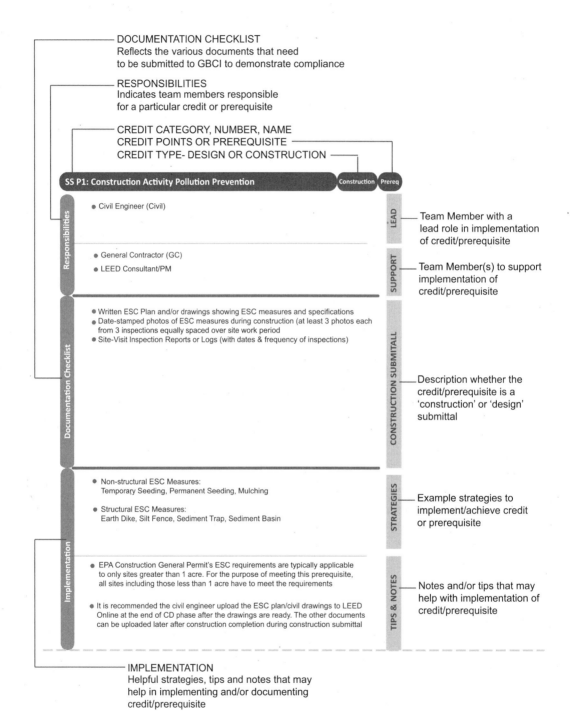

DOCUMENTATION CHECKLIST
Reflects the various documents that need
to be submitted to GBCI to demonstrate compliance

RESPONSIBILITIES
Indicates team members responsible
for a particular credit or prerequisite

CREDIT CATEGORY, NUMBER, NAME
CREDIT POINTS OR PREREQUISITE
CREDIT TYPE- DESIGN OR CONSTRUCTION

SS P1: Construction Activity Pollution Prevention   Construction Prereq

Responsibilities

● Civil Engineer (Civil)

LEAD — Team Member with a
lead role in implementation
of credit/prerequisite

● General Contractor (GC)
● LEED Consultant/PM

SUPPORT — Team Member(s) to support
implementation of
credit/prerequisite

Documentation Checklist

● Written ESC Plan and/or drawings showing ESC measures and specifications
● Date-stamped photos of ESC measures during construction (at least 3 photos each
from 3 inspections equally spaced over site work period
● Site-Visit Inspection Reports or Logs (with dates & frequency of inspections)

CONSTRUCTION SUBMITALL — Description whether the
credit/prerequisite is a
'construction' or 'design'
submittal

Implementation

● Non-structural ESC Measures:
Temporary Seeding, Permanent Seeding, Mulching

● Structural ESC Measures:
Earth Dike, Silt Fence, Sediment Trap, Sediment Basin

STRATEGIES — Example strategies to
implement/achieve credit
or prerequisite

● EPA Construction General Permit's ESC requirements are typically applicable
to only sites greater than 1 acre. For the purpose of meeting this prerequisite,
all sites including those less than 1 acre have to meet the requirements

● It is recommended the civil engineer upload the ESC plan/civil drawings to LEED
Online at the end of CD phase after the drawings are ready. The other documents
can be uploaded later after construction completion during construction submittal

TIPS & NOTES — Notes and/or tips that may
help with implementation of
credit/prerequisite

IMPLEMENTATION
Helpful strategies, tips and notes that may
help in implementing and/or documenting
credit/prerequisite

Figure 2   Description of page 2 of credit implementation worksheet.

CREDIT IMPLEMENTATION—

SUSTAINABLE SITES

Introduction

This section provides a credit-by-credit framework that may be followed by project team members to implement the requirements of credits in the *Sustainable Sites* category of the LEED-NC v3 rating system. It also identifies the responsible team members, includes a documentation checklist, and provides tips and notes for each credit.

The *Sustainable Sites* category of credits provides guidance on how to select, design, and manage project sites in an environmentally responsible manner. It specifically encourages utilization of previously developed land, design of regionally appropriate landscape, and the use of best management practices for stormwater quality and quantity control. The category also rewards smart transportation choices as well as reduction in erosion, light pollution, heat island effect, and construction activity pollution. The implementation of these credits involves conducting an investigation of the site and performing calculations related to parking and stormwater (described in Chapter 4). Up to 26 points may be achieved in this category. This section provides implementation worksheets for the following credits:

SS Prereq 1, Construction Activity Pollution Prevention

SS Credit 1, Site Selection

SS Credit 2, Development Density and Community Connectivity

SS Credit 3, Brownfield Redevelopment

SS Credit 4.1, Alternative Transportation—Public Transportation Access

SS Credit 4.2, Alternative Transportation—Bicycle Storage and Changing Rooms

SS Credit 4.3, Alternative Transportation—Low-Emitting and Fuel-Efficient Vehicles

SS Credit 4.4, Alternative Transportation—Parking Capacity

SS Credit 5.1, Site Development—Protect or Restore Habitat

SS Credit 5.2, Site Development—Maximize Open Space

SS Credit 6.1, Stormwater Design—Quantity Control

SS Credit 6.2, Stormwater Design—Quality Control

SS Credit 7.1, Heat Island Effect—Non-roof

SS Credit 7.2, Heat Island Effect—Roof

SS Credit 8, Light Pollution Reduction

Registration

Stage I

Pre-SD

Design Application

Stage II

SD LEED Phase

- Compare local code with 2003 EPA Construction General Permit for Erosion and Sedimentation Control (ESC) requirements.
- Choose the more stringent code and develop erosion and sedimentation control measures accordingly.

Civil   SD

DD LEED Phase

- Identify erosion prone areas on-site and outline soil stabilization measures.
- Develop an erosion control plan with details.

Civil   DD

CD LEED Phase

- Incorporate erosion and sedimentation control measures in construction drawings and specifications.

Civil   CD

Construction Application

Stage III

Constn. LEED Review

- Implement ESC plan per civil drawings and specifications during construction until project completion.
- Document on-site measures via photographs throughout construction.
- After site is stabilized, remove any temporary ESC measures.

GC

- Document ESC measures via site inspection reports (monthly inspections).
- Complete online credit form.
- Upload civil drawings and specifications describing ESC measures.
- Upload photographs (provided by contractor), inspection logs/reports.

Civil   Construction

- Review online form and documentation; submit to GBCI.

LEED PM

O&M

Responsibilities

LEAD

- Civil Engineer (Civil)

SUPPORT

- General Contractor (GC)
- LEED Consultant/PM

Documentation Checklist

CONSTRUCTION SUBMITALL

- Written ESC Plan and/or drawings showing ESC measures and specifications
- Date-stamped photos of ESC measures during construction (at least 3 photos each from 3 inspections equally spaced over site work period)
- Site visit inspection reports or logs (with dates and frequency of inspections)

Implementation

STRATEGIES

- Non-structural ESC Measures:
  Temporary Seeding, Permanent Seeding, Mulching

- Structural ESC Measures:
  Earth Dike, Silt Fence, Sediment Trap, Sediment Basin

TIPS & NOTES

- EPA Construction General Permit's ESC requirements are typically applicable only to sites greater than 1 acre. For the purpose of meeting this prerequisite, all sites including those less than 1 acre have to meet the requirements.

- It is recommended the civil engineer upload the ESC plan/civil drawings to LEED Online at the end of CD phase after the drawings are ready. The other documents can be uploaded later after construction is complete during the construction submittal.

Registration

Stage I — SD LEED Phase *(Owner / LEED PM — Pre-SD)*

- If site has not been chosen, then owner can use characteristics outlined in SS C1 to select the site.
- If site has been selected, then the LEED PM should research/obtain information on site's environmental characteristics from site surveys, local zoning plans, Google Earth, and Civil/Environmental expert.
- Determine if credit can be achieved or if further information is required.

Design Application

Stage II — SD LEED Phase *(LEED PM — SD)*

- Determine if site meets all of the following criteria:
  - Not a prime farmland
  - Site elevation is not lower than 5 feet above the elevation of a 100-year flood plain as determined by FEMA
  - Not a land identified as habitat for endangered species
  - Not within 100 feet of wetlands
  - Not within 50 feet of a water body
  - Not a public parkland prior to acquisition
- Determine if credit can be achieved or if there are any special circumstances.

Stage II — DD LEED Phase *(Civil / LEED PM — DD)*

- Prepare documentation to support any *special circumstances* that will allow the site to meet the intent of the credit.

CD LEED Phase *(LEED PM — CD)*

- Complete online credit form.
- Upload any documentation, such as site surveys, zoning information, etc., to support credit compliance.

- Review online credit form and documentation; submit to GBCI.

Construction Application

Stage III — Constn. LEED Review *(Construction)*

(O&M)

Responsibilities

LEAD

- LEED PM

SUPPORT

- Civil Engineer (Civil)
- Landscape Architect
- Owner/Developer

Documentation Checklist

DESIGN SUBMITALL

- Online credit form
- Documentation to support special circumstances regarding site selection criteria

Implementation

STRATEGIES

TIPS & NOTES

- Small man-made ponds—like stormwater retention ponds, or ponds created for recreational purposes—are exempt from the "not within 50 feet of a water body" requirement.

- Man-made wetlands and water bodies created to restore natural habitats and ecological systems are not exempt from this requirement.

- In addition to a site survey, local City zoning/building department websites are a great resource for gathering information regarding a site's characteristics.

Registration

Stage I — SD LEED Phase — *Pre-SD* (LEED PM)

- Calculate development density of neighborhood and project.
  If > 60,000 sf/acre, then pursue Option 1.
- Review site vicinity using Google Earth and local zoning websites to determine community services with pedestrian access and within 1/2 mile radius of site
  If >10 services and residential neighborhood with 10 units/acre found, then pursue Option 2.

Design Application

Stage II — SD LEED Phase — *SD* (LEED PM)

- Determine Option applicable to project.
- Prepare site vicinty drawing showing development density radius or community connectivity, depending on option chosen.

Stage II — DD LEED Phase — *DD* (LEED PM)

- Complete online credit form.
- Update calculations, if needed.
- Finalize site vicinty drawing showing development density radius or community connectivity, depending on option chosen.

CD LEED Phase — *CD* (LEED PM)

- Review online credit form documentation; submit to GBCI.

Construction Application

Stage III — Constn. LEED Review — *Construction*

O&M

Responsibilities

- LEED PM

- Architect
- Civil Engineer
- Owner/Developer

Documentation Checklist

- Online credit form
- Option 1:
  - Site vicinity plan showing development density radius
  - Record area of buildings within the density radius along with their site areas in the credit form
- Option 2:
  - Site vicinity plan showing community services and residential neighborhood
  - Record service name and type in the online credit form
- For both options, narrative explaining any special circumstances

Implementation

- If site has not been selected, then owners may choose sites based on neighborhood, infrastructure, community services, and transportation network around site.

Registration

Stage I

Pre-SD — LEED PM / Owner

- If site has not been selected, owners should investigate potential benefits and disadvantages of developing on brownfield sites; engage an environmental consultant to conduct an Environmental Site Assessment (ESA) to identify contaminants and recommend remediation measures.
- If site has been selected, then LEED PM should find out whether the site has been classified as a brownfield or documented as contaminated according to Phase II ESA.

Design Application

Stage II

SD LEED Phase

SD — Env. Consultant

- Prepare documentation describing the site contamination and remediation efforts undertaken.

DD LEED Phase

DD — LEED PM

CD LEED Phase

CD — LEED PM

- Obtain documentation from owner/environmental consultant that describes contamination and remediation efforts undertaken.
- Complete online credit form and upload documentation to LEED Online.

- Review online credit form and documentation; submit to GBCI.

Construction Application

Stage III

Constn. LEED Review

Construction

- General contractor may need to incorporate remediation activities into construction schedule if the site has not already been remediated when development work begins.

O&M

Responsibilities

- LEED PM

- Environmental Consultant (Env. Consultant)
- Owner/Developer

Documentation Checklist

- Online credit form
- Document describing site contamination and remediation efforts
- Narrative explaining any special circumstances

Implementation

- If site has not been chosen, then owners should weigh the pros and cons of developing on a contaminated/brownfield site and select accordingly. Some pros are: existing infrastructure, often less expensive, and incentives available. Some cons are: time and cost of remediation, perceived health risks by future occupants, and liability of contaminant migration to neighbors.

- Good resource to get more information about brownfields: http://www.epa.gov/brownfields/

Registration

Stage I

LEED PM | Pre-SD

- Research site vicinity and local transportation options (using site area maps, Google Earth, and local transit authority maps) to determine if
  - Option 1: project is within 1/2 mile walking distance of an existing, planned, or funded rail station;
  - Option 2: project is within 1/4 mile walking distance of 1 or more stops for 2 or more bus lines.

Design Application

Stage II

SD LEED Phase

LEED PM / Owner | SD

- Determine option applicable for the project.
- Prepare site vicinity plan showing location of rail stations or bus stops with labeled walking paths depending on option applicable.
- For planned/funded stops and stations, owner or civil engineer must obtain verification from local authority about the proposed stops or stations.

DD LEED Phase

LEED PM | DD

CD LEED Phase

LEED PM | CD

- Depending on applicable option, provide information regarding rail stations or bus stops and bus lines, and their distance from the main building entrance in the credit form.
- Upload the site vicinity plan and submit to GBCI.

Construction Application

Stage III

Constn. LEED Review

Construction

O&M

Responsibilities

- LEED PM

- Architect
- Owner/Developer

Documentation Checklist

- Online credit form—list rail station or bus stop distances and routes
- Site Vicinity Plan (to scale) labeled with location of rail stations or bus stops and walking path from project building entrance to station or stops
- Train route or bus route maps and schedules from local transit authority
- Narrative explaining any special circumstances

Implementation

- If site has not been chosen already, then owners may select a site with convenient access to existing transportation networks.
- If no bus stops are within the 1/4 mile walking distance, consider initiating conversation with local transit authorities for possible addition of stops closer to project site.
- Consider providing shuttle service for occupants from project site to nearest transit center.

- Exemplary Performance (1 point) can be achieved for this credit by:
  - Option 1: Developing a comprehensive transportation management plan that demonstrates a quantifiable reduction in personal automobile use.
  - Option 2: Doubling the transit ridership by locating the project within 1/2 mile of 2 rail stations or 1/4 mile of 2 or more bus stops with 4 or more lines. Combination of both is allowed and frequency of service must be at least 200 transit rides per day, total, at these stops.

Registration

Stage I

Pre-SD

LEED PM

- Gather data from building owners/management on total number of occupants —full-time staff, part-time staff, peak transients, and residents.
- Determine full time equivalent (FTE) occupants.

Design Application

Stage II

SD LEED Phase

SD

LEED PM

- Determine number of bike racks and/or shower and changing facilities required:
  - Commercial project: bike racks = 5% of total building users and shower and changing facilities = 0.5% FTE occupants
  - Residential project: only bike racks = 15% of total building users

Architect

- Perform research on bike rack storage products.
- Incorporate bike rack storage spaces in plans.
- Incorporate shower and changing rooms in plans.

DD LEED Phase

DD

LEED PM

- Review drawings for accuracy of number, type, and location of bike racks and shower/changing facilities.

Architect

- Revise drawings, if needed.

CD LEED Phase

CD

Architect

- Revise drawings, if needed.

LEED PM

- Review and upload drawings showing location of bike rack and/or shower spaces.
- Complete online credit form and submit to GBCI.

Construction Application

Stage III

Constn. LEED Review

Construction

O&M

- Encourage bicycle ridership among occupants by means of incentives, educational resources, etc. (recommended).

Responsibilities

- LEED PM

- Architect

Documentation Checklist

- Online credit form information
- Architectural drawings showing location of bike racks and/or shower facilities
- Bike rack details on drawings or manufacturer's cut sheets
- Narrative explaining approach to design or special circumstances (optional)

Implementation

- Number of space-saving bike storage solutions available in market; research for appropriate solution.
- Consider providing wall mounted bike racks in covered parking garages if there are site constraints.

- Exemplary Performance (1 point) can be achieved for this credit by developing a comprehensive transportation management plan that demonstrates a quantifiable reduction in personal automobile use. If exemplary point is awarded for SS C4.1, then exemplary point for SS C4.2 cannot be claimed.

LEAD

SUPPORT

DESIGN SUBMITALL

STRATEGIES

TIPS & NOTES

Registration

Stage I — Pre-SD — LEED PM

- Discuss options with owner and architect:
  - Option 1: Preferred parking for 5% of total parking capacity
  - Option 2: Alternative-fuel fueling stations for 3% of total parking capacity
  - Option 3: Low-emitting and fuel effiecint vehicles (FEV's) for 3% of Full-Time Equivalent (FTE) occupants
  - Option 4: Low-emitting or fuel efficient vehicle sharing program
- Determine option that project will pursue.

Design Application

Stage II

SD LEED Phase — SD — LEED PM / Architect/ Owner

- Calculate number of preferred parking spaces, fueling stations or FEV's required, depending on option chosen.

- Opt 1: Designate parking spaces in drawings; provide signage/striping details.
- Opt 2: Incorporate fueling stations in drawings.
- Opt 3: Owner to initiate procurement of vehicles; designate preferred parking.
- Opt 4: Owner to procure shared vehicle; designate preferred parking in dwgs.

DD LEED Phase — DD — LEED PM / Architect

- Opt 3: Obtain information on purchased vehicles—make, model and fuel type
- Opt 4: • Owner to prepare contractual agreement for vehicle-sharing
  - Prepare narrative explaining vehicle-sharing program
  - Estimate number of customers served by the program
- All options: Review drawings for accuracy of number and location of preferred parking spaces as well as signage details or fueling stations per option chosen.
- All options: Prepare management policy regarding preferred parking for FEVs.

CD LEED Phase — CD — Architect / LEED PM

- Revise drawings, if needed.

- Review and upload drawings and documents with preferred parking locations, signage details, fueling stations, or vehicle type based on option chosen.
- Complete online credit form and submit to GBCI.

Construction Application

Stage III

Constn. LEED Review — Construction

— O&M

- Communicate and administer the preferred parking policy to building occupants by providing the policy along with a list of qualifying vehicles, parking stickers, discounted parking, etc.

Responsibilities

- LEED PM

- Architect
- Owner/Developer

Documentation Checklist

- Online credit form
- Opt 1: • Drawings showing location of preferred parking spaces and signage or striping details, or if discounted parking is being provided, then written decription about the discount program
- Opt 2: • Drawings showing location of alternative-fuel fueling stations
  - Information on the type of alternative fuel station along with manufacturer, model, and fueling capacity per station
- Opt 3: • Drawings showing location of preferred parking spaces and signage or striping details
  - Information about purchased vehicles—make, model, and fuel type
- Opt 4: • Drawings showing location of parking for shared vehicles with pedestrian path from parking to entrance highlighted
  - Information about shared vehicles—make, model, and fuel type
  - Copy of contractual agreement with the shared vehicle program
  - Information about the program, its administration, and estimated number of users
- Narrative explaining approach to design or special circumstances (optional)

Implementation

- Exemplary Performance (1 point) can be achieved for this credit by developing a comprehensive transportation management plan that demonstrates a quantifiable reduction in personal automobile use.
  If exemplary point is awarded for SS C4.1, then exemplary point for SS C4.3 cannot be claimed.

Registration

Stage I — Pre-SD

- Gather information on parking requirements according to local zoning ordinance.
- Discuss with owner and architect the proposed parking capacity for the project and any opportunities in sizing the parking capacity to meet but not exceed the local zoning requirements.
- Discuss opportunities to provide carpool spaces or programs for encouraging shared vehicle use.

LEED PM

Design Application

Stage II

SD LEED Phase — SD — LEED PM / Architect

- Determine option to be pursued and incorporate requirements in drawings:
  - Non residential, Opt 1: Ensure parking capacity = local zoning requirements and designate preferred parking spaces for carpools/vanpool for 5% of parking
  - Non residential, Opt 2: If parking provided is for < 5% of FTE occupants, then designate preferred parking for carpools/vanpools for 5% of total parking spaces
  - Non residential, Opt 3: Provide no new parking
  - Residential, Opt 1: Parking capacity = local zoning requirements and institute infrastructure/programs to facilitate shared vehicle use
  - Residential, Opt 2: Provide no *new* parking

DD LEED Phase — DD — LEED PM / Architect

- Review drawings to see if parking capacity is according to the applicable project type and option and if details on the preferred parking spaces have been included.
- Owner to pursue any zoning negotiations over parking requirements, if required.
- If pursuing Option 1 for residential projects, then coordinate with building management to develop a program for promoting shared vehicle use such as car share services, ride boards, and shuttle services to mass transit.

CD LEED Phase — CD — Architect / LEED PM

- Revise drawings, if needed.

- Review and upload drawings and documents with preferred parking locations, signage details and shared-vehicle use program, based on option chosen.
- Complete online credit form and submit to GBCI.

Construction Application

Stage III

Constn. LEED Review — Construction

O&M

- Administer preferred parking policies or any programs to encourage shared-vehicle use by occupants.

Responsibilities

- LEED PM

- Architect
- Civil Engineer
- Owner/Developer

Documentation Checklist

- Online credit form
- Case 1 Non-Residential Projects
  - Opt 1: • Drawings with parking data showing parking provided <= parking required
    - Drawings showing location of preferred parking spaces and signage or striping details, or if discounted parking is being provided, then written decription about the discount program
    - Building management's program/policy for preferred parking
  - Opt 2: • Description/drawings with parking data showing parking provided is for less than 5% of FTE occupants
    - Drawings showing location of preferred parking spaces and signage or striping details, or if discounted parking is being provided, then written decription about the discount program
  - Opt 3: • Description/drawings showing that no *new* parking has been provided
- Case 2 Residential Projects
  - Opt 1: • Drawings with parking data showing parking provided <= parking required
    - Document/brochures describing shared-vehicle usage support programs
  - Opt 2: • Description/drawings showing that no *new* parking has been provided
- Case 3 Mixed Use Projects
  - Opt 1: • If commercial area < 10% of total area, then follow case 2 options, or if commercial area > 10%, then follow both case 1 and case 2 requirements for respective areas
  - Opt 2: • Description/drawings showing that no *new* parking has been provided

Implementation

- Exemplary Performance (1 point) can be achieved for this credit by developing a comprehensive transportation management plan that demonstrates a quantifiable reduction in personal automobile use. If exemplary point is awarded for SS C4.1, then exemplary point for SS C4.3 cannot be claimed.

Registration

Stage I — LEED PM — Pre-SD

- Identify if project site is a greenfield site or has been previously developed and determine option applicable to project:
  - Option 1: Greenfield sites—limit site disturbance by defining boundaries of disturbance from various building and hardscape components.
  - Option 2: Previously developed sites—restore 50% of site (excluding building footprint) or 20% of total site area, whichever is greater, with native or adapted vegetation.

Design Application

Stage II

SD LEED Phase — Civil / Landscape — SD

- Option 1:
  - Prepare site plan showing limits of site disturbance as per credit requirement.

- Option 2:
  - Design to minimize building footprint and impervious areas (architect).
  - Perfom calculations to determine area of site that needs to be restored with native and adapted vegetation (landscape architect).
  - Include native/adapted vegetation in the design (landscape architect).

DD LEED Phase — LEED PM — DD

- Option 1:
  - Review site plan showing limits of disturbance to verify that the defined limits are according to the credit's requirement.
- Option 2:
  - Review landscape drawings to see if the area restored with native or adapted vegetation is according to the credit's requirement.

CD LEED Phase — Landscape / Civil — CD

- Revise drawings as needed.
- Complete online credit form.
- Prepare construction drawings and documents showing the limits of site disturbance or native/adapted vegetation area based on option applicable.

Construction Application

Stage III

Constn. LEED Review — Landscape / Civil / GC — Construction

- Option 1:
  - Implement the site disturbance plan, by clearly identifying construction entrances, inspecting boundaries and fences, and ensuring protected areas are not encroached upon.
  - Document measures by means of photographs.
  - Update civil drawings, if needed.

- Option 2:
  - Update drawings based on any changes that were made during construction.

  - For both options, upload final drawings (civil or landscape) as well as photographs to LEED Online and submit to GBCI.

O&M

Responsibilities

- Case 1: Civil Engineer
- Case 2: Landscape Architect

- Architect
- General Contractor (GC)
- LEED PM

Documentation Checklist

- Online credit form
- Opt 1: • Site plan showing disturbance boundaries
  - • Photographs showing implementation during construction
- Opt 2: • Landscape site plan showing protected or restored site area
  - • Drawings showing list of native and/or adapted plant species
- Narrative explaining approach to design or special circumstances (recommended)

Implementation

- Exemplary Performance (1 point) can be achieved for this credit by restoring or protecting a minimum of 75% of the site (excluding building footprint) or 30% of total site area (including building footprint), whichever is greater, with natural or adapted vegetation.

- It is recommended that exhibit drawings highlighting the limits of disturbance or native vegetation area be prepared for this credit, in addition to the construction drawings.

Registration

Stage I

Pre-SD

LEED PM

- Find out if there are any open space requirements according to local zoning and determine case applicable to project:
  Case 1: Sites with local zoning open space requirements
  Case 2: Sites with no local zoning open space requirements
  Case 3: Sites with zoning ordinances but no open space requirements

Design Application

Stage II

SD LEED Phase

SD

Landscape / Civil

- Perform calculations to determine the open space area that will be needed to meet the requirement of the applicable case:
  Case 1: Open space area provided is 25% greater than open space area required
  Case 2: Open space area provided = area of building footprint
  Case 3: Open space area provided = 20% of project site area
- Design to reduce overall building footprint to meet the open space area requirement (architect).
- Design to reduce development footprint—compact parking, roads, hardscape, etc.—to meet the credit requirement (civil engineer).
- Design to increase vegetated open space (landscape architect).

DD LEED Phase

DD

Landscape　LEED PM

- Review drawings and open space area calculations to ensure they are according to credit requirements.

- Update open space area calculations, if needed, based on changes in design.

CD LEED Phase

CD

LEED PM　Landscape　Landscape

- Revise drawings as needed.
- Complete online credit form.
- Prepare drawings showing the vegetated open space area and upload to LEED Online.

- Review online documentation and submit to GBCI.

Construction Application

Stage III

Constn. LEED Review

Construction

O&M

Responsibilities

- Civil Engineer

- Architect
- Landscape Architect
- LEED PM

Documentation Checklist

- Online credit form
- Landscape drawings—site plan showing vegetated open space area
- Exhibit site plan, with vegetated open space area highlighted (recommended)
- Drawing/document showing open space area calculations

Implementation

- According to the *LEED Reference Guide*, open space area is defined as "the property area minus the development footprint." It must also be vegetated and pervious.

- Projects earning SS C2 can include vegetated roofs, pedestrian oriented hardscapes, and wetlands as part of the open space area.

- Exemplary Performance (1 point) can be achieved for this credit by:
  Case 1: Open space area provided is 50% greater than open space area required
  Case 2: Open space provided = twice the area of building footprint
  Case 3: Open space area provided = 40% of site area

- It is recommended that exhibit drawings highlighting the vegetated open space area be prepared for this credit, in addition to the construction drawings.

LEAD

SUPPORT

DESIGN SUBMITALL

STRATEGIES

TIPS & NOTES

Registration

Stage I — Pre-SD — Civil

- Determine case applicable to project:
  Case 1: Sites with existing imperviousness 50% or less
  Case 2: Sites with existing imperviousness greater than 50%
- Explore opportunities in design to decrease impervious area of site and reduce runoff volumes.

Design Application

Stage II

SD LEED Phase — SD — Landscape / Civil

- Incorporate stormwater management strategies to reduce impervious area; consider pervious paving, infiltration swales, stormwater harvesting, etc.
- Perform calculations to determine if the proposed design meets credit requirements of the applicable case:
  Case 1, Opt 1: Post-development peak discharge rate and quantity should not exceed predevelopment peak discharge rate and quantity.
  Opt 2: Rate and quantity of post-developement stormwater runoff should be below critical values of receiving waterways.
  Case 2, Post-development peak discharge rate and quantity should be at least 25% less than predevelopment peak discharge rate and quantity.

DD LEED Phase — DD — LEED PM

- Review drawings and stormwater calculations to ensure they are according to credit requirements.

— Civil

- Coordinate with landscape architect in stormwater design strategies.
- Update calculations, if needed, based on changes in design.

CD LEED Phase — CD — Civil

- Revise drawings as needed.
- Complete online credit form.
- Prepare drawings showing the stormwater management plan and upload to LEED Online.

— LEED PM

- Review online documentation and submit to GBCI.

Construction Application

Stage III

Constn. LEED Review — Construction

- Contractor to implement stormwater design strategies identified in the design.

— O&M

Responsibilities

- Civil Engineer

- Architect
- Landscape Architect
- LEED PM

Documentation Checklist

- Online credit form—provide rate and quantity calculations
- Civil drawings—plan showing stormwater management strategies
- Narrative describing stormwater management strategies and percentage of rainfall each strategy is designed to handle

Implementation

- Some strategies to minimize stormwater runoff volume are: pervious paving materials, rainwater harvesting for reuse in irrigation or indoor water use, infiltration swales and retention ponds, vegetated filter strips, and vegetated roofs.

- Exemplary Performance (1 point) can be achieved for this credit by documenting a comprehensive approach to capture and treat rainwater beyond the requirements of the credit.

Registration

Stage I

Civil | Pre-SD

- Explore opportunities in stormwater design to reduce impervious cover and promote infiltration, capturing, and treating stormwater.
- Determine applicable local codes and permits required.

Design Application

Stage II

SD LEED Phase

Landscape / Civil | SD

- Incorporate stormwater management strategies to reduce impervious area and promote infiltration; consider pervious paving, infiltration swales, and stormwater harvesting.
- Determine the non-structural and structural measures needed to capture and treat stormwater according to the credit's requirements. The proposed measures must be able to treat 90% of annual average rainfall and be capable of removing 80% of the post-development total suspended solids (TSS).

DD LEED Phase

LEED PM | DD

- Review drawings and stormwater best management practices to ensure they are according to credit requirements.

Civil

- Coordinate with landscape architect and plumbing engineer in development of stormwater design strategies to include any reuse of water for irrigation or indoor purposes.

CD LEED Phase

Civil | CD

- Finalize stormwater management systems; revise drawings as needed.
- Complete online credit form.
- Prepare drawings showing the stormwater management plan and upload to LEED Online.

LEED PM

- Review online documentation and submit to GBCI.

Construction Application

Stage III

Constn. LEED Review

Construction

- Contractor to implement stormwater design strategies identified in the design.

O&M

Responsibilities

- Civil Engineer

- Architect
- Landscape Architect
- LEED PM

Documentation Checklist

- Online credit form—list and describe structural and non-structural stormwater measures used, percentage of annual rainfall treated by each measure, and estimated TSS removal range
- Civil drawings—plans showing stormwater management strategies
- Narrative describing stormwater management strategies
- Additional performance monitoring data/documentation to support any structural best management practice's pollutant removal performance

Implementation

- Some non-structural strategies to increase pervious areas are: pervious paving materials, infiltration swales, retention ponds, vegetated filter strips, etc.

- Some structural measures to increase pervious area and treat stormwater are: rainwater cisterns, manhole treatment devices, ponds, continuous deflection separation units, etc.

- Exemplary Performance (1 point) can be achieved for this credit by documenting a comprehensive approach to capture and treat rainwater beyond the requirements of the credit.

Registration

Stage I — Pre-SD — Architect

- Explore opportunities in site design to minimize hardscape surfaces, provide shaded or underground parking.
- Choose appropriate option
  Opt 1: Provide a combination of shaded paving, high solar reflective index (SRI) paving, or open grid paving for 50% of the site hardscape area.
  Opt 2: Place minimum of 50% of parking spaces under cover.

Design Application

Stage II

SD LEED Phase — SD — Architect

- Incorporate strategies to reflect the requirements of the option chosen
  Opt 1: Coordinate with landscape and civil in the incorporation of vegetation, open grid paving, high SRI paving, etc.
  Opt 2: Provide parking under cover, underground or in a separate parking garage.

DD LEED Phase — DD — Architect

- Incorporate into specifications any requirements pertaining to paving or other hardscape materials.
- Perform calculations to determine if credit requirements will be met.

— LEED PM

- Review drawings and calculations to ensure they are according to credit requirements.

CD LEED Phase — CD — Architect

- Finalize site plan, parking drawings, and specifications for construction.
- Complete online credit form.
- Upload required drawings demonstrating compliance with option chosen.

— LEED PM

- Review online documentation and calculations.

Construction Application

Stage III

Constr. LEED Review — Construction — GC

- Contractor to execute site plan according to construction drawings.

— Architect

- Update drawings and calculations to reflect any changes that may have occurred during construction and upload to LEED Online.

— LEED PM

- Review online documentation and calculations; submit to GBCI.

O&M

Responsibilities

- Architect

- Civil Engineer
- Landscape
- LEED PM

Documentation Checklist

- Online credit form—list information about SRI value of various compliant surfaces and number of covered and uncovered parking spaces
- Opt 1: Site plan drawing showing non-roof hardscape areas and areas that count towards achievement of credit (recommended to be prepared as an exhibit)
- Opt 2: Parking drawings with covered and uncovered parking areas highlighted (recommended to be prepared as an exhibit)
- Manufacturer's data on SRI values of paving materials, architectural shading devices, etc.

Implementation

- Exemplary Performance (1 point) can be achieved for this credit by:
  Opt 1: 100% of non-roof hardscape area shaded or constructed with high SRI
        paving or open-grid paving
  Opt 2: 100% of on-site parking is located under cover

Registration

Stage I — Architect / Pre-SD
- Explore opportunities in design to provide a cool roof (with high reflectance) or a vegetated roof.
- Determine cost implications, if any, associated with providing cool roofs, vegetated roofs, or combination of both.

Design Application

Stage II

SD LEED Phase — Architect / SD
- Research available products for cool roofs and vegetated roofing solutions.
- Choose appropriate option
  Opt 1: Use roofing materials with SRI=78 for low-sloped roof and SRI=29 for steep-sloped roof for a minimum of 75% of the roof surface.
  Opt 2: Install vegetated roof on at least 50% of roof area.
  Opt 3: Install combination of above options.
- Coordinate with landscape architect or vegetated roofing consultant, if vegetated roofs will be provided.
- Provide roof SRI information to energy modeler for input into the energy model.

DD LEED Phase — Architect / DD
- Incorporate roof specfications and details in drawings.
- Perform calculations to determine if credit requirements will be met.
- Coordinate with structural engineer on roof loads if vegetated roof is being provided.

LEED PM
- Review drawings and calculations to ensure they are according to credit requirements.

CD LEED Phase — Architect / CD
- Finalize roof plan drawings, details, and specifications for construction.
- Complete online credit form.
- Upload required drawings demonstrating compliance with option chosen.

LEED PM
- Review online documentation and calculations; submit to GBCI.

Construction Application

Stage III

Constn. LEED Review — GC / Construction
- Contractor to procure roof materials as specified and/or install vegetated roof according to installation drawings.

O&M
- Regular cleaning, inspection, and maintenance of high reflectance roof as well as vegetated roof system is needed.

Responsibilities

- Architect

- Civil Engineer
- Landscape Architect
- LEED PM
- Structural Engineer
- Vegetated Roofing Consultant

Documentation Checklist

- Online credit form—list roofing products used with their emittance and reflectance percentages, SRI values, and slope of roof; input roof areas to demonstrate compliance
- Roof plan drawings showing all compliant SRI roof areas and/or vegetated roof areas
- Manufacturers' data/specifications showing product characteristics

Implementation

- Cool Roof Rating Council's website lists a number of available roofing solutions with their manufacturers and SRI data (Website: www.coolroofs.org).

- Exemplary Performance (1 point) can be achieved for this credit by demonstrating 100% of the project's roof area consists of a vegetated roof system.

- Area of roof covered by mechanical equipment, solar energy panels, and appurtenances are not included in the total roof area calculations. Subtract this area from total roof area before determining the roof area that is compliant with the credit.

- Ensure energy model reflects the specified cool roof's characteristics.

Registration

Stage I

Lighting — Pre-SD

- Determine local regulations, if any, for lighting.
- Determine exterior lighting zone classification for the project.

Design Application

Stage II

SD LEED Phase

Lighting — SD

- Explore options to determine compliance path for interior lighting:
  - Opt 1: Provide automatic controls to reduce input power by at least 50% of all non-emergency interior lights with direct line of sight to openings in the envelope.
  - Opt 2: Provide automatic shielding of all openings in the envelope with a direct line of sight to non-emergency interior luminaires.
- Research available exterior lighting product options.
- Design exterior lighting according to credit requirements.
- Prepare a site lighting simulation to determine compliance.
- Coordinate with architect, landscape architect, and electrical engineer.

DD LEED Phase

Lighting — DD

- Develop interior and exterior lighting designs according to credit requirements and lighting simulation.
- Perform lighting power density (LPD) calculations to determine if exterior lighting requirements will be met.

LEED PM

- Review drawings and simulation to see if they meet credit requirements.

CD LEED Phase

Lighting — CD

- Update site lighting simulation, if needed.
- Finalize lighting layout drawings, details, and specifications for construction.
- Coordinate with architect, landscape architect, and electrical engineer.
- Complete online credit form.
- Upload required drawings demonstrating compliance.

LEED PM

- Review online documentation and calculations; submit to GBCI.

Construction Application

Stage III

Constn. LEED Review

CxA / GC — Construction

- Commission lighting systems according to commissioning plan to ensure lighting systems and controls are installed and operated correctly.

O&M

Responsibilities

- Lighting Consultant (Lighting)

- Electrical Engineer
- Landscape Architect
- LEED PM
- Architect

Documentation Checklist

- Online credit form
- Interior Lighting: Option 1
  - Interior lighting drawings showing location of automatic controls and light fixtures they serve
  - Drawings, specifications, or building operation plan describing the sequence of operation for lighting

- Interior Lighting: Option 2
  - Drawings showing location, type, and specfications of automatic shading devices
  - Drawings, specifications, or building operation plan describing the sequence of operation for lighting
  - Manufacturers' data/specifications showing automatic shading device characteristics

- Exterior Lighting
  - Zone classification for project site
  - Site photometric plan to show compliance with light tresspass requirements
  - List of alll exterior lighting fixtures and their manufacturers' data
  - Narrative describing the light trespass analysis procedure

Implementation

- Note that interior lighting requirements have changed significantly from LEED–NC v2.2. Some projects, such as mid-rise, multifamily buildings, might now find it difficult to meet this credit's requirement of incorporating automatic controls and/or automatic shading devices. Both elements are tough and cost prohibitive to implement in typical residential settings.

Sample Documents

The following pages present excerpts from documents that were submitted to USGBC by the teams of the case study projects (discussed in Chapter 8) for some credits in the *Sustainable Sites* category. Sample documents for the following credits have been included.

SS Credit 2, Development Density and Community Connectivity

SS Credit 4.1, Alternative Transportation—Public Transportation Access

SS Credit 4.3, Alternative Transportation—Low-Emitting and Fuel-Efficient Vehicles

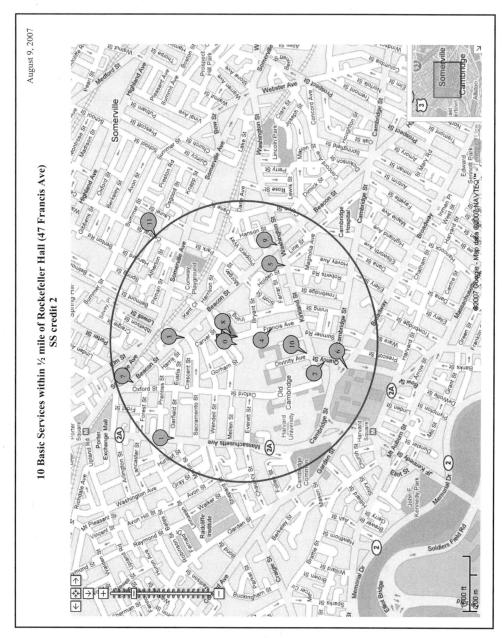

August 9, 2007

10 Basic Services within ½ mile of Rockefeller Hall (47 Francis Ave)
SS credit 2

Sample Document: SS Credit 2, Development Density and Community Connectivity (page 1 of 2).

Credit: Harvard University, Office of Sustainability.

August 9, 2007

10 Basic Services within ½ mile of Rockefeller Hall (47 Francis Ave)
SS credit 2

Basic Services Legend:

0	Rockefeller Hall – 47 Francis Ave, Cambridge, MA
1	Cambridge Trust Co - more info » 1720 Massachusetts Ave, Cambridge, MA (617) 503-4080 - 0.4 mi NW
2	**Church** of the New Jerusalem - more info » 50 Quincy St, Cambridge, MA (617) 864-4552 - 0.3 mi S
3	Shaw's Supermarket - more info » 275 Beacon St, Somerville, MA (617) 354-7023 - 0.2 mi N
4	Oxford Street **Day Care** Co-Op - more info » 25 Francis Ave, Cambridge, MA (617) 547-3175 - 0.1 mi S
5	Kirkland Cleaners - more info » 86 Kirkland St, Cambridge, MA (617) 876-3262 - 1 review - 0.3 mi SE
6	**Fire** Department: Cambridge - more info » 491 Broadway, Cambridge, MA (617) 349-4900 - 0.4 mi S
7	Beacon Street Laundromat - more info » 358 Beacon St, Somerville, MA (617) 864-8654 - 0.4 mi NW
8	Andover-Harvard Theology **Library** - more info » 45 Francis Ave, Cambridge, MA (617) 495-5770 - 0.1 mi SE
9	Bright Dental - more info » 435 Washington St # B, Somerville, MA (617) 491-2829 - 0.3 mi SE
10	Harvard University Peabody Museum - more info » 11 Divinity Ave, Cambridge, MA (617) 496-1027 - 9 reviews - 0.3 mi S
11	½ mile radius line

Note: There is pedestrian access between Rockefeller Hall and all of the above listed services via sidewalks.

Sample Document: SS Credit 2, Development Density and Community Connectivity (page 2 of 2).
Credit: Harvard University, Office of Sustainability.

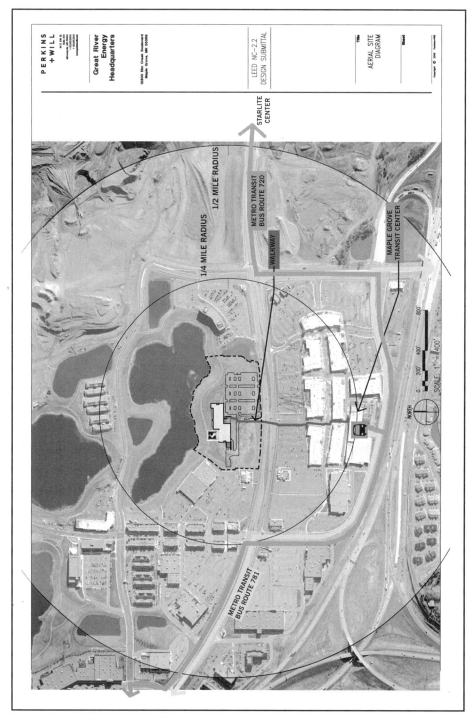

Sample Document: SS Credit 4.1, Alternative Transportation—Public Transportation Access.

Credit: Perkins + Will.

August 13, 2007

Harvard Divinity School - Rockefeller Hall
SS credit 4.3 – Alternative Transportation: Alternative Fueled Vehicles
Narrative

To meet the alternative fuel vehicle requirements for SS Credit 4.3, Harvard Divinity School (HDS) would like to refer to SS Credit 4.3 CIR Ruling dated 1/26/2005, which states that Zipcars can be used to supply alternative fuel vehicles if the project also achieves credit SSc4.1 AND Zipcars are counted as 8 privately owned vehicles.

If one Zipcar replaces 8 privately owned cars, a single Zipcar is able to serve 3.7% of our population. We have one Zipcar under contract to achieve this point. Since the existing Letter Template doesn't reflect this calculation method, we have developed an alternative Letter Template that incorporates rulings from the CIR's. This template is included on page 2 of this submittal.

There are actually multiple Zipcars on the Harvard campus. For the purpose of this credit we are claiming one of them, the Toyota Prius (named Plymouth) parked near the corner of Garden St. and Fernald Dr (28 Fernald Dr.) The Prius exceeds the LEED requirement of a minimum ACEEE green score of 40. See a screen shot of this vehicle on page 3 below, confirming the type of vehicle and its location. This Zipcar is parked within walking distance of Rockefeller Hall, and also adjacent to one of the campus shuttle stops at Moors Hall, as shown on the map on page 4. The convenient shuttle and the straight-forward walking route with pedestrian sidewalks make this a preferred parking location. For a list of all of Harvard's zipcars, and which LEED projects they are allocated to, see page 6.

A copy of our current agreement with Zipcar is included on Page 7-16. This agreement was signed on April 3, 2006 for a three year period that will end in April of 2009, at which time Harvard will likely renew this successful program. Highlighted in the agreement, paragraph 2.4 states that all of Harvard's Zipcars will become hybrid by January 2007. On page 17, find an email from Harvard's Zipcar contact Dan Curtin dated 7/03/07 which confirms the intention of Zipcar to keep Harvard's zipcars as hybrids.

We hope this demonstrates the school's commitment reducing single occupant vehicle use to and from campus.

Contents:

Rockefeller Hall SSc4.3
Page 1 of 18

Sample Document: SS Credit 4.3, Alternative Transportation—Low-Emitting and Fuel-Efficient Vehicles (page 1 of 6).
Credit: Harvard University, Office of Sustainability.

SS Credit 4.3: Alternative Transportation, Alternative Fuel Vehicles

I, _____ Jesse Foote _____ , declare that alternative fuel vehicles (through a contract with Zipcar) will be provided for 3% of building occupants and preferred parking will be provided for these vehicles. I also declare that we have met the requirements for Sustainable Sites Credit 4.1 - Public Transportation.

I have provided the following to support the declaration:

☐   Signed 2-year contract for hybrid Zipcars, highlighting relevant details

☐   Customers/car calculation below demonstrating that at least 3% of building occupants will be provided with an alternative fuel vehicle (assume each Zipcar replaces 8 privately owned vehicles)

☐   Site drawing or parking plan highlighting preferred parking for alternative fuel vehicles

Number of regular building occupants assumed:	217
Number of alternative fuel vehicles provided:	1
Percentage of regular building occupants provided with an alternative fuel vehicle:	3.69%

SS Cr 4.3 (1 point): Alternative Transportation, Alternative Fuel Vehicles

Points Documented
1

Name:	Jesse Foote
Organization:	Harvard Green Campus Initiative
Role in Project:	Green Building Consultant
Signature:	*Jesse Foote*
Date:	13-Aug-07

Created July 1, 200?

Rockefeller Hall SSc4.3
Page 2 of 18

Sample Document: SS Credit 4.3, Alternative Transportation—Low-Emitting and Fuel-Efficient Vehicles (page 2 of 6).
Credit: Harvard University, Office of Sustainability.

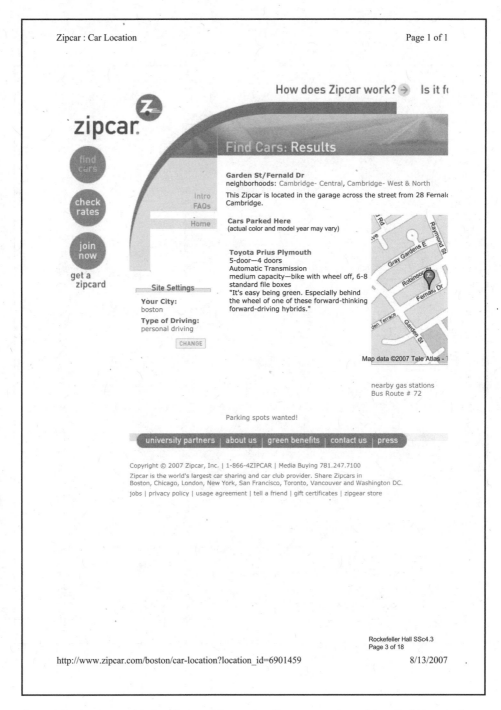

Zipcar : Car Location   Page 1 of 1

How does Zipcar work? →   Is it fc

zipcar

find cars

check rates

join now

get a zipcard

Find Cars: Results

Intro
FAQs

Home

Garden St/Fernald Dr
neighborhoods: Cambridge- Central, Cambridge- West & North
This Zipcar is located in the garage across the street from 28 Fernalc
Cambridge.

Cars Parked Here
(actual color and model year may vary)

Toyota Prius Plymouth
5-door—4 doors
Automatic Transmission
medium capacity—bike with wheel off, 6-8
standard file boxes
"It's easy being green. Especially behind
the wheel of one of these forward-thinking
forward-driving hybrids."

Site Settings

Your City:
boston

Type of Driving:
personal driving

CHANGE

Map data ©2007 Tele Atlas - ͺ

nearby gas stations
Bus Route # 72

Parking spots wanted!

university partners | about us | green benefits | contact us | press

Copyright © 2007 Zipcar, Inc. | 1-866-4ZIPCAR | Media Buying 781.247.7100
Zipcar is the world's largest car sharing and car club provider. Share Zipcars in
Boston, Chicago, London, New York, San Francisco, Toronto, Vancouver and Washington DC.
jobs | privacy policy | usage agreement | tell a friend | gift certificates | zipgear store

Rockefeller Hall SSc4.3
Page 3 of 18

http://www.zipcar.com/boston/car-location?location_id=6901459   8/13/2007

**Sample Document: SS Credit 4.3, Alternative Transportation—Low-Emitting and
Fuel-Efficient Vehicles (page 3 of 6).**

Credit: Harvard University, Office of Sustainability.

ZipCar Parking Location, within walking Distance and on bus routes from Rockefeller Hall (47 Francis Ave)
SS credit 4.3

August 13, 2007

Rockefeller Hall SSc4.3
Page 4 of 18

Sample Document: SS Credit 4.3, Alternative Transportation—Low-Emitting and Fuel-Efficient Vehicles (page 4 of 6).

Credit: Harvard University, Office of Sustainability.

August 13, 2007

ZipCar Parking Location, within walking Distance and on bus routes from Rockefeller Hall (47 Francis Ave)

SS credit 4.3

Basic Services Legend:

0	Rockefeller Hall – 47 Francis Ave, Cambridge, MA
1	Prius Plymouth – Garden St / Fernald Dr
2	Campus Shuttle Stop – Annenberg Hall
3	Campus Shuttle Stop – Law School
4	Campus Shuttle Stop – Moore Hall

Rockefeller Hall SSc4.3
Page 5 of 18

Sample Document: SS Credit 4.3, Alternative Transportation—Low-Emitting and Fuel-Efficient Vehicles (page 5 of 6)

Credit: Harvard University, Office of Sustainability.

August 13, 2007

ZipCar Allocation for LEED SS Cr. 4.3, Alternative Fueled Vehicles

Vehicle	Location	Project Name	FTE Count	% of FTE Count
Prius Pat	Peabody Terrace	Blackstone	142	5.6%
Prius Priscilla	111 Western Ave. near HBS	Hamilton Hall	74	10.8%
Prius Portland	219 Western Ave. near HBS	Wyss Hall		
Prius Pittsburgh	Agassiz/North Hall	60 Oxford Street	140	5.7%
Prius Parnassus	Mt Auburn/Banks St.	90 Mt. Auburn Street	86	9.3%
Prius Plymouth	Garden St./Fernald Dr.	Rockefeller Hall	217	3.7%
Prius Paulsen	One Western Ave.	10 Akron St.	215	3.7%
Escape Hybrid Etta	Harvard Square/Dunster St.	5 Cowperthwaite St.	201	4.0%

Rockefeller Hall SSc4.3
Page 6 of 18

Sample Document: SS Credit 4.3, Alternative Transportation—Low-Emitting and Fuel-Efficient Vehicles (page 6 of 6).
Credit: Harvard University, Office of Sustainability.

CREDIT IMPLEMENTATION—
WATER EFFICIENCY

Introduction

This section provides a credit-by-credit framework to implement the requirements of credits in the *Water Efficiency* category of the LEED-NC v3 rating system. It also identifies responsible team members, includes a documentation checklist, and provides tips and notes for each credit.

The consumption of public water supply by Americans and the world in general is continually increasing. Several reports and research show that "the United States uses about 400 billion gallons (1.5 trillion liters) of water per day. Of this, the vast majority is used for thermoelectric power generation (48%) and irrigation on farms (34%). Water use in and around buildings, from both public water supplies and well water, accounts for about 47 billion gallons (180 billion liters) per day, or 12% of U.S. water use."[1] The use of large volumes of water not only increases maintenance and life-cycle costs for building operations but also increases consumers' costs for additional municipal supply and treatment facilities. On the other hand, if buildings use water efficiently, significant reduction may be achieved in supply costs and sewage volume. Water shortages are expected to become more and more common in the coming decades. A 2003 report by the U.S. General Accounting Office reported that 36 states are likely to experience water shortages by 2013.[2] The practice of water conservation measures can help improve both environmental and human well-being, which may be affected when reservoirs and groundwater aquifers are depleted because lower water levels can concentrate both natural contaminants and human pollutants.[3]

The credits and prerequisites in the *Water Efficiency* category outline water reduction targets and encourage the use of various strategies, such as high-efficiency fixtures, rainwater reuse, etc., to reduce the amount of potable water consumption in buildings, thereby minimizing the strain on municipal water supply sources and systems. It also encourages the use of efficient irrigation and landscape design strategies to reduce the amount of water utilized for irrigation purposes. While some strategies, such as using high-efficiency fixtures, may have minimal impact on project costs, there are others such as, on-site waste treatment, that may require significant up-front investment. A cost-effective and comprehensive water efficiency plan must be prepared to address all issues identified in this category of credits. A maximum of 10 points may be achieved in this category.

This section includes implementation worksheets for the following credits:

WE Prereq 1, Water Use Reduction—20% Reduction

WE Credit 1, Water Efficient Landscaping

WE Credit 2, Innovative Wastewater Technologies

WE Credit 3, Water Use Reduction

NOTES

[1] Alex Wilson, "Water: Doing More With Less," http://www.buildinggreen.com/auth/article. cfm/2008/2/3/Water-Doing-More-With-Less/ (accessed July 16, 2010).
[2] USGBC, *LEED Reference Guide for Green Building Design and Construction* (USGBC, 2009).
[3] Ibid.

Registration

Stage I — Pre-SD — LEED PM

- Establish goal of 20% water use reduction from baseline and explore opportunities to achieve this.
- Determine total number of occupants (full-time equivalents, visitors, students, residents, etc.)

Design Application

SD LEED Phase — SD — Plumb. Eng.

- Research high efficiency water fixture product options.
- Perform preliminary water use calculations, using LEED occupancy counts to determine optimal flow rates of fixtures.
- Investigate all options to meet the prerequisite; consider high efficiency or dual flush toliets, low-flow fixtures, rainwater reuse for sewage, etc.
- Coordinate with architect, landscape architect, civil engineer, and owner to determine appropriate fixtures and strategies.

Stage II

DD LEED Phase — DD — Plumb. Eng.

- Incorporate chosen fixtures and strategies into drawings.
- Ensure specifications clearly identify the maximum flow and flush rates of fixtures.
- Update calculations, if needed.

LEED PM

- Review drawings and calculations to ensure prerequisite requirements will be met.

CD LEED Phase — CD — Plumb. Eng.

- Update calculations, if needed.
- Finalize plumbing drawings, details, and specifications for construction.
- Coordinate with architect, landscape architect, and civil engineer.
- Complete online credit form.
- Upload required drawings demonstrating compliance.

LEED PM

- Review online documentation and calculations; submit to GBCI.

Construction Application

Stage III

Constn. LEED Review — Construction — GC

- Contractor to procure and install fixtures according to specified flow/flush rates.

O&M

Responsibilities

- Plumbing Engineer (Plumb. Eng.)

- Civil Engineer
- Landscape Architect
- LEED PM
- Architect
- Owner

Documentation Checklist

- Online credit form
- Plumbing drawings with fixture schedule showing type, flow/flush rates, location, and name of manufacturer
- Manufacturers' data showing water consumption rates and model of each fixture

Implementation

- Baseline is calculated based on Energy Policy Act (EPAct) 1992 Standards:

	Commercial	Residential
Toilets	1.6 gallons per flush (gpf)	1.6 gpf
Urinals	1.0 gpf	NA
Lavatory Faucets:		
private applications	2.2 gallons per minute (gpm)	2.2 gpm
other applications	0.5 gallons per minute (gpm)	NA
metering faucets	0.25 gallons per minute (gpm)	NA
Prerinse Spray Valve	<1.6 gpm	NA
Showerheads	NA	2.5 gpm

- The U.S. Environmental Protection Agency (EPA) has a label called *WaterSense* for fixtures that are efficient and conserve water. When specifying products, look for the *WaterSense* label on manufacturers' data sheets. More information on *WaterSense* products is available at: www.epa.gov/watersense.

Registration

Stage I — Landscape — Pre-SD

- Investigate options to reduce use of water for landscaping.
- Investigate availability of municipally supplied non-potable water to site for irrigation purposes.
- Discuss with owner, civil engineer, and architect options to use captured rainwater and/or treated greywater for irrigation purposes.

Design Application

Stage II

SD LEED Phase — Landscape — SD

- Research available product options for irrigation systems, rainwater sensors, rainwater capture and reuse solutions, etc.
- Incorporate strategies in landscape design that reduce overall water use for landscape; consider native/adapted vegetation, group plants with similar water needs, reduce continuous turf areas, use efficient irrigation systems, and supplement potable water use with non-potable water sources.
- Perform baseline and design irrigation calculations to determine percentage reduction of water use.

DD LEED Phase — LEED PM / Landscape — DD

- Develop landscape drawings to reflect chosen strategies.
- Ensure specifications clearly identify the type of irrigation system and species of all plantings.
- Update calculations as needed.

- Review drawings and calculations.

CD LEED Phase — LEED PM / Landscape — CD

- Update calculations as needed.
- Finalize landscape drawings, details, and specifications for construction.
- Coordinate with architect, plumbing engineer, and civil engineer in implementation of any rainwater or greywater reuse systems.
- Upload required drawings demonstrating compliance.

- Review online documentation and calculations; submit to GBCI.

Construction Application

Stage III

Constn. LEED Review — Construction

O&M

- Landscape Architect

- Civil Engineer
- LEED PM
- Architect
- Plumbing Engineer

- Online credit form—calculations showing baseline and design irrigation consumption and percentage reduction
- Landscape drawings showing planting plan and schedule as well as irrigation plan with specifications
- Manufacturers' data showing any special irrigation system used (if applicable)

Registration

Stage I

Plumb. Eng. — Pre-SD

- Investigate availability of municipally supplied non-potable water to site.
- Research applicable codes and local laws applicable to treatment of wastewater on-site.
- Discuss with owner, landscape architect, and civil engineer options to achieve this credit.

Design Application

Stage II

SD LEED Phase

Plumb. Eng. — SD

- Research available high efficiency flush fixture product options.
- Investigate options of using captured rainwater, recycled greywater, or treated wastewater for building sewage conveyance.
- Perform preliminary water use calculations, using LEED occupancy counts, to determine optimal flow rates of fixtures and/or volume of non-potable water required.
- Coordinate with architect, landscape architect, civil engineer, and owner to determine appropriate fixtures and strategies.
- Determine appropriate option:
  Opt 1: Reduce potable water use for sewage conveyance by 50%
  Opt 2: Treat 50% of wastewater on-site to tertiary standards

DD LEED Phase

Plumb. Eng. — DD

- Design to incorporate chosen fixtures and/or treatment systems.
- Ensure specifications clearly identify flush rates of fixtures.
- Update calculations as needed.

LEED PM

- Review drawings and calculations to determine whether credit requirements will be met.

CD LEED Phase

Plumb. Eng. — CD

- Update calculations, if needed.
- Finalize plumbing drawings, details, and specifications for construction.
- Coordinate with architect, landscape architect, and civil engineer.
- Complete online credit form.
- Upload required drawings demonstrating compliance.

LEED PM

- Review online documentation and calculations; submit to GBCI.

Construction Application

Stage III

Constn. LEED Review

GC — Construction

- Contractor to procure fixtures with specified flush rates and/or install rainwater or greywater treatment and reuse systems according to specifications.

O&M

Responsibilities

- Plumbing Engineer (Plumb. Eng.)

- Civil Engineer
- Landscape Architect
- LEED PM
- Architect
- Owner

Documentation Checklist

- Online credit form—list type and number of occupants, calculations describing reduction in potable water use consumption, volume of captured rainwater and/or treated wastewater
- Plumbing drawings with fixture schedule showing type, flush rates, location, and manufacturer
- Drawings/details describing rainwater and/or graywater systems or on-site wastewater treatment systems and their capacity
- Manufacturers' data showing water consumption rates and model of fixtures or systems used in the project

Implementation

- WE P1 and WE C3 refer to reducing potable water consumption for all indoor water use needs and include both flow (faucets, showerheads) and flush (toilets, urinals) fixtures. WE C2 refers to reducing potable water consumption for only indoor sewage conveyance and counts only flush fixtures—toilets and urinals.

Registration

Stage I — Pre-SD — LEED PM

- Establish goal of reducing water use beyond the 20% reduction required for WE P1.
- Explore opportunities to achieve this.
- Determine total number of occupants (full-time equivalents, visitors, students, residents, etc.)

Design Application

Stage II

SD LEED Phase — SD — Plumb. Eng.

- Research available high efficiency water fixture product options.
- Perform preliminary water use calculations using LEED occupancy counts to determine optimal flow rates of fixtures to achieve credit.
- Investigate all options to increase water reduction percentage; consider high efficiency or dual flush toliets, low flow fixtures, rainwater reuse for sewage, waterless urinals, etc.
- Investigate cost implications of options to decide the best strategies.
- Coordinate with architect, landscape architect, civil engineer, and owner to determine appropriate fixtures and strategies.

DD LEED Phase — DD — Plumb. Eng.

- Incorporate chosen fixtures and strategies into drawings.
- Ensure specifications clearly identify the maximum flow and flush rates of fixtures.
- Update calculations, if needed.

LEED PM

- Review drawings and calculations to determine whether credit requirements will be met.

CD LEED Phase — CD — Plumb. Eng.

- Update calculations, if needed.
- Finalize plumbing drawings, details, and specifications for construction.
- Coordinate with architect, landscape architect, and civil engineer.
- Complete online credit form.
- Upload required drawings demonstrating compliance.

LEED PM

- Review online documentation and calculations; submit to GBCI.

Construction Application

Stage III — Constn. LEED Review — Construction — GC

- Contractor to procure fixtures according to specified flow/flush rates.

O&M

LEAD

SUPPORT

Responsibilities

- Plumbing Engineer (Plumb. Eng.)

- Civil Engineer
- Landscape Architect
- LEED PM
- Architect
- Owner

Documentation Checklist

DESIGN SUBMITALL

- Online credit form
- Plumbing drawings with fixture schedule showing type, flow/flush rates, location, and name of manufacturer
- Manufacturers' data showing water consumption rates and model of each fixture

Implementation

STRATEGIES

- Baseline is calculated based on Energy Policy Act (EPAct) 1992 Standards:

	Commercial	Residential
Toilets	1.6 gallons per flush (gpf)	1.6 gpf
Urinals	1.0 gpf	NA
Lavatory Faucets:		
Private applications	2.2 gallons per minute (gpm)	2.2 gpm
Other applications	0.5 gallons per minute (gpm)	NA
Metering faucets	0.25 gallons per minute (gpm)	NA
Prerinse Spray Valves	<1.6 gpm	NA
Showerheads	NA	2.5 gpm

- The U.S. Environmental Protection Agency (EPA) has a label called *WaterSense* for fixtures that are efficient and conserve water. When specifying products, look for the *WaterSense* label on manufacturers' data sheets. More information on *WaterSense* products is available at: www.epa.gov/watersense.

TIPS & NOTES

Sample Documents

The following pages present excerpts from documents submitted to USGBC by the teams of the case study projects (discussed in Chapter 8). Sample documents for the following *Water Efficiency* credit have been included.

WE Credit 2, Innovative Wastewater Technologies

M6   Water Cycle

Minnesota is known for it's abundance of water and an important project goal is to demonstrate good stewardship of those resources.

Data Type: Design Year		
1) Precipitation Managed On-site:		69%
2) How much potable water is used indoors and outdoors? Cooking water and dishwasher process water not included—information not available	929,624 gal/yr	
3) Total Percentage of Capture Rainwater Used:	(29%) 265,420 gal/yr	
4) Total Water Used Indoors:	558,307 gal/yr	
5) Total Water Outdoors:	371,317 gal/yr	
6) Comparative Water Savings Summary		
Total Annual Water Saved from Baseline: (Indoor & Outdoor):	1,641,453 gal/yr	64%
Annual Outdoor Water Savings via Native & Adapted Plantings:	1,327,967 gal/yr	78%
Annual Indoor Water Savings via Fixture Efficiency:	313,486 gal/yr	36%
Annual Indoor Potable Water Savings w/ Rainwater Harvesting:	578,904 gal/yr	66%

Sample Document: WE Credit 2, Innovative Wastewater Technologies (page 1 of 2).
Credit: Perkins + Will and Dunham Associates.

Rainwater Harvesting for Maple Grove, Minnesota

Month	Collection Area (sq ft) ×	Collection Efficiency ×	Mean Precipitation × (inches) ×	0.6233 gal/in =	Rainwater Volume (gal)	Amount of rainwater needed for toilets/urinals	Amount of rainwater used for toilets/urinals (gal/mo)	Amount of rainwater overflow to rain garden (gal/mo)
January	51335	0.25	0.86	0.6233	6879	348422 / 12 = 29035 gal/month	6879	0
February	51335	0.25	0.84	0.6233	6719	348422 / 12 = 29035 gal/month	6719	0
March	51335	0.25	1.61	0.6233	12879	348422 / 12 = 29035 gal/month	12879	0
April	51335	0.75	2.21	0.6233	53035	348422 / 12 = 29035 gal/month	29035	24,000
May	51335	0.75	3.43	0.6233	82313	348422 / 12 = 29035 gal/month	29035	53,278
June	51335	0.75	4.22	0.6233	101271	348422 / 12 = 29035 gal/month	29035	72,236
July	51335	0.75	3.62	0.6233	86872	348422 / 12 = 29035 gal/month	29035	57,837
August	51335	0.75	3.43	0.6233	82313	348422 / 12 = 29035 gal/month	29035	53,278
September	51335	0.75	2.91	0.6233	69834	348422 / 12 = 29035 gal/month	29035	40,799
October	51335	0.75	2.07	0.6233	49676	348422 / 12 = 29035 gal/month	29035	20,641
November	51335	0.25	1.47	0.6233	11759	348422 / 12 = 29035 gal/month	28259	*
December	51335	0.25	0.93	0.6233	7439	348422 / 12 = 29035 gal/month	7439	0
					570,988	Total rainwater reuse for Toilets/Urinals (Jan – Dec,) (gallons)	265,420	

Tank Size is 20,000 gallons (16,500 usable capacity)
* At the beginning of November the tank is assumed to be full to provide rain water until the tank reaches the low level limit.

Sample Document: WE Credit 2, Innovative Wastewater Technologies (page 2 of 2).
Credit: *Perkins + Will and Dunham Associates.*

CREDIT IMPLEMENTATION— ENERGY AND ATMOSPHERE

Introduction

This section provides a credit-by-credit framework to implement the requirements of credits in the *Energy and Atmosphere* category of the LEED-NC v3 rating system. It also identifies responsible team members, includes a documentation checklist, and provides tips and notes for each credit.

Buildings consume approximately 39% of the energy and 74% of the electricity produced annually in the United States. Electricity is often generated by burning fossil fuels, which releases carbon dioxide and other greenhouse gases that contribute to climate change.[1] There is significant evidence to show that the temperatures of the earth are increasing at levels that may lead to several alarming problems, such as rise in sea levels, occurrence of natural disasters, changes in weather patterns, etc. It is imperative, therefore, for buildings not only to reduce their energy use but also to use more benign forms of energy. The less energy a building consumes, the fewer greenhouse gases are emitted from energy production. Improved energy performance also leads to significant reduction in operational costs for building owners.

The *Energy and Atmosphere* category of credits and prerequisites sets standards for minimum energy efficiency of buildings and encourages the use of various strategies to improve their energy performance. It also rewards monitoring energy use and supplementing it with on-site renewable energy systems. Up to 35 points can be achieved by implementing various credit requirements in this category. This section includes implementation worksheets for the following credits.

EA Prereq 1, Fundamental Commissioning of Building Energy Systems
EA Prereq 2, Minimum Energy Performance
EA Prereq 3, Fundamental Refrigerant Management
EA Credit 1, Optimize Energy Performance

EA Credit 2, On-Site Renewable Energy

EA Credit 3, Enhanced Commissioning

EA Credit 4, Enhanced Refrigerant Management

EA Credit 5, Measurement and Verification

EA Credit 6, Green Power

NOTE

[1] USGBC, *LEED Reference Guide for Green Building Design and Construction* (USGBC, 2009).

Registration

Stage I — CxA, Owner — Pre-SD

- Owner to designate Commissioning Authority (CxA).
- Owner and CxA to prepare owner's project requirements (OPR)
  OPR should include: owner and user requirements, environmental and sustainability goals, energy efficiency goals, indoor environmental quality requirements, equipment and system expectations, and building occupant and operations personnel requirements.

Design Application

Stage II

SD LEED Phase — CxA, MEP, Architect — SD

- Design team and CxA to prepare basis of design (BOD) documents based on OPR; BOD includes: primary design assumptions, applicable codes and standards, and narrative descriptions of systems to be commissioned (HVAC, lighting, etc.).
- CxA to review OPR and BOD for clarity and completeness.

DD LEED Phase — CxA — DD

- Develop and implement commissioning plan; include in plan:
  Cx Overview—objectives, project information, systems to be commissioned
  Cx Team—team roles, communication protocol, meetings, and coordination
  Cx Process—documenting OPR, preparing BOD, developing functional test procedures, verifying systems performance, reporting deficiencies and resolution, and accepting building systems.

CD LEED Phase — CxA, Design Team — CD

- Incorporate commissioning requirements into construction documents and/or specifications.

Construction Application

Stage III

Constn. LEED Review — GC, CxA — Construction

- During construction, verify installation and performance of commissioned systems:
  - Perform prefunctional inspections
  - Perform systems performance testing
  - Evaluate results; compare with OPR and BOD
- At substantial completion, CxA to complete a summary commissioning report with summary of the Cx process and results, history of system deficiencies and resolution, system performance test results, and confirmation that systems meet the OPR, BOD and contract documents.
- Complete, and upload all required documents to LEED Online.

- LEED PM to review all uploaded documentation and submit to GBCI. — LEED PM

O&M

Responsibilities

- Commissioning Authortiy (CxA)

- Architect
- MEP
- Owner
- LEED PM

Documentation Checklist

- Online credit form—list of systems commissioned and confirmation that CxA has documented experience of previously working on at least 2 building projects
- Owner's project requirements
- Basis of design document
- Commissioning specifications
- Commissioning plan
- Commissioning summary report

Implementation

- For the purpose of LEED fundamental Cx, only the following systems need to be commissioned: HVAC&R sytems and associated controls, lighting and daylighting controls, domestic hot water systems, and renewable energy systems. Additional items—envelope, water treatment systems, etc.—may be added but are not required for meeting this prerequisite.

- The LEED PM should serve as facilitator of the Cx process, ensuring all Cx activities take place and there is communication between the CxA, team, and owner.

Registration

Stage I

Pre-SD   LEED PM, Owner

- Determine compliance path applicable to project:
  Opt 1: Performance path—whole building simulation.
  Opt 2: Prescriptive path—use ASHRAE Advanced Energy Design Guides applicable for small office and retail buildings under 20,000 sf or for small warehouses and self storage buildings under 50,000 sf.
  Opt 3: Prescriptive path—use Advanced Buildings Core Performance Guide applicable for buildings under 100,000 sf; NA for healthcare, warehouse, or labs.

Design Application

Stage II

SD LEED Phase

SD   EM, Design Team

- Option 1
  - Owner to designate energy simulation expert (could be a third-party consultant or the mechanical engineer or LEED consultant).
  - Energy modeler to create preliminary model and make energy efficiency recommendations to design team.
  - Design team to discuss and incorporate recommendations.
  - MEP team to design according to applicable mandatory provisions of ASHRAE 90.1-2007.
- Options 2 and 3
  - Design team to incorporate prescriptive measures into design.

DD LEED Phase

DD   EM

- Update energy model based on design development drawings.
- LEED PM to review model results and drawings to verify accuracy of assumptions, and identify any potential inconsistencies.

CD LEED Phase

CD   LEED PM, EM

- Update energy model inputs based on final construction documents.
- Prepare documentation required to demonstrate compliance with prerequisite and upload to LEED Online.

- LEED PM to review all uploaded documentation and submit to GBCI.

Construction Application

Stage III

Constn. LEED Review

Construction

O&M

Responsibilities

- Energy Simulation Expert/Energy Modeler (EM)

- Architect
- MEP Engineers
- Lighting Consultant
- Owner
- LEED PM

Documentation Checklist

- Online credit form—list climate zone for project location, compliance path chosen, and information requested related to energy model inputs or prescriptive measures followed
- ASHRAE 90.1 2007 mandatory provisions compliance forms
- Option 1
  - Energy simulation annual performance rating reports of both baseline and design case
  - Energy efficiency features included in the design with differences from baseline listed

- Options 2 and 3
  - Drawings and specifications highlighting the prescriptive measures for building envelope, HVAC, and service water heating

Implementation

- Many energy simulation software packages are available to demonstrate compliance with this prerequisite; commonly used examples are eQuest, VisualDOE, EnergyPlus, and IES-VE.

- During early design stages, use energy modeling to inform decisions related to building envelope design, orientation, and massing of building. As design progresses, use model to make detailed decisions related to building energy systems.

- For energy simulation option, the performance rating method outlined in Appendix G of ASHRAE 90.1-2007 should be used. The energy cost budget method (Section 11) should NOT be used.

- Building performance according to the performace rating method is measured by the annual energy cost and NOT annual energy used for both baseline and proposed cases.

Registration

Stage I

Pre-SD — LEED PM, Owner

- Establish requirement of zero use of chlorofluorocarbon-based (CFC) refrigerants in the new base building HVAC&R systems.
- If existing base building systems with CFC-based refrigerants are being reused, then owner to establish a CFC phaseout conversion plan prior to project completion.

Design Application

Stage II

SD LEED Phase

SD — Mech.

- Mechanical Engineer to confirm no CFC-based refrigerants will be used in base building systems.
- For CFC phaseout, perform economic analysis of conversion and/or replacement and develop phaseout plan.

DD LEED Phase

DD

CD LEED Phase

CD — LEED PM, Mech.

- Specify appropriate refrigerants and incorporate in drawings and specifications.
- Complete required documentation and upload to LEED Online.

- LEED PM to review all uploaded documentation and submit to GBCI.

Construction Application

Stage III

Constn. LEED Review

Construction

O&M

Responsibilities

- Mechanical Engineer (Mech.)

- Architect
- Owner
- LEED PM

Documentation Checklist

- Online credit form
- Mechanical drawings showing the type of refrigerant used
- Manufacturers' documentation demonstrating type of refrigerant used
- For major renovations, phaseout plan and implementation documentation

Implementation

- It is standard practice in new buildings to install equipment that does not use CFCs, so this prerequisite is fairly easy to meet in the case of new construction.

- For major renovations where CFC phaseout is not feasible, a third party audit, demonstrating that it is not economically feasible (simple payback for replacement or conversion longer than 10 years) to replace or convert the system, can be used as an alternate compliance approach.

- For projects connecting with existing chilled water systems, either the system should be CFC free, or the project team must demonstrate commitment to phasing out CFC-based refrigerants no later than 5 years after project is completed.

- For projects connecting to district energy systems, use technical guidance provided by USGBC under registered project tools.

Registration	**Stage I**	• Determine compliance path applicable to project: Opt 1: Performance path—whole building simulation Opt 2: Prescriptive path—use ASHRAE Advanced Energy Design Guides applicable for small office and retail buildings under 20,000 sf or for small warehouses and self storage buildings under 50,000 sf Opt 3: Prescriptive path—use Advanced Buildings Core Performance Guide applicable for buildings under 100,000 sf; (NA for healthcare, warehouse, or labs)	LEED PM, Owner	**Pre-SD**

Design Application	**Stage II** — SD LEED Phase	• Option 1
		  • Owner to designate energy simulation expert (could be a third-party consultant or the mechanical engineer or LEED consultant).
		  • Energy modeler to create preliminary model and make energy efficiency recommendations to design team to optimize energy performance of building.
		  • Design team to discuss and establish energy optimization goals.
		  • Design team to incorporate recommendations in design.
		• Option 2 and 3
		  • Design team to identify prescriptive measures that will help optimize building performance and incorporate in design.
			EM, Design Team	**SD**

	DD LEED Phase	• Opt 1: Update energy model based on design development drawings. • Opt 2 and 3: Review and ensure prescriptive measures are included in drawings. • LEED PM to review model results and drawings to verify accuracy of assumptions, and identify any potential inconsistencies.	LED PM, Team, EM	**DD**
	CD LEED Phase	• Opt 1: Prepare detailed energy model with inputs based on final CDs. • Opt 2 & 3: Incorporate all prescriptive measures into CDs. • Prepare documentation required to demonstrate compliance with credit and upload to LEED Online. • LEED PM to review all uploaded documentation and submit to GBCI.	LEED PM, EM	**CD**
Construction Application	**Stage III** — Constn. LEED Review			**Construction**
				O&M

Responsibilities

- Energy Simulation Expert/Energy Modeler (EM)

- Architect
- Mechanical Electrical Plumbing Engineers (MEP)
- Lighting Consultant
- Owner
- LEED PM

Documentation Checklist

- Online credit form—list climate zone for project location, compliance path chosen, and information requested related to energy model inputs or prescriptive measures followed

- Option 1
  - Energy simulation annual performance rating reports of both baseline and design case
  - Energy efficiency features included in the design with differences from baseline listed

- Option 2 and 3
  - Drawings and specifications highlighting the prescriptive measures for building envelope, HVAC, and service water heating

Implementation

- Many energy simulation software packages are available to demonstrate compliance with this credit; commonly used examples are eQuest, VisualDOE, EnergyPlus, and IES-VE.

- During early design stages, use energy modeling to inform decisions related to building envelope design, orientation, and massing of building. As design progresses, use model to make detailed decisions related to building energy systems.

- Use simulation option (Opt 1) to take advantage of number of points (up to 19) available under the option. Prescriptive options (Opt 2 & 3) only allow up to 3 points.

- For energy simulation option, the performance rating method outlined in Appendix G of ASHRAE 90.1-2007 should be used. The energy cost budget method (Section11) should NOT be used.

- Building performance according to the performace rating method is measured by the annual energy cost and NOT annual energy used for both baseline and proposed cases.

Registration

Stage I

LEED PM *Pre-SD*

- Research any incentives and/or rebates that may be available for the project location for installation of on-site renewable energy systems, like solar, wind, etc.
- Research different renewable energy systems to identify optimal solution for the project site.
- Discuss options with owners and project team.

Design Application

Stage II

SD LEED Phase

MEP, RE Cons. *SD*

- Use estimated annual energy costs from building energy simulation model to calculate the percentage of energy costs to be offset with renewable energy.
- Calculate size of renewable system required (engage a renewable energy consultant, if required).
- Perform cost benefit analysis of system proposed; take into account incentives and rebates and operational benefits.
- Determine optimum size, type, and design of system.
- Coordinate with architect and owners on location of system.

DD LEED Phase

MEP, RE Cons. *DD*

- Incorporate system design, layout, and specifications in drawings.

CD LEED Phase

MEP, RE Cons. *LEED PM* *CD*

- Update calculations based on refined energy model energy costs.
- Refine, coordinate, and finalize renewable system in construction drawings and specifications.
- Complete required documentation and upload to LEED Online.

- LEED PM to review all uploaded documentation and submit to GBCI.

Construction Application

Stage III

Constn. LEED Review

CxA, GC *Construction*

- Commission renewable energy systems according to Cx Plan to ensure proper installation and operation.

O&M

Responsibilities

● Mechanical Engineer (Mech.) or Renewable Energy Consultant (RE Con.)

● Architect
● Owner
● LEED PM

Documentation Checklist

● Online credit form—on-site renewable energy system type, total annual energy generation, and back up energy sources
● Calculations showing energy generated from renewable energy source
● Drawings and/or specifications showing the layout, location, and details of renewable energy system
● Manufacturers' documentation showing the system characteristics
● Documentation of any incentives received

Implementation

● Eligible systems—photovoltaic, wind energy, solar thermal, bio-fuel based electrical, geothermal heating, geothermal electric, low-impact hydroelectric power, and wave and tidal power systems.

● Ineligible systems—passive solar or daylighting strategies and geo-exchange systems.

● The Database of State Incentives for Renewable Energy website is a very good resource to identify both federal and state incentives for renewable energy systems: www.dsireusa.org.

● Consider offsetting the energy cost of common areas such as corridors, garages, common amenities etc., with renewable energy systems. These areas are typically paid for by the owner/manager of the property and shifting these areas to a renewable source, can bring down operational costs significantly.

Registration

Stage I (CxA, Owner — Pre-SD)

- Owner to designate Commissioning Authority (CxA).
- Owner and CxA to prepare owner's project requirements (OPR). OPR to include: Owner and user requirements, environmental and sustainability goals, energy efficiency goals, indoor environmental quality requirements, equipment and system expectations, and building occupant and operations personnel requirements

Design Application

Stage II

SD LEED Phase (CxA, MEP, Architect — SD)

- Design team and CxA to prepare basis of design (BOD) documents based on OPR.BOD to include: Primary design assumptions, applicable codes and standards, and narrative descriptions of systems to be commissioned (HVAC, lighting, etc.)
- CxA to review OPR and BOD for clarity and completeness.

DD LEED Phase (DD)

- Develop and implement commissioning plan; include in plan:
  Cx Overview—objectives, project information, and systems to be commissioned
  Cx Team—team roles, communication protocol, meetings, and coordination
  Cx Process—documenting OPR, preparing BOD, developing functional test procedures, verifying systems performance, reporting deficiencies and resolution, and accepting building systems

CD LEED Phase (CxA, MEP — CD)

- Incorporate commissioning requirements into construction documents and/or specifications.
- CxA to conduct review of mid-construction documents to ensure that the owner's project requirements are satisfactorily addressed in the drawings.

Construction Application

Stage III

Constn. LEED Review (GC, CxA — Construction)

- During equipment procurement and installation, CxA should review contractor submittals applicable to systems being commissioned.
- During construction, verify installation and performance of commissioned systems.
- Develop systems manual for commissioned systems; include for each system: final BOD, system single-line diagrams, as-built sequence of operations, operating instructions, recommended schedules for maintenance requirements, and retesting and for calibrating sensors.
- Verify that requirements for training operating personnel and building occupants have been completed.
- At substantial completion, CxA to complete a summary commissioning report.
- Complete and upload all required documents to LEED Online.
- LEED PM to review all uploaded documentation and submit to GBCI. (LEED PM)

(CxA, MEP — O&M)

- CxA to review building operation within 10 months after substantial completion and address any system deficiencies or operational issues.

Responsibilities

- Commissioning Authority (CxA)

- Architect
- MEP
- Owner
- LEED PM

Documentation Checklist

- Online credit form—list of systems commissioned and confirmation that CxA has documented experience of working on at least 2 building projects
- Owner's project requirements
- Basis of design document
- Commissioning specifications
- Commissioning plan
- Commissioning summary report
- Written schedule of building operator training
- Sytems manual
- Comissioning design review comments, design team responses, and confirmation of comments addressed

Implementation

- For the purpose of LEED Cx, only the following systems need to be commissioned: HVAC&R sytems and associated controls, lighting and daylighting controls, domestic hot water systems, and renewable energy systems. Additional items—envelope, water treatment systems, etc.—may be added but are not required for meeting this prerequisite.

- The LEED PM should serve as facilitator of the Cx procss, ensuring all Cx activities take place and there is communication between the CxA, team, and owners.

LEAD

SUPPORT

CONSTRUCTION SUBMITTAL

STRATEGIES

TIPS & NOTES

Registration

Stage I

Pre-SD

Design Application

Stage II

SD LEED Phase

- Determine applicable option:
  Option 1: Do not use refrigerants.
  Option 2: Select refrigerants that minimize the emission of compounds that contribute to ozone depletion and global warming (maximum emission threshold calculated according to formula provided).
- If pursuing option 2, perform calculations to determine the refrigerants to be specified.
- Select fire suppression systems that do not have CFCs, HCFCs, or halons.

Mech.

SD

DD LEED Phase

DD

CD LEED Phase

- Specify appropriate refrigerants and fire suppression systems, and incorporate in drawings and specifications.
- Complete required documentation and upload to LEED Online.

- LEED PM to review all uploaded documentation and submit to GBCI for Design Review.

LEED PM, Mech.

CD

Construction Application

Stage III

Constn. LEED Review

Construction

O&M

Responsibilities

- Mechanical Engineer (Mech.)

- Architect
- Owner
- LEED PM

Documentation Checklist

- Online credit form—list base building systems containing refrigerants and the type of refrigerant along with its ozone depleting potential (ODP) and global warming potential (GWP)
- Mechanical drawings and/or specifications showing the type of refrigerant used
- Manufacturers' documentation demonstrating type of refrigerant used
- Manufacturers' data indicating halons, CFCs, or HCFCs are not in fire-suppression systems

Implementation

- Using natural ventilation with no active cooling systems can help achieve this credit via Option 1 of not using any refrigerants.

- Using HVAC&R systems with natural refrigerants like water, ammonia, and carbon dioxide is another approach to achieve this credit via Option 1.

- Select refrigerants with low ODP and GWP.

- Any cooling equipment that contains less than 0.5 pounds of refrigerant is not subject to the requirements of this credit. Examples of such equipment are small HVAC units, standard refrigerators, or small water coolers.

- This credit is usually easier to achieve in projects such as office, hotel or high-rise multifamily buildings, which typically use R-134A or R-22 as refrigerants. It is uncommon for mid-rise multifamily buildings to achieve this credit that typically use split DX systems with R-410A refrigerant, which has higher calculated emissions than the allowable threshold set by the credit.

LEAD

SUPPORT

DESIGN SUBMITTAL

STRATEGIES

TIPS & NOTES

Registration

Stage I

Pre-SD — LEED PM, Owner

- Identify the team member who will be responsible for developing and implementing the measurement and verification (M&V) plan. The mechanical engineer on the team may do this, or third-party providers may be identified.

Design Application

Stage II

SD LEED Phase — **SD** — MEP, MV Cons.

- Decide on M&V path to be followed:
  Opt 1—Option D of the International Performance and Verification protocol (IPMVP)
  Opt 2—Option B of the IPMVP
- Develop M&V plan for the project based on option chosen.

DD LEED Phase — **DD** — MEP, MV Cons.

- MEP engineers to incorporate meters or other measurement systems in drawings and specifications.

CD LEED Phase — **CD** — MEP, MV Cons.

- MEP to finalize location and type of meters and/or other measurement systems in drawings and specifications.
- Review meter placement in drawings.

Construction Application

Stage III

Constn. LEED Review — **Construction** — MV Cons., MEP, GC

- Contractor to install meters and/or other measurement systems as specified.
- Verify proper installation and operation of meters.
- Upload required documents to LEED Online.

LEED PM

- LEED PM to review all uploaded documentation and submit to GBCI for Construction Review.

O&M

- After occupancy, begin M&V activities based on option chosen and continue for at least 1 year. Log data, recalibrate base energy model or estimated energy use, report on energy savings, and provide suggestions for continuous improvement.

Responsibilities

● MEP Engineer (MEP) or M&V Consultant (MV Cons.)

● Architect
● Owner
● LEED PM

Documentation Checklist

● Online credit form
● An IPMVP compliant measurement and verification plan as per the option chosen
● Drawings and/or specifications showing the layout, location, and details of meters and/or other measurement systems
● Manufacturers' documentation regarding meters or other measurement device characteristics

Implementation

Registration

Stage I

Pre-SD — LEED PM, Owner

- Discuss with owners green power options available to the project site—utility provided green power or Renewable Energy Certificates (RECs).

Design Application

Stage II

SD LEED Phase

SD — LEED PM, Owner, EM

- Perform calculations to determine amount of green power required:
  Opt 1: Use electricity consumption estimated from the energy model
  Opt 2: Use U.S. Department of Energy's Commercial Buildings Energy
             Consumption Survey to determine estimated electricity use
- Procure cost estimates for both utility-supplied green power and RECs for the estimated green power quantity.
- Discuss with owners to determine the appropriate green power choice.

DD LEED Phase

DD

CD LEED Phase

CD — LEED PM

- Owner to sign a 2-year contract for purchase of renewable energy certified by Green-e or equivalent, for minimum of 35% of building's estimated annual electricity consumption (either from utility or via purchase of RECs). This can happen any time before submission to GBCI for Construction Review.

Construction Application

Stage III

Constn. LEED Review

Construction — LEED PM

- Obtain contract between green power provider and owners.
- Complete online documentation required.
- Submit to GBCI for Construction Review.

O&M

Responsibilities

- LEED PM

- Energy Simulation Expert (EM)
- MEP Engineers
- Owner

Documentation Checklist

- Online credit form—list estimated electricity consumption and percentage of renewable energy purchased
- 2-year contract showing purchase of renewable energy certified by Green-e or equivalent

Implementation

- Green-e certified renewable energy provider list can be found at: www.green-e.org

- Green Power provided by local utility companies may be more expensive than purchasing RECs. Compare costs before making decisions.

- RECs are widely available in all U.S. states at competitive prices. However, prices fluctuate with time, so it is important to lock in rates and enter into contract as early as possible to avoid future increases. Rates can be fixed in advance even before building is operational.

- Exemplary Performance (1 point) can be achieved by purchasing 100% of electricity from renewable sources.

LEAD

SUPPORT

CONSTRUCTION SUBMITTAL

STRATEGIES

TIPS & NOTES

Sample Documents

The following pages present excerpts from documents submitted to USGBC by the teams of the case study projects (discussed in Chapter 8) for credits in the *Energy and Atmosphere* category. Sample documents for the following credits have been included.

EA Credit 1, Optimize Energy Performance
EA Credit 2, On-Site Renewable Energy
EA Credit 6, Green Power

Mechanical System Description and Modeling Methodology

The mechanical system consists of six major components as shown on the diagram below. All components use some type of heat pump, either water-to-water for the displacement unit or water-to-air for the other units. The water source for all of the heat pumps is a lake. This is a complex mechanical system which DOE2.1E is not capable of modeling without Exceptional Calculation methods. There are several elements of Exceptional Calculations that were needed for this simulation. A detailed description of each of the primary components of the mechanical system is provided on the following pages, and that is followed by a general summary list of the exceptional calculations.

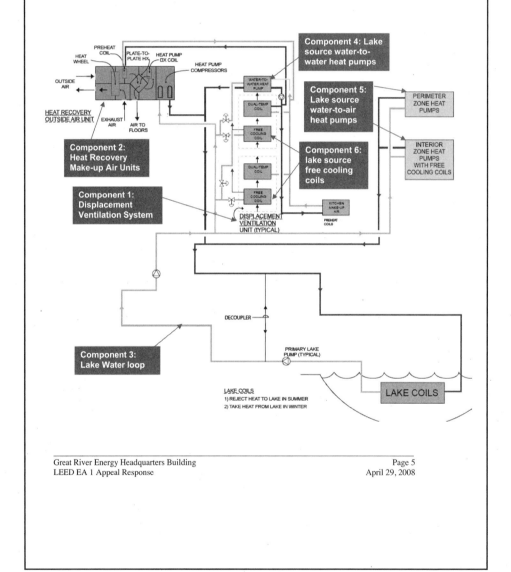

Great River Energy Headquarters Building
LEED EA 1 Appeal Response

Page 5
April 29, 2008

Sample Document: EA Credit 1, Optimize Energy Performance (page 1 of 3).

Credit: Dunham Associates and Weidt Group.

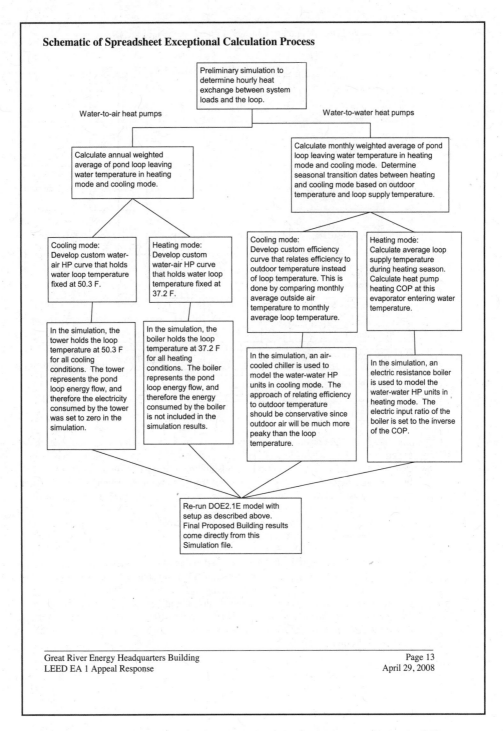

Schematic of Spreadsheet Exceptional Calculation Process

Preliminary simulation to determine hourly heat exchange between system loads and the loop.

Water-to-air heat pumps

Water-to-water heat pumps

Calculate annual weighted average of pond loop leaving water temperature in heating mode and cooling mode.

Calculate monthly weighted average of pond loop leaving water temperature in heating mode and cooling mode. Determine seasonal transition dates between heating and cooling mode based on outdoor temperature and loop supply temperature.

Cooling mode:
Develop custom water-air HP curve that holds water loop temperature fixed at 50.3 F.

Heating mode:
Develop custom water-air HP curve that holds water loop temperature fixed at 37.2 F.

Cooling mode:
Develop custom efficiency curve that relates efficiency to outdoor temperature instead of loop temperature. This is done by comparing monthly average outside air temperature to monthly average loop temperature.

Heating mode:
Calculate average loop supply temperature during heating season. Calculate heat pump heating COP at this evaporator entering water temperature.

In the simulation, the tower holds the loop temperature at 50.3 F for all cooling conditions. The tower represents the pond loop energy flow, and therefore the electricity consumed by the tower was set to zero in the simulation.

In the simulation, the boiler holds the loop temperature at 37.2 F for all heating conditions. The boiler represents the pond loop energy flow, and therefore the energy consumed by the boiler is not included in the simulation results.

In the simulation, an air-cooled chiller is used to model the water-water HP units in cooling mode. The approach of relating efficiency to outdoor temperature should be conservative since outdoor air will be much more peaky than the loop temperature.

In the simulation, an electric resistance boiler is used to model the water-water HP units in heating mode. The electric input ratio of the boiler is set to the inverse of the COP.

Re-run DOE2.1E model with setup as described above. Final Proposed Building results come directly from this Simulation file.

Great River Energy Headquarters Building
LEED EA 1 Appeal Response

Page 13
April 29, 2008

Sample Document: EA Credit 1, Optimize Energy Performance (page 2 of 3).
Credit: Dunham Associates and Weidt Group.

A summary in kBtu/SF and total energy dollars is shown in the table below.

Model #	Model Description	Itemized kBtu/SF							Total Cost	
		Heating	Cooling	Fan/Pump	DHW	Lights	Equipment	Total		
1	Appendix G Baseline	45.4	6.5	13.1	1.4	15.3	22.7	104.4	$ 351,343	
2	Proposed Architectural and Lighting Design Interim Model	47.5	6.0	11.8	1.4	8.0	22.7	97.3	$ 326,959	
3	Proposed Architectural and Lighting and HVAC Model	9.2	5.3	23.4	1.4	8.0	22.7	70.0	$ 221,071	
	Diifference Model 2 minus Model 3 (Exceptional Calculation 2)	38.3	0.7	(11.6)	-	(0.0)	-	27.4	$ 105,888	
4	Final Proposed Design Model (Model 3 with water side economizer Exceptional Calculation 1)	9.2	2.1	23.4	1.4	8.0	22.7	66.8	$ 206,336	
	Total Savings (Model 1 minus 4)	36.2	4.4	(10.3)	-	7.3	-	37.6	$ 145,007	41%
	Onsite Renewable Savings								$ 29,593	
	Grand Total								$ 174,600	50%

The energy design consultant for the project has over 25 years of energy simulation experience using a wide range of programs. Over that time we have developed deep experience using energy simulation models and have compared these results to whole building metered energy performance. Over the years we have developed and tested a number of custom functions inside DOE2.1E to compare strategies and systems that cannot be directly modeled. Also, our modeling philosophy has been to develop methods that allow us to error on the conservative side of estimating energy savings so client expectations are meet after the building goes into operation.

We have had the opportunity to have modeled and studied the metered energy performance of a building project with similar HVAC system, daylighting control, and climate characteristics that has been operating for over 6 years. During that time we have reviewed sub metering by end-use data and have compared these results to our original design model. The actual building has been operating 10 to 15 % better than our design model results, and is operating in the 27,000 to 29,000 BTU/SF range, far lower than the performance simulated for this project.

Sample Document: EA Credit 1, Optimize Energy Performance (page 3 of 3).
Credit: Dunham Associates and Weidt Group.

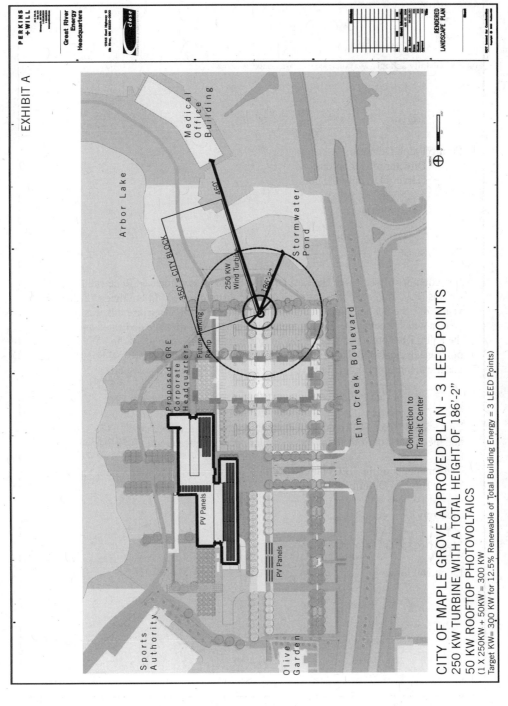

CITY OF MAPLE GROVE APPROVED PLAN - 3 LEED POINTS
250 KW TURBINE WITH A TOTAL HEIGHT OF 186'-2"
50 KW ROOFTOP PHOTOVOLTAICS
(1 X 250KW + 50KW = 300 KW
Target KW= 300 KW for 12.5% Renewable of Total Building Energy = 3 LEED Points)

Sample Document: EA Credit 2, On-Site Renewable Energy.
Credit: Perkins + Will

Harvard Divinity School   45 FRANCIS AVENUE, CAMBRIDGE, MA 02138

To: USGBC

From: Ralph DeFlorio
        Director of Operations
        Harvard Divinity School

Re: REC purchases

Dear USGBC,

For a number of years, the Harvard Divinity School has purchased Renewable Energy Credits (RECs) to cover the entire electrical consumption of our buildings. For fiscal year 2009, the School purchased a total of 1,350,000 kWh. Because this was the fiscal year in which Rockefeller Hall completed construction, we have dedicated a portion of the FY09 REC purchase specifically to Rockefeller. The amount of the FY09 REC purchase dedicated to Rockefeller Hall was 584,209 kWh. This is equal to 70% of the electrical consumption predicted by the energy model over a 2 year period:

$$417{,}292 \text{ kWh / year } * \ 70\% \ * \ 2 \text{ years } = \ 584{,}209 \text{ kWh}$$

This exceeds the LEED requirement for Energy & Atmosphere credit 6 (35% for 2 years) and meets the requirement for the exemplary performance credit (70% for two years).

The remainder of the RECs purchased for FY09 were assigned to the other buildings used by the Harvard Divinity School.

Sincerely,

Ralph DeFlorio

Sample Document: EA Credit 6, Green Power.
Credit: Harvard Divinity School

CREDIT IMPLEMENTATION— MATERIALS AND RESOURCES

Introduction

This section provides a credit-by-credit framework to implement the requirements of credits in the *Materials and Resources* category of the LEED-NC v3 rating system. It also identifies responsible team members, includes a documentation checklist, and provides tips and notes for each credit.

Building construction and operations generate a tremendous amount of waste. The extraction, processing, transportation, use, and disposal of building materials can have significant impact on the environment, as they pollute water and air, destroy native habitats, and deplete natural resources. Procuring materials in an environmentally responsible manner, minimizing waste, and recycling materials can significantly reduce the negative impacts associated with the production and transport of building materials.[1]

The *Materials and Resources* category of credits focuses on reducing the amount of waste that buildings generate during both construction and operation cycles. It encourages the reuse of existing building materials that are available on-site or sourced from other locations. It also gives credit to the selection of sustainable materials that incorporate recycled content, are made from rapidly renewable materials, and are sourced regionally. In addition, the category rewards the utilization of Forest Stewardship Council (FSC) certified wood and minimization of landfill disposal by recycling, reusing, or donating construction and demolition waste generated on-site. The intent behind these strategies is to extend the life cycle of existing building stock, conserve resources, reduce waste, and limit the environmental impacts resulting from extraction, production, and transportation of virgin materials. Up to 14 points can be achieved in this category. This section includes implementation worksheets for the following credits.

MR Prereq 1, Storage and Collection of Recyclables

MR Credit 1.1, Building Reuse—Maintain Existing Walls, Floors, and Roof

MR Credit 1.2, Building Reuse—Maintain 50% of Interior Non-Structural Elements

MR Credit 2, Construction Waste Management

MR Credit 3, Materials Reuse

MR Credit 4, Recycled Content

MR Credit 5, Regional Materials

MR Credit 6, Rapidly Renewable Materials

MR Credit 7, Certified Wood

NOTE

[1] USGBC, *LEED Reference Guide for Green Building Design and Construction* (USGBC, 2009).

Registration

Stage I

Architect | Pre-SD

- Establish area(s) for collection and storage of recyclables in the building and include them in the building program.

Design Application

Stage II

SD LEED Phase

Architect, Bldg. Mgmt. | SD

- Obtain input from local recycling/waste hauler regarding their preferred method of recycling (on-site or off-site separation), items eligible for recycling, and preferred location for collection from site.
- Discuss with building management the logistics of occupant recycling—number, type and location of bins, space for recyclables on every floor, collection by custodial staff, etc.
- Incorporate into design drawings space(s) designated for collection of recyclables based on input from local hauler and building management.
- At a minimum, the following items should be recycled: paper, plastics, metals, glass, and cardboard.

DD LEED Phase

LEED PM, Bldg. Mgmt. | DD

- Review drawings to determine if space(s) for collection and storage of recyclables are according to prerequisite's requirements.
- Work with building management to develop an occupant recycling program that outlines benefits of recycling to occupants and educates the custodial staff on proper recycling procedures.

CD LEED Phase

Architect | LEED PM | CD

- Finalize sizes and locations of recyling areas in drawings and upload documentation to LEED Online.

- LEED PM to review uploaded documentation and submit to GBCI for Design Review.

Construction Application

Stage III

Constn. LEED Review

Construction

O&M

353

Responsibilities

- Architect

- Building Management (Bldg. Mgmt.)
- Owner
- LEED PM

Documentation Checklist

- Online credit form—list of items that will be recycled and total building area
- Drawings—site plan and/or floor plan drawings that show all recycling storage areas
- Building management's policy/program outlining the details of the recycling procedures

Implementation

Stage I

- If site has not been selected, consider utilizing existing building.
- If existing building is being renovated, then consider reusing existing building elements and the cost savings associated with it.

Architect, Owner | Pre-SD

Design Application

Stage II

SD LEED Phase

- Inventory existing conditions and develop a floor plan showing existing structural components, walls, windows, doors, etc.
- Identify components that will be reused and establish that they can be reused and maintained.
- Design to reuse as many existing building components as possible.
- Calculate the surface area of all existing structural components and the area of components that will be reused to determine percentage of building reuse.

Architect | SD

DD LEED Phase

DD

CD LEED Phase

- Finalize drawings and specifications, clearly outlining components that will be reused, how they will be treated, and any special measures during construction needed to preserve them.
- Update building reuse areas and percentage calculations.

LEED PM, Architect | CD

Construction Application

Stage III

Constn. LEED Review

- Contractor to implement measures to preserve and reuse existing building components identified in the construction drawings and specifications.
- Architect to update drawings and calculations based on any changes that may have occurred during construction.
- Upload required drawings and documentation to LEED Online.

Architect, GC | Construction

- LEED PM to review uploaded documentation and submit to GBCI for Construction Review.

LEED PM

O&M

Responsibilities

- Architect

- Building Management
- Owner
- LEED PM
- General Contractor (GC)

Documentation Checklist

- Online credit form—list areas of total new, existing, and reused building components
- Floor plan drawings highlighting components that are new, existing, and reused
- Narrative explaining any elements that were excluded from calculations

Implementation

- Measurements for areas of components (in square feet) should be taken similar to the way they are typically taken when preparing bid for construction of a building:
  - Structural floor and roof: count area of each component
  - Exterior walls: count area of exterior wall minus exterior doors and windows
  - Interior structural walls: count area of one side of existing wall element

- Non-structural roofing material, window assemblies, structural and envelope materials that are deemed structurally unsound and hazardous should be excluded from calculations.

- If existing building components are reused but the reuse percentage is not enough to garner a point under this credit, then projects can apply the portion reused towards the achievement of MR Credit 2—Construction Waste Management.

Stage I

Architect, Owner — Pre-SD

- If existing building is being renovated, consider reusing existing building's interior non-structural elements and materials.

Design Application

Stage II

SD LEED Phase

Architect — SD

- Inventory existing conditions and develop a floor plan showing interior non-structural elements that can be reused and will be retained in their existing location or moved to another location.
- Design to reuse as many existing interior building elements as possible.
- Calculate the surface area (sf) of all (existing and new) interior non-structural elements and determine reuse percentage by dividing total area of retained interior non-structural elements with the total area of interior non-structural elements.

DD LEED Phase

DD

CD LEED Phase

LEED PM, Architect — CD

- Finalize drawings and specifications, clearly outlining elements that will be reused, how they will be treated, and any special measures during construction needed to preserve them.
- Update reuse areas and percentage calculations.

Construction Application

Stage III

Constn. LEED Review

Architect, GC — Construction

- Contractor to implement measures to preserve and reuse existing interior elements identified in the construction drawings and specifications.
- Architect to update drawings and calculations based on any changes that may have occurred during construction.
- Upload required drawings and documentation to LEED Online.

LEED PM

- LEED PM to review uploaded documentation and submit to GBCI for Construction Review.

O&M

Responsibilities

- Architect

- Building Management
- Owner
- LEED PM
- General Contractor

Documentation Checklist

- Online credit form—list areas of total interior non-structural elements area and area of interior non-structural elements reused/retained
- Floor plan drawings highlighting interior non-structural elements that are new, existing, and reused
- Narrative explaining approach to building elements reuse (recommended)

Implementation

- Measurements for areas of components should be taken similar to the way they are typically taken when preparing bids for flooring, ceiling or painting:
  - Finished ceilings and flooring: count area of each element
  - Interior non-structural walls: count finished area between ceiling and floor on both sides
  - Interior finishes of structural and party walls: count area of one side
  - Interior doors: count surface area once
  - Interior casework: calculate visible surface area

- If existing building components are reused but the reuse percentage is not enough to garner a point under this credit, then projects can apply the portion reused towards the achievement of MR Credit 2—Construction Waste Management.

Registration

Stage I

Pre-SD — LEED PM

- Discuss with owner and general contractor requirements of credit— to recycle demolition and construction waste—and establish a target recycling goal.

Design Application

Stage II

SD LEED Phase

SD — LEED PM, GC

- Identify construction waste recycling vendors for the project location and their recycling protocol—whether they take co-mingled or source separated waste, how recycling rates will be tracked, etc.

DD LEED Phase

DD — LEED PM, GC

- LEED PM or architect to develop a draft construction waste management plan in consultation with contractor and recycling vendor (plan can be part of specifications or drawings).

CD LEED Phase

CD — LEED PM, GC

- Contractor to finalize construction waste management plan and address any issues or concerns.

Construction Application

Stage III

Constn. LEED Review

Construction — GC

- Implement measures outlined in the construction waste management plan.
- Designate clearly labeled containers and/or areas for collecting recyclable waste.
- Organize a pre-construction LEED kick-off meeting to explain waste management procedures to subcontractors.
- Track construction waste diversion rates by means of tickets/receipts collected from recycling hauler, and submit progress to LEED PM and/or architect.
- At end of construction, compile all progress data, calculate final diversion rate and upload required documentation to LEED Online.

LEED PM

- LEED PM to review uploaded documentation and submit to GBCI for Construction Review.

O&M

Responsibilities

- General Contractor (GC)

- Architect
- Owner
- LEED PM

Documentation Checklist

- Online credit form—list construction waste generated by type, quantities of each type that were diverted and landfilled, total percentage of waste diverted from landfill
- Construction waste management plan outlining the approach, materials diverted, implementation protocols, and parties responsible for implementing the plan
- Receipts, tickets, logs, from hauler as back-up information (recommended)

Implementation

- Calculations of waste recycled and landfilled can be done either by weight or volume, but must be consistent throughout.
- There are multiple options for construction waste recycling. Materials (both recyclable and non-recyclable) can be mixed together on-site and separated off-site by the recycling hauler. This type of recycling is known as commingled recycling. The other option is to separate materials on-site into recyclable and non-recyclable waste, which is known as source separated recycling. Costs vary for each option, and in some areas, both options may not be available. Choose option according to location, availability and budget.
- Hazardous waste should be excluded from calculations.

- Investigate with suppliers/manufacturers to determine whether they will take back waste generated on-site from their products. For instance, many carpet manufacturers offer to take back scrap carpet waste generated on-site to recycle in their facilties. This would count towards the diversion rate.

Registration

Stage I

Pre-SD — Architect, Owner

- Assess opportunities to reuse materials available on-site or procure salvaged materials from other locations/sites.
- If any materials will be reused from existing building, then identify those items before demolition begins.

Design Application

Stage II

SD LEED Phase

SD — GC, Architect

- Using preliminary project budget, estimate the dollar amount of materials that should be reused and/or salvaged to obtain the required percentage for the credit (minimum 5% of total materials cost).
- Research available salvaged materials for the project and coordinate with contractor to identify materials that will help meet the minimum percentage requirement.
- Incorporate salvaged or reused materials in building design.

DD LEED Phase

DD

CD LEED Phase

CD — GC, Architect

- Finalize drawings and specifications, indicating materials that will be reused/salvaged and their procurement sources..

Construction Application

Stage III

Constn. LEED Review

Construction — GC

- Procure materials according to specifications and drawings.
- Keep track of costs incurred for procurement of off-site salvaged materials and replacement value of any on-site materials that were reused.
- Retain receipts/records of transactions showing costs of materials.
- After construction is complete, perform calculations to determine percentage of reused materials.
- Upload documentation to LEED Online.

LEED PM

- LEED PM to review uploaded documentation and submit to GBCI for Construction Review.

O&M

Responsibilities

- General Contractor (GC)

- Architect
- Owner
- LEED PM

Documentation Checklist

- Online credit form—list of reused and salvaged materials and their corresponding costs and total material cost (excluding labor and equipment)
- Forms or spreadsheets used to keep track of material costs (recommended)

Implementation

- Use line item budget or schedule of values per CSI Division to keep track of costs of materials (excluding labor and equipment) for credits MR C3 to MR C7.
- Only materials in CSI Masterformat 2004 Divisions 3 to 10, Sections 31.60.00, 32.10.00, 32.30.00, and 32.90.00 should be included.
- Exclude mechanical, electrical, and plumbing components, appliances, and equipment.
- Furniture and furnishings should be excluded, unless they are included consistently in MR C3 to MR C7.

- An exemplary performance point (1 point) is available when the value of reused or salvaged materials used in the project is 15% or more of the total materials cost.

Registration / Stage I		Pre-SD

Design Application

Stage II

SD LEED Phase

- Using preliminary project budget, identify items that will contribute towards the achievement of the credit (minimum recycled content value of 10% of total materials cost); establish target recycled content value goals.
- Research available materials and coordinate with contractor to identify materials that will help meet the target goal.
- Incorporate materials with recycled content in building design.

Architec, GC, LEED PM — SD

DD LEED Phase

- Develop *green* specifications, indicating requirements for procuring materials with recycled content. Introduce a separate section on *LEED Requirements*. Include submittal expectations from contractor and provide forms and spreadsheets for contractor and subcontractors to keep track of information.

Architect, LEED PM — DD

CD LEED Phase

- Finalize drawings and specifications indicating requirements for procuring materials with recycled content.

Architect — CD

Construction Application

Stage III

Constn. LEED Review

- Procure materials according to specifications and drawings.
- Organize a pre-construction kick-off meeting to explain LEED credit expectations to subcontractors (LEED PM may assist with this).
- Keep track of information received from subcontractors.
- Compile costs of materials with recycled content as well as total costs of materials for applicable CSI Divisions and submit progress documentation to architect or LEED PM.
- After construction is complete, perform calculations to determine final percentage of recycled content value.
- Upload documentation to LEED Online.
- LEED PM to review uploaded documentation and submit to GBCI for Construction Review.

GC — Construction

LEED PM

O&M

Responsibilities

- General Contractor (GC)

- Architect
- Owner
- LEED PM

Documentation Checklist

- Online credit form—list of product names, manufacturers' names, costs, and percentage post-consumer content and pre-consumer content
- Manufacturers' data documenting recycled content percentages
- Spreadsheet/schedule of values used to keep track of material costs (recommended)

Implementation

- Use line item budget or schedule of values according to CSI Division to keep track of costs of materials (excluding labor and equipment). A single spreadsheet can be created to keep track of materials for credits MR C3 to MR C7.

- Include only materials in CSI Masterformat 2004 Divisions 3 to 10, Sections 31.60.00, 32.10.00, 32.30.00, and 32.90.00.

- Exclude mechanical, electrical, as well as plumbing components, appliances, and equipment.

- Furniture and furnishings should be excluded, unless they are included consistently in MR C3 to MR C7.

- An exemplary performance point (1 point) can be achieved if the total recycled content value is 30% or more.

- For products that are an assembly of multiple materials (like windows), recycled content value can be calculated by weight of the material with recycled content in the assembly.

Registration

Stage I

Design Application

Stage II

SD LEED Phase

- Using preliminary project budget, identify items that will help to contribute towards the achievement of the credit (minimum 10% of total materials cost); establish target regional material value goals.
- Research available materials, and coordinate with contractor to identify materials that will help meet the target goal.
- Incorporate into design materials that have been extracted, harvested, as well as manufactured within 500 miles of the project site.

Architect, GC, LEED PM

SD

DD LEED Phase

- Develop *green* specifications, indicating requirements for procuring regional materials. Introduce a separate section on *LEED Requirements*. Include submittal expectations from contractor and provide forms and spreadsheets for contractor and subcontractors to keep track of information.

Architect, LEED PM

DD

CD LEED Phase

- Finalize drawings and specifications, indicating requirements for procuring regional materials.

Architect

CD

Construction Application

Stage III

Constn. LEED Review

- Procure materials according to specifications and drawings.
- Organize a pre-construction kick-off meeting to explain LEED credit expectations to subcontractors (LEED PM may assist with this).
- Keep track of information received from subcontractors.
- Compile costs of regional materials and total costs of materials. for applicable Divisions and submit progress documentation to architect or LEED PM.
- After construction is complete, perform calculations to determine final percentage of regional materials.
- Upload documentation to LEED Online.
- LEED PM to review uploaded documentation and submit to GBCI for Construction Review.

GC

Construction

LEED PM

O&M

Responsibilities

- General Contractor (GC)

- Architect
- Owner
- LEED PM

LEAD

SUPPORT

Documentation Checklist

- Online credit form—list of product names, manufacturers' names, costs, distance between project and manufacturer, and distance between project and extraction site
- Manufacturers' data documenting material origin and manufacture within 500 mile radius of site
- Spreadsheet/schedule of values used to keep track of material costs (recommended)

CONSTRUCTION SUBMITTAL

Implementation

- Use line item budget or schedule of values according to CSI Division to keep track of costs of materials (excluding labor and equipment). A single spreadsheet can be created to keep track of materials for credits MR C3 to MR C7.

- Include only materials in CSI Masterformat 2004 Divisions 3 to 10, Sections 31.60.00, 32.10.00, 32.30.00, and 32.90.00.

- Exclude mechanical, electrical, and plumbing components, appliances, and equipment.

- Furniture and furnishings should be excluded, unless they are included consistently in MR C3 to MR C7.

- An exemplary performance point (1 point) can be achieved if the total value of regionally harvested, extracted, and manufactured materials is 30% or more.

- For products that are an assembly of multiple materials (like windows), break down the components by weight and use percentage of weight that is regional to calculate the dollar value of the product that contributes towards the credit.

STRATEGIES

TIPS & NOTES

Registration

Stage I

Design Application

Stage II

SD LEED Phase

- Using preliminary project budget, identify items that will help to contribute towards the achievement of the credit (minimum 2.5% of total materials cost).
- Research available materials and coordinate with contractor to identify materials that will help meet the target goal.
- Incorporate rapidly renewable materials into design.

Architect, GC, LEED PM

SD

DD LEED Phase

- Develop *green* specifications, indicating requirements for procuring rapidly renewable materials. Introduce a separate section on *LEED Requirements*. Include submittal expectations from contractor and provide forms and spreadsheets for contractor and subcontractors to keep track of information.

Architect, LEED PM

DD

CD LEED Phase

- Finalize drawings and specifications indicating requirements for procuring rapidly renewable materials.

Architect

CD

Construction Application

Stage III

Constn. LEED Review

- Procure materials according to specifications and drawings.
- Organize a pre-construction kick-off meeting to explain LEED credit expectations to subcontractors (LEED PM may assist with this).
- Keep track of information received from subcontractors.
- Compile costs of rapidly renewable materials and total costs of materials for applicable CSI Divisions and submit progress documentation to architect or LEED PM.
- After construction is complete, perform calculations to determine final percentage of rapidly renewable materials.
- Upload documentation to LEED Online.
- LEED PM to review uploaded documentation and submit to GBCI for Construction Review.

GC

Construction

LEED PM

O&M

Responsibilities

- General Contractor (GC)

- Architect
- Owner
- LEED PM

Documentation Checklist

- Online credit form—list of product names, manufacturers' names, costs, the percentage of each product that is rapidly renewable (by weight), and compliant value
- Manufacturers' data documenting rapidly renewable criteria
- Spreadsheet/schedule of values used to keep track of material costs (recommended)

Implementation

- Use line item budget or schedule of values according to CSI Division to keep track of costs of materials (excluding labor and equipment). A single spreadsheet can be created to keep track of materials for credits MR C3 to MR C7.

- Include only materials in CSI Masterformat 2004 Divisions 3 to 10, Sections 31.60.00, 32.10.00, 32.30.00, and 32.90.00. Exclude mechanical, electrical,plumbing components, appliances and equipment. Furniture & furnishings should be excluded, unless they are included consistently in MR C3 to MR C7.

- Common examples of rapidly renewable materials are bamboo flooring, cotton batt insulation, linoleum flooring, wheatboard cabinetry, as well as coir, jute, and cork flooring.

- An exemplary performance point (1 point) can be achieved if the total value of rapidly renewable materials is 5% or more.

- For products that are an assembly of multiple materials (like windows), break down the components by weight and use percentage of weight that is rapidly renewable to calculate the dollar value of the product that contributes towards the credit.

Registration

Stage I — SD LEED Phase
- Assess the feasibility of incorporating FSC-certified wood products in the building; discuss costs, availability, and schedule impact associated with procuring certified wood.

Architect, Owner — **Pre-SD**

Design Application

Stage II — SD LEED Phase
- Contact local vendors, suppliers, and manufacturers of FSC-certified products and discuss availability and costs.
- Incorporate FSC-certified wood products into design.

Architect, GC, LEED PM — **SD**

Stage II — DD LEED Phase
- Develop *green* specifications, indicating requirements for procuring FSC-certified wood products. Introduce a separate section on *LEED Requirements*. Include submittal expectations from contractor and provide forms and spreadsheets for contractor and subcontractors to keep track of information. Include list of vendors/sources from where FSC-certified wood can be procured and chain of custody information required from vendors.

Architect, LEED PM — **DD**

CD LEED Phase
- Finalize drawings and specifications, indicating requirements for procuring FSC-certified wood.

Architect — **CD**

Construction Application

Stage III — Constn. LEED Review
- Procure materials according to specifications and drawings.
- Obtain and retain chain of custody certificates and invoices from vendors.
- Compile costs of all wood products and products that are FSC-certified.
- After construction is complete, perform calculations to determine final percentage of certified wood.
- Upload documentation to LEED Online.

GC — **Construction**

- LEED PM to review uploaded documentation and submit to GBCI for Construction Review

LEED PM — **O&M**

Responsibilities

- General Contractor (GC)

- Architect
- Owner
- LEED PM

Documentation Checklist

- Online credit form—list of product names, manufacturers' names, costs, and percentage of certified wood
- Copies of vendor invoices for each certified wood product
- Manufacturers' data documenting FSC-Certification or chain of custody (COC)
- Spreadsheet/schedule of values used to keep track of material costs (recommended)

Implementation

- Use line item budget or schedule of values according to CSI Division to keep track of costs of materials (excluding labor and equipment). A single spreadsheet can be created to keep track of materials for credits MR C3 to MR C7.

- Note that this credit is calculated based on percentage of FSC-certified wood costs to total new wood costs. It is not based on total materials cost of the project like credits MR C3 to MR C6. This credit is also applicable only to *new* wood permanently installed in the project. Reused or salvaged wood products will not count towards this credit.

- An exemplary performance point (1 point) can be achieved if the total value of FSC-certified wood is 95% or more of the total new wood in the project.

- This credit is easier to achieve in projects with limited wood use. For instance, an office building with mostly steel and concrete will have limited wood-based products, and therefore, procuring FSC-certified wood for 50% of total wood used in the project is financially viable. However, in an all wood framed building, procuring 50% FSC-certified wood may be cost prohibitive and might also impact construction schedule.

Sample Documents

The following pages present excerpts from some *Materials and Resources* credit documents that were submitted to USGBC by the teams of the case study projects discussed in Chapter 8. Sample documents for the following credit have been included.

MR Credit 2, Construction Waste Management

46 Blackstone Street
4th Floor - South
Cambridge, MA 02139
green.harvard.edu

HARVARD UNIVERSITY DIVINITY SCHOOL
ROCKEFELLER HALL

MR Credit 2 – Construction Waste Management

USGBC REVIEW - CONSTRUCTION APPLICATION

The LEED Submittal Template has been provided stating that the project has diverted 456.03 tons (90.42 %) of on-site generated construction waste from landfill. Calculations have been provided to document the waste types and receiving agencies for recycled materials. A narrative has been provided describing the project's Construction Waste Management Plan. However, it appears that the "mixed debris" was sent to a commingled sorting facility. It is unclear whether the amounts of "mixed debris" were calculated taking into account the average annual recycling rate of the sorting facility or whether these are the actual measurements of the amount of the mixed debris diverted. The LEED-NC v2.1 CIR Ruling dated 12/02/2005 indicates that "the average annual recycling rate for the specific sorting facility is acceptable as long as the facility's method of recording and calculating the recycling rate is regulated by a local or state government authority."

USGBC TECHNICAL ADVICE

Please provide a narrative to clarify how the amounts of diverted "mixed debris" were determined. Provide the annual recycling rate for the specific sorting facility along with confirmation that it is regulated by a local or state government, or documentation showing the amount of diverted materials sorted from the "mixed debris." This could include copies of the weight tickets, volumes, etc.

RESPONSE

We apologize for not providing more detailed information in the original submittal. On the original submittal, we combined the total amount (tons) of diverted/recycled materials by each hauler. If a hauler processed more than one type of material, the material description listed the total combined tons of materials as "mixed debris".

Based on your technical advice, we have revised the MRc2 LEED Letter Template to break out, listing individually, all materials originally combined into the general description of "mixed debris". However, for two of the sorting facilities, Jet-A-Way Inc and Waste Management, documentation was not kept showing the breakdown of how "mixed debris" was recycled. For this reason, the tonnage of construction waste sent to those two facilities is counted as 100% landfilled on the LEED template (the amount of recycled material from those two facilities was removed from the "Diverted" section of the template and added to the "landfill" section.) Even with this change, the project achieves a waste diversion rate of 84.952%, and still achieves both MRc2 points.

In addition to revising the Template, we have attached documentation to support the revisions made.

Thank you for considering this re-submittal,

Jesse Foote

Jesse Foote
Manager, Green Campus Building Services

Sample Document: MR Credit 2, Construction Waste Management (page 1 of 3).

Credit: Harvard University, Office of Sustainability and Shawmut Construction.

CERTIFICATE
OF DONATION AND RECYCLING

Awarded to

Harvard University Divinity School

It is hereby certified that the above mentioned successfully donated 8.58 tons of surplus furnishings on the date in June, 2007. The materials collected were recycled through donation for re-use and shipped to domestic & international relief efforts.

__6-27-2007__	*Stacey Clark*
DATE	INSTITUTION RECYCLING NETWORK

Sample Document: MR Credit 2, Construction Waste Management (page 2 of 3).
Credit: Harvard University, Office of Sustainability and Shawmut Construction.

CONSTRUCTION WASTE MANAGEMENT
MONTHLY PROJECT PROGRESS REPORT

SHAWMUT DESIGN

Project Title:	Harvard-Rockefeller Hall		Shawmut Superintendent	
Project Size:	28,000 Square Feet		CWM Contractor:	Waste Solutions, Inc.
Project Address:	45 Francis Street, Cambridge, MA		Month/Year:	September 2007-May 2008
Project Number:	70262			

Container or ticket number	Haul Date	Total Tonnage (sum of concrete, metal, wood, other and residual)	Concrete Tonnage (diverted)	Metal Tonnage (diverted)	Wood Tonnage (diverted)	Gypsum Tonnage	Paper and Card-board Products Tonnage (diverted)	Total Diverted Tonnage	Residual or Trash Tonnage (landfilled)	% Recycled/ Diverted Material	Name and location material was sent to. (If Other Recycled Material please also include description of material (IE Clean Green, Drywall, Glass, Etc.)	Comments. If landfilled, please explain why material was not diverted. If large amount appears as "Other" please explain contents.
197459	9/24/2007	4.04	0.44	0.48	2.22	0.00	0.08	3.23	0.81	80.00%	Stoughton Recycling	
198361	10/2/2007	4.15	0.42	0.42	2.70	0.00	0.21	3.74	0.42	90.00%	Pond View Recycling	
199929	10/15/2007	7.56	0.83	0.91	4.16	0.00	0.15	6.05	1.51	80.00%	Stoughton Recycling	
199603	10/29/2007	6.63	0.66	0.66	3.98	0.00	0.33	5.64	0.99	85.00%	LL&S Wood Recycling	
200441	11/6/2007	6.85	0.75	0.82	3.77	0.00	0.14	5.48	1.37	80.00%	Stoughton Recycling	
202886	11/16/2007	7.13	0.78	0.86	3.92	0.00	0.14	5.70	1.43	80.00%	Stoughton Recycling	
201802	11/28/2007	3.28	0.36	0.39	1.80	0.00	0.07	2.62	0.66	80.00%	Stoughton Recycling	
203009	12/7/2007	12.75	1.40	1.53	7.01	0.00	0.26	10.20	2.55	80.00%	Stoughton Recycling	
204358	12/13/2007	6.82	0.75	0.82	3.75	0.00	0.14	5.46	1.36	80.00%	Stoughton Recycling	
204122	12/18/2007	5.79	0.58	0.58	3.47	0.00	0.29	4.92	0.87	85.00%	LL&S Wood Recycling	
204477	12/28/2007	10.06	1.11	1.21	5.53	0.00	0.20	8.05	2.01	80.00%	Stoughton Recycling	
203544	1/7/2008	4.50	0.50	0.54	2.48	0.00	0.09	3.60	0.90	80.00%	Stoughton Recycling	
203484	1/11/2008	5.06	0.56	0.61	2.78	0.00	0.10	4.05	1.01	80.00%	Stoughton Recycling	
206526	1/18/2008	8.35	0.92	1.00	4.59	0.00	0.17	6.68	1.67	80.00%	Stoughton Recycling	
199187	1/23/2008	3.71	0.41	0.45	2.04	0.00	0.07	2.97	0.74	80.00%	Stoughton Recycling	
206597	1/29/2008	3.15	0.35	0.38	1.73	0.00	0.06	2.52	0.63	80.00%	Stoughton Recycling	
206445	2/4/2008	3.74	0.41	0.45	2.06	0.00	0.07	2.99	0.75	80.00%	Stoughton Recycling	
205140	2/8/2008	3.56	0.39	0.43	1.96	0.00	0.07	2.85	0.71	80.00%	Stoughton Recycling	
208826	2/16/2008	6.84	0.75	0.82	3.76	0.00	0.14	5.47	1.37	80.00%	Stoughton Recycling	
208566	2/22/2008	2.60	0.26	0.26	1.56	0.00	0.13	2.21	0.39	85.00%	LL&S Wood Recycling	
207518	2/29/2008	4.21	0.46	0.51	2.32	0.00	0.08	3.37	0.84	80.00%	Stoughton Recycling	
207826	3/7/2008	4.55	0.50	0.55	2.50	0.00	0.09	3.64	0.91	80.00%	Stoughton Recycling	
209711	3/13/2008	4.95	0.54	0.59	2.72	0.00	0.10	3.96	0.99	80.00%	Stoughton Recycling	
210457	3/19/2008	4.14	0.46	0.50	2.28	0.00	0.08	3.31	0.83	80.00%	Stoughton Recycling	
210713	3/26/2008	9.11	1.00	1.09	5.01	0.00	0.18	7.29	1.82	80.00%	Stoughton Recycling	
211263	4/2/2008	5.09	0.56	0.61	2.80	0.00	0.10	4.07	1.02	80.00%	Stoughton Recycling	
211219	4/8/2008	8.63	0.95	1.04	4.75	0.00	0.17	6.90	1.73	80.00%	Stoughton Recycling	
209491	4/11/2008	3.21	0.35	0.39	1.77	0.00	0.06	2.57	0.64	80.00%	Stoughton Recycling	
211246	4/16/2008	4.25	0.47	0.51	2.34	0.00	0.09	3.40	0.85	80.00%	Stoughton Recycling	
211724	4/23/2008	7.49	0.82	0.90	4.12	0.00	0.15	5.99	1.50	80.00%	Stoughton Recycling	
212061	5/1/2008	8.49	0.93	1.02	4.67	0.00	0.17	6.79	1.70	80.00%	Stoughton Recycling	
211132	5/9/2008	8.66	0.95	1.04	4.76	0.00	0.17	6.93	1.73	80.00%	Stoughton Recycling	
214561	5/15/2008	5.56	0.56	0.56	3.34	0.00	0.28	4.73	0.83	85.00%	LL&S Wood Recycling	
213933	5/21/2008	3.37	0.34	0.34	2.02	0.00	0.17	2.86	0.51	85.00%	LL&S Wood Recycling	
215863	5/23/2008	4.00	0.44	0.48	2.20	0.00	0.08	3.20	0.80	80.00%	Stoughton Recycling	
215920	5/27/2008	7.70	0.85	0.92	4.24	0.00	0.15	6.16	1.54	80.00%	Stoughton Recycling	
214129	5/29/2008	3.07	0.34	0.37	1.69	0.00	0.06	2.46	0.61	80.00%	Stoughton Recycling	

© Waste Solutions Inc.

6/27/2008

Sample Document: MR Credit 2, Construction Waste Management (page 3 of 3).
Credit: Harvard University, Office of Sustainability and Shawmut Construction.

Construction Waste Management Plan

Project: Great River Energy (GRE), Maple Grove, MN
Company: McGough Construction
Designated Recycling Coordinator: Amanda Nonnemacher, McGough Construction

Waste Management Goals:

➤ This project will recycle or salvage for reuse a **minimum of 75%** by weight of the waste generated on-site.

➤ Each construction trailer and break area will be provided with a recycling bin for paper and a recycling bin for cans and bottles.

Communication Plan:

➤ Waste prevention and recycling activities will be discussed at each site safety/LEED meeting.

➤ As each new subcontractor comes on-site, they will be presented with a copy of the Waste Management Plan and given a tour of the recycling areas.

➤ The subcontractors are responsible for separating and recycling/disposing of their own materials. Subcontractors will be expected to make sure their crews comply with the Waste Management Plan.

➤ All recycling containers will be clearly labeled.

➤ Lists of acceptable/unacceptable materials will be posted throughout the site.

➤ Shamrock Recycling will provide all dumpsters and oversight of waste management progress. Shamrock will submit monthly Waste Management Progress Reports showing waste recycled and total waste generated.

Expected Project Materials, Disposal, and Handling:

The following chart identifies waste materials expected on this project, their disposal method, and handling procedures.

Material	Disposal Method	Handling Procedure
Concrete	Recycle – Rehbein Companies	Deposit in concrete bin
Clean Wood Scrap	Scraps reused for formwork, etc. Remaining recycled – Environmental Wood Supply	Stack reusable pieces in boneyard. Place unusable clean wood in wood recycling dumpster
Scrap Metal	Recycle – SCI Recycling Service	Deposit all metals in metal dumpster
Drywall	Recycle – Shamrock Recycling	Deposit all drywall in drywall dumpster
Cardboard & Paper	Recycle – Pioneer Paper	Break down and deposit all in cardboard and paper dumpster
Steel and Aluminum Cans & Plastic and Glass Bottles	Recycle – Recycle America Alliance	Deposit in steel and aluminum cans & plastic and glass bottles dumpster
Asphalt at Existing Road	Recycle – Rehbein Companies	Hauled off-site for recycling
Construction Debris	Recycle – Shamrock Recycling	Dispose in construction debris dumpster. Sorted at recycling facility.
Carpet	Recycle- Shaw Carpet	All carpet scraps sent back to Shaw
Access Floor Panels	Recycle- SCI Recycling Service	Deposit in access floor panel Dumpster

Sample Document MR Credit 2, Construction Waste Management Plan.

Credit: McGough Construction

CREDIT IMPLEMENTATION— INDOOR ENVIRONMENTAL QUALITY

Introduction

The section provides a credit-by-credit framework to implement the requirements of credits in the *Indoor Environmental Quality* category of the LEED-NC v3 rating system. It also identifies responsible team members, includes a documentation checklist, and provides tips and notes for each credit.

This category of credits is written with the intent of improving the overall indoor environmental quality of buildings. The EQ credits reward projects that provide improved ventilation, monitor outdoor air, are designed with occupant thermal comfort in mind, and provide individual controllability of lighting and thermal comfort systems. It also encourages the use of strategies to maximize daylighting and access to views in interior spaces. This section includes implementation worksheets for the following credits.

EQ Prereq 1, Minimum Indoor Air Quality Performance

EQ Prereq 2, Environmental Tobacco Smoke (ETS) Control

EQ Credit 1, Outdoor Air Delivery Monitoring

EQ Credit 2, Increased Ventilation

EQ Credit 3.1, Construction IAQ Management Plan—During Construction

EQ Credit 3.2, Construction IAQ Management Plan—Before Occupancy

EQ Credit 4.1, Low-Emitting Materials—Adhesives and Sealants

EQ Credit 4.2, Low-Emitting Materials—Paints and Coatings

EQ Credit 4.3, Low-Emitting Materials—Flooring Systems

EQ Credit 4.4, Low-Emitting Materials—Composite Wood and Agrifiber Products

EQ Credit 5, Indoor Chemical and Pollutant Source Control

EQ Credit 6.1, Controllability of Systems—Lighting

EQ Credit 6.2, Controllability of Systems—Thermal Comfort

EQ Credit 7.1, Thermal Comfort—Design

EQ Credit 7.2, Thermal Comfort—Verification

EQ Credit 8.1, Daylight and Views—Daylight

EQ Credit 8.2, Daylight and Views—Views

Registration

Stage I — Mech. — Pre-SD

- Discuss with owner and/or building management any special ventilation requirements based on building type, occupant density, zone categories, and occupant needs.

Design Application

Stage II

SD LEED Phase — Mech. — SD

- Design mechanical ventilation system according to requirements of Sections 4 through 7 of ASHRAE standard 62.1-2007 or local code, whichever is more stringent.
- Design natural ventilation system according to the requirements of Section 5.1 of of ASHRAE standard 62.1-2007.
- Coordinate with architect in the design and sizing of operable windows to ensure natural ventilation requirements are met.

DD LEED Phase — Mech. — DD

- Incorporate ventilation rate procedure and/or natural ventilation calculations in drawings.

CD LEED Phase — LEED PM, Mech. — CD

- Finalize drawings and specifications showing the ventilation system design, components, calculations, and specifications.
- Upload required documentation to LEED Online.

- LEED PM to review uploaded documentation and submit to GBCI for Design Review.

Construction Application

Stage III

Constn. LEED Review — Construction

— O&M

- Mechanical Engineer (Mech.)

- Architect
- Owner
- LEED PM

- Online credit form
- Drawings or documents demonstrating compliance with ventilation requirements of ASHRAE 62.1-2007

- Multifamily buildings that typically use natural ventilation for residential units may need to pay special attention to the design of the units and their windows (size, location, etc.) to ensure that the natural ventilation requirements outlined in Section 5.1 of ASHRAE 62.1-2007 are met.

Registration

Stage I — *Pre-SD* — LEED PM, Architect

- Discuss with owner and building management their policy with respect to smoking for the building and site—whether the building and property will be non-smoking or whether designated smoking areas will be provided.

Design Application

Stage II

SD LEED Phase — *SD* — Architect, Mech.

- Determine applicable compliance path and option for the project.
- Incorporate requirements into design according to applicable case and option:
  Case 1, Opt 1: Provide signage to prohibit smoking in the building and within 25 feet of entries, outdoor air intakes, and operable windows
  Case 1, Opt 2: Provide signage prohibiting smoking in the building except in those areas designated as *smoking rooms*. Design designated smoking rooms to contain, capture, and remove ETS from building, with direct exhaust to outdoors
  Case 2: Provide signage as Case 1, Opt 1 and design residential units according to prerequisite's requirements. Weatherstrip exterior doors and doors leading to common hallways and seal penetrations in walls.

DD LEED Phase — *DD* — Architect, Mech.

- Incorporate signage details, exhaust designs, penetration sealing details, and other details into the design drawings.
- LEED PM to work with building management to develop a written policy regarding ETS control.

CD LEED Phase — *CD* — LEED PM, Architect

- Finalize drawings and specifications, reflecting the requirements of the applicable option, for implementation by construction team.
- Complete required documentation and upload to LEED Online.

- LEED PM to finalize ETS policy document with building management.
- LEED PM to review uploaded documentation and submit to GBCI for Design Review.

Construction Application

Stage III — *Constn. LEED Review* — *Construction* — GC, Third-party tester

- Contractor to ensure proper sealing of smoking rooms, as identified in the construction drawings.
- For Case 1, Opt 1: Verify performance of smoking rooms by testing the differential pressure between the smoking room with respect to adjacent area and in each vertical chase, with the doors of the smoking room closed.
- For Case 2: Conduct blower door test to demonstrate acceptable sealing of the residential units.

O&M

Responsibilities

- Architect

- Mechanical Engineer (Mech.)
- Owner
- Building Management (Bldg. Mgmt.)
- LEED PM

Documentation Checklist

- Online credit form
- Drawings demonstrating implementation of smoking policy—designated rooms, signage details, ventilation information, weatherstripping, prohibited areas, etc.
- Record of testing data for smoking rooms or residential units to demonstrate no transfer of ETS to adjacent spaces
- Document describing the ETS policy for the building

Implementation

- Residential buildings that need to demonstrate acceptable sealing by means of a blower door test can use the progressive sampling methodology defined in Chapter 4 (Compliance through Quality Construction) of the Residential Manual for Compliance with California's 2001 Energy Efficiency Standards.

Registration

Stage I — LEED PM — Pre-SD

- Discuss with owner, building facility managers, and mechanical engineer the feasibility of providing carbon dioxide monitors and outdoor airflow measurement devices according to the credit requirements.

Design Application

Stage II

SD LEED Phase — Mech. — SD

- Calculate number of carbon dioxide monitors needed, based on the credit's requirement of providing monitors in densely occupied spaces (25 people or more per 1000 sf) and between 3 and 6 feet above the floor.
- Research options for outdoor airflow measurement devices and calculate number required to measure outdoor airflow for non-densely occupied spaces.
- Ensure device is capable of measuring airflow with an accuracy of plus or minus 15% of the design minimum outdoor air rate as defined by ASHRAE 62.1-2007.
- Identify locations of carbon dioxide monitors and outdoor airflow measurement devices for mechanically ventilated spaces and carbon dioxide monitors in naturally ventilated spaces.

DD LEED Phase — Mech. — DD

- Update type, numbers, and location of monitors as design progresses, if needed.

CD LEED Phase — LEED PM, Mech. — CD

- Finalize drawings and specifications, reflecting the location, type, number of carbon dioxide monitors and outdoor airflow measurement devices.
- Complete all required documentation and upload to LEED Online.
- LEED PM to review uploaded documentation and submit to GBCI for Design Review.

Construction Application

Stage III

Constn. LEED Review — CxA, GC — Construction

- Contractor to ensure proper installation of carbon dioxide monitors and airflow measurement devices according to drawings and specifications.
- CxA to commission monitors and airflow measurement devices as part of ventilation system commissioning to ensure proper installation and operation.

Bldg. Mgmt. — O&M

- Operate and maintain monitors; implement corrective action if required.

Responsibilities

- Mechanical Engineer (Mech.)

- Architect
- Owner
- Building Management (Bldg. Mgmt.)
- LEED PM

Documentation Checklist

- Online credit form
- Drawings showing carbon dioxide and airflow monitors—number, type, location, schedules, etc.
- Manufacturer's data describing the monitors' performance characteristics

Implementation

Registration

Stage I · Pre-SD · Mech.

- Discuss with owner and/or building management any special ventilation requirements based on building type, occupant density, zone categories, and occupant needs.
- Discuss any issues related to increasing ventilation rates beyond ASHRAE 62.1-2007.

Design Application

Stage II

SD LEED Phase · SD · Mech., Architect

- Case 1: Design mechanical ventilation system to increase breathing zone outdoor air ventilation rates by at least 30% above the minimum rates required by ASHRAE 62.1-2007, as determined by EQ P1.
- Case 2: Design natural ventilation system to meet the recommendations set forth in the Carbon Trust *Good Practice Guide 237*.
- Coordinate with architect in the design and sizing of operable windows to ensure natural ventilation requirements are met.

DD LEED Phase · DD · Mech.

- Incorporate ventilation rate procedure calculations for mechanically ventilated spaces, and/or diagrams/calculations/analytic models for naturally ventilated spaces, in drawings.

CD LEED Phase · CD · LEED PM, Mech.

- Finalize drawings, showing the ventilation system design, components, calculations, and specifications.
- Upload required documentation to LEED Online.
- LEED PM to review uploaded documentation and submit to GBCI for Design Review.

Construction Application

Stage III · Constn. LEED Review · Construction · O&M

- Mechanical Engineer (Mech.)

- Architect
- Owner
- LEED PM

- Online credit form
- Case 1:
  Completed ASHRAE 62.1-2007 calculator or a local version of the calculator that is at least as stringent as the ASHRAE version
- Case 2 Opt 1:
  Diagrams and calculations to show that the design of the natural ventilation systems meets the recommendations set forth in *Natural Ventilation in Non-Domestic Buildings* (CIBSE Applications Manual 10, 2005).
- Case 2 Opt 2:
  Summary report from a macroscopic, multi-zone, analytic model to predict that room-by-room airflows will effectively naturally ventilate, defined as providing the minimum ventilation rates required by ASHRAE 62.1-2007, for at least 90% of occupied spaces
- Any additional documents to support any alternative compliance path approach (Optional)
- Any additional documentation to support any special circumstances (Optional)

- For mechanical ventilation, use ASHRAE User Manual and EQ P1 calculators available for download from LEED Online.

- For natural ventilation, use *Natural Ventilation in Non-Domestic Buildings* (Chartered Institution of Building Services Engineers (CIBSE) Applications Manual 10, 2005) to determine opening sizes for operable windows, trickle vents, and louvers. Or use a macroscopic, multizone analytic model to predict room-by room airflow rates.

- Natural Ventilation may not be an effective strategy in all project scenarios or locations. Evaluate potential indoor air quality issues that may arise because of traffic exhaust, polluting industries, or waste management sites around the building. Use the flow diagram process of the CIBSE Applications Manual 10, to determine whether natural ventilation is an effective strategy for the project. Diagram is also available in the *LEED Reference Guide*.

- For mechanical ventilation, increasing outdoor air rates may impact the design and size of the HVAC system. This may lead to greater energy use and an increase in operating costs. Evaluate all impacts before designing for increased ventilation rates.

Registration

Stage I

Pre-SD

Design Application

Stage II

SD LEED Phase

- Discuss with contractor, architect, and owner the various measures to be implemented on-site during construction to manage indoor air quality. Measures include, but are not limited to, protecting HVAC systems and ducts, using low-VOC paints, regular housekeeping, replacing filtration media prior to occupancy, etc.

LEED PM

SD

DD LEED Phase

- Incorporate in specifications a section on indoor air quality (IAQ) management, identifying responsibilities of the contractor. A sample IAQ plan as well as forms and checklists to record implementation methods may be provided. LEED PM may assist architect with this item.

Architect, LEED PM

DD

CD LEED Phase

- Contractor to finalize IAQ management plan according to project schedule and sequence of installation.

LEED PM, GC

CD

Construction Application

Stage III

Constn. LEED Review

- Provide copies of IAQ plan to subcontractors and also post on project site.
- Inform subcontractors of the IAQ management procedures to be followed.
- Implement measures identified in the IAQ management plan—protect HVAC, install low VOC materials, isolate work areas, use MERV 8 filters, as well as perform housekeeping and schedule installation of materials with impact on IAQ in mind.
- Document measures implemented via photographs and inspection checklists.
- Submit progress documentation to architect or LEED PM.
- Compile all photographs and inspection logs and upload to LEED Online along with other required documentation.
- LEED PM to review uploaded documentation and submit to GBCI for Construction Review.

GC

LEED PM

Construction

O&M

Responsibilities

- General Contractor (GC)

- Architect
- Owner
- LEED PM

Documentation Checklist

- Online credit form
- Construction indoor air quality management plan document showing measures implemented during construction
- Photographs documenting implementation of each of the measures during construction with dates and description of activity
- Inspection reports and/or checklists with corrective action taken

Implementation

- The credit references the Sheet Metal and Air Conditioning National Contractors Association (SMACNA) IAQ Guidelines for Occupied Buildings Under Construction, 2nd edition, 2007, ANSI/SMACNA 008-2008 (Chapter 3). Use these SMACNA guidelines to develop the IAQ management plan.

- If used during construction, permanently installed air handlers are required to have filtration media with a minimum efficiency reporting value (MERV) of 8. These filters should be replaced prior to occupancy, and there is no specific MERV rating requirement for the new filters. However, if a project is also targeting EQ C5, then the new filtration media should have a minimum MERV rating of 13.

- It will be helpful to designate an individual from the contractor's team to be the LEED champion overseeing and documenting all LEED-related credit implementation procedures on-site.

Registration

Stage I / Pre-SD

Design Application

Stage II

SD LEED Phase
- Discuss with contractor, architect, and owner the options available to meet the credit requirements:
  Opt 1: Perform building flush-out
  Opt 2: Perform indoor air quality (IAQ) testing
- Evaluate impacts of both options on schedule and costs; determine appropriate option.
- If Opt 1 is chosen, then incorporate building flush-out period in construction schedule.
- If Opt 2 is chosen, engage services of third-party testing agency.

GC, Owner, LEED PM / SD

DD LEED Phase
- Incorporate in specifications a section on indoor air quality (IAQ) management, clearly identifying the responsibilities of the contractor. A sample IAQ plan as well as forms and checklists to record implementation methods may be provided. LEED PM may assist architect with this item.

Architect, LEED PM / DD

CD LEED Phase
- Contractor to finalize IAQ management plan, highlighting the procedures for building flush-out or IAQ testing depending on option chosen.

LEED PM, GC / CD

Construction Application

Stage III

Constn. LEED Review
- Replace any filtration media from air handlers that were used during construction.
- Opt 1: Perform building flush-out according to credit requirements.
- Opt 1: Document dates, occupancy, outdoor air delivery rates, internal temperatures, and any other special considerations during flush-out procedure.
- Opt 2: Coordinate with third-party testing agency to perform IAQ testing according to the sampling protocol identified in the credit requirements. If a test sample does not meet the requirement, then flush-out that space with outdoor air and retest the space.
- Opt 2: Obtain copy of testing report showing details about the test samples, contaminants tested, and any corrective measures taken.
- Upload all documentation to LEED Online.
- LEED PM to review uploaded documentation and submit to GBCI.

LEED PM, GC / Construction

O&M

Responsibilities

- General Contractor (GC)

- Architect
- Owner
- LEED PM

Documentation Checklist

- Online credit form
- Construction indoor air quality management plan highlighting procedures followed before occupancy
- Option 1: Flush-out procedure:
  - Record of dates, occupancy, outdoor air delivery rates, internal temperature, humidity maintained during implementation of flush-out
- Option 2: IAQ testing
  - Testing report verifying that all required contaminants are accounted for and reported in the correct unit of measure. Also include any failed samples, corrective action, and retesting results.

Implementation

- Both options need to be implemented after installation of all interior finishes.

- For the flush-out option, it may take upto 28 days to reach the volume of air required by the credit (14,000 cu feet per square foot of floor area).

- For the IAQ testing option to be successful, ensure that there are no product substitutions and all interior finishes—paints, coatings, adhesives, sealants, carpets, etc.—are low-emitting.

Registration

Stage I

SD LEED Phase

Pre-SD

Design Application

Stage II

SD LEED Phase

- Research and select interior finish products according to the credit requirements:
  EQ C4.1: Adhesives and sealants not to exceed specified VOC limits
  EQ C4.2: Paints and coatings not to exceed specified VOC limits
  EQ C4.3: Carpets to be CRI certified and hard surface flooring to be
  FloorScore compliant
  EQ C4.4: Composite wood and agrifiber products must not contain added
  urea-formaldehyde resins

Architect, Interior Designer

SD

DD LEED Phase

- Incorporate requirements for all items covered under these credits into the specifications manual. Include standards to be met, VOC limits to be followed, and information to be provided by subcontractors.

Architect, LEED PM

DD

CD LEED Phase

- Finalize specifications, clearly identifying all credit requirements.
- Prepare an *instructions to bidders or subcontractors* document, clearly highlighting information required from subcontractors, and a schedule for submission of information.

LEED PM, Architect

CD

Construction Application

Stage III

Constn. LEED Review

- Organize a pre-construction kick-off meeting to discuss the requirements of these credits and submittal expectations from subcontractors. (LEED PM may assist with this.)
- Distribute lists and post allowable VOC limit levels for all adhesives and sealants to be used on-site.
- Ensure products are procured according to specifications without substitutions.
- Compile information provided by subcontractors.
- Upload all documentation to LEED Online.

LEED PM, GC

Construction

- LEED PM to review uploaded documentation and submit to GBCI.

O&M

Responsibilities

- General Contractor (GC)

- Architect
- Interior Designer
- LEED PM

Documentation Checklist

- Online credit forms
- EQ C4.1
  - List of each indoor adhesive, sealant, sealant primer, and aerosol adhesive used in the project along with manufacturer's name, product name, and VOC data (g/L)
- EQ C4.2
  - List of each indoor paint and coating used in the project along with manufacturer's name, product name, and specific VOC data (g/L)
- EQ C4.3
  - List of each carpet, carpet cushion, and carpet adhesive along with manufacturer's product name, and specific VOC data (g/L)
  - List of each hardsurface flooring product, tile setting ashesive, finish, and grout name, along with manufacturer's name, product name and specific VOC data (g/L)
- EQ C4.4
  - List of each composite wood and agrifiber product installed in the building interior along with manufacturer's data or cut sheet, indicating the product does not have any added urea-formaldehyde resins
  - For all credits: Manufacturer's cut sheets, MSD sheets, and test reports to support information

Implementation

- All these credits are applicable only to products that are applied on-site and used in the interior of the weatherproofing system. This includes all interior floors, walls, ceilings, interior furnishings, suspended ceiling systems and materials above those suspended systems, ventilation system components, materials inside wall cavities, ceiling cavities, floor cavities, horizontal or vertical chases and caulking materials for windows, as well as ceiling or wall insulation.

- Subcontractors play a critical role in the implementation of these credits; they should:
  - procure materials according to the specifications
  - bring attention to any item that is not available or exceeds the allowable VOC limit
  - seek approval from general contractor for any substitutions
  - provide data requested for each product: product, name, manufacturer's name, VOC data (g/L) of the product, and cut sheets or MSD sheets as supporting documents

- Note that there is no percentage of materials that needs to meet these requirements. All applicable products installed inside the building must meet the low-emitting criteria set forth in these credits.

Registration

Stage I

Pre-SD

Design Application

Stage II

SD LEED Phase

- Decide on type and locations of permanent entryways systems.
- Incorporate these systems at regularly used exterior entrances; ensure at least 10 feet of space is available from building entrance to the interior.
- Identify locations of areas where chemicals or high-volume copy, fax, and print equipment will be used. Consider stacking or locating them close to each other to minimize exhaust ductwork; provide enclosed rooms.
- Locate chemical storage and mixing areas away from occupant work areas.

Architect, Mech. · *SD*

DD LEED Phase

- Architect to ensure physical isolation of chemical and equipment rooms by providing deck to deck partitions or sealed gypsum board partitions.
- Mechanical engineer to include in drawings MERV 13 filters for the building ventilation systems; provide exhaust systems for chemical and equipment rooms; and provide separate drainage piping for rooms where chemicals are mixed and disposed of.

Architect, Mech. · *DD*

CD LEED Phase

- Finalize drawings and specifications with all the credit requirements—entryway system, isolation and exhaust of equipment rooms, MERV 13 filters, and proper disposal of chemical contaminants.
- Upload all required documentation to LEED Online.

- LEED PM to review uploaded documentation and submit to GBCI for Design Review.

LEED PM, Architect · *CD*

Construction Application

Stage III

Constn. LEED Review

Construction

O&M

Responsibilities

LEAD

- Architect

SUPPORT

- Mechanical Engineer (Mech.)
- Interior Designer
- LEED PM

Documentation Checklist

DESIGN SUBMITTAL

- Online credit forms
- Drawings showing location, size, and type of entryway system; table listing all entryway systems; entryway system specifications (manufacturer's data or cut sheets)
- Drawings showing locations of exhaust systems and rooms requiring separation or isolation measures
- Mechanical schedules or documentation showing MERV 13 filters for air handling units
- Manufacturer's data or cut sheets of the MERV 13 filters

Implementation

STRATEGIES

TIPS & NOTES

- Mid-rise, multifamily buildings and other projects that use low-capacity packaged air handling systems may not be able to use MERV 13 filters because the size of the filters may create a pressure drop across the system. This credit may not be practically feasible in such scenarios.

- Convenience printers, fax, and copy machines are not required to be isolated; the credit is applicable only to high-volume equipment rooms.

Registration

Stage I

Pre-SD

Design Application

Stage II

SD LEED Phase

- Design lighting layout and controls to meet occupant needs and tasks.
- Design to provide individual control of lighting to 90% of the building occupants.
- Design to provide controls in shared multi-occupant spaces to enable adjustments that meet group needs.

Architect, Lighting   SD

DD LEED Phase

- Coordinate with electrical engineer for power and circuitry requirements.

Lighting, Electrical   DD

CD LEED Phase

- Finalize lighting layout, type, and controls in drawings and specifications; perform calculations to ensure 90% of the total individual occupant spaces have lighting control.
- Complete required documentation and upload to LEED Online.

- LEED PM to review uploaded documentation and submit to GBCI.

LEED PM, Lighting   CD

Construction Application

Stage III

Constn. LEED Review

- Contractor to install fixtures and controls according to drawings and specifications.
- CxA to verify proper calibration, installation, and operation during the commissioning process.

CxA, GC   Construction

O&M

Responsibilities

- Lighting Consultant (Lighting)

- Electrical Engineer (Electrical)
- Architect
- Interior Designer
- LEED PM

Documentation Checklist

- Online credit form—list space name, type of space (shared or individual), lighting control type, and description; also provide total quantity of spaces—individual, shared classroom, etc.
- Floor plan drawings with interior layout of individual and shared work areas, showing location, zoning, and type of lighting controls

Implementation

- Although multilevel lighting or dimmers are recommended for meeting individual task needs because they provide maximum level of flexibility, for the purpose of achieving this credit, a simple on-off switch control is sufficient to satisfy the credit requirement.

Registration

Stage I   Pre-SD

Design Application

Stage II

SD LEED Phase

- Evaluate options of providing maximum individual control over thermal comfort for occupants. Consider providing individual thermostats, diffusers, or radiant panels, or using an underfloor air distribution system.
- Consider incorporating operable windows as a means of providing individual thermal comfort to occupants close to building perimeter.

Architect, Mech.   SD

DD LEED Phase

- Identify types of thermal comfort controls that will be provided and show their locations in the drawings. Coordinate with architect and building management to ensure that the controls are at appropriate locations and can be easily operated by future building occupants.

Architect, Mech.   DD

CD LEED Phase

- Finalize drawings showing location, number, and type of thermal comfort control systems and the areas they serve.
- Complete required documentation and upload to LEED Online.

- LEED PM to review uploaded documentation and submit to GBCI.

LEED PM, MEch.   CD

Construction Application

Stage III

Constn. LEED Review

- Contractor to install controls according to drawings and specifications.
- CxA to verify proper calibration, installation and operation during the commissioning process.

CxA, GC   Construction

O&M

Responsibilities

LEAD

- Mechanical Engineer (Mech.)

SUPPORT

- Electrical Engineer
- Architect
- Interior Designer
- LEED PM

Documentation Checklist

DESIGN SUBMITTAL

- Online credit form—list space name, type of space (shared or individual), and thermal comfort control type and description; also provide total quantity of spaces—individual, shared, classroom, etc.
- Floor plan drawings with interior layout of individual and shared work areas showing location, zoning, and type of thermal comfort controls

Implementation

STRATEGIES

TIPS & NOTES

- For residential projects, programmable thermostats and operable windows in each residential unit would help satisfy the individual thermal comfort control requirement.

Registration

Stage I

Pre-SD   CxA, Owner

- Incorporate thermal comfort requirements in owner's project requirements (OPR) to demonstrate that the building HVAC systems and envelope meet the requirements of ASHRAE 55-2004, Thermal Environmental Conditions for Human Occupancy.

Design Application

Stage II

SD LEED Phase   SD   Mech., Architect

- Incorporate requirements of ASHRAE 55-2004 in the basis of design (BOD) document.
- Design building envelope and HVAC systems based on the thermal comfort criteria set forth by ASHRAE 55-2004; evaluate both natural and mechanical ventilation options to decide best approach.

DD LEED Phase   DD   Mech.

- Develop drawings to include all components of the HVAC system design.

CD LEED Phase   CD   LEED PM, Mech.

- Finalize drawings showing the mechanical design.
- Use Section 6.1.1 of ASHRAE 55-2004 to prepare a document demonstrating compliance with the credit.
- Complete required documentation and upload to LEED Online.

- LEED PM to review uploaded documentation and submit to GBCI.

Construction Application

Stage III

Constn. LEED Review   Construction

O&M

Responsibilities

- Mechanical Engineer (Mech.)

- Electrical Engineer
- Architect
- Interior Designer
- LEED PM

Documentation Checklist

- Online credit form—list thermal comfort criteria assumptions
- Documentation with inputs and results of calculations or simulations. Include worst case design outdoor conditions and worst case predicted indoor conditions for each month. Show predicted worst case indoor conditions for each month on Figure 5.3 of ASHRAE 55.
- Documentation with PMV/PPD calculation; ASHRAE comfort tool results; and/or psychrometric comfort zone chart from Standard 55
- Any other drawings that will show compliance with credit

Implementation

LEAD

SUPPORT

DESIGN SUBMITTAL

STRATEGIES

TIPS & NOTES

Registration

Stage I

Pre-SD

Design Application

Stage II

SD LEED Phase

- Discuss with owner and/or building management the feasibility of administering a building occupant survey after occupancy and determine the party responsible for preparing the survey and administering it.
- Discuss the feasibility of including permanent monitoring systems to ensure comfort criteria as determined by EQ C7.1 are maintained.

LEED PM, Owner   SD

DD LEED Phase

- Develop a preliminary thermal comfort survey based on the requirements of ASHRAE 55-2004 and also a plan for corrective action.
- Coordinate with architect, MEP team, and building management to prepare the survey.
- Choose appropriate monitoring systems and incorporate in design.

LEED PM   DD

CD LEED Phase

- Finalize thermal comfort survey and plan for corrective action with the building management; decide on schedule for administration of the survey.
- Finalize type and locations of permanent monitoring systems in drawings and specifications.

- Upload required documents, obtain owner signature on credit form, and submit to GBCI for Design Review.

LEED PM, Owner   CD

Construction Application

Stage III

Constn. LEED Review

Construction

- After 6–18 months of occupancy, building management or third-party consultants should administer the survey, and if required, implement corrective measures.

O&M

Responsibilities

- LEED PM or Mech.—to prepare survey
- Building Management—to administer survey

- Mechanical Engineer (Mech.)
- Architect
- Owner

Documentation Checklist

- Online credit form
- Copy of thermal comfort survey along with information on how and when it will be administered
- Plan for corrective action if more than 20% of the building occupants are dissatisfied with the thermal comfort in the building
- Drawings showing location of permanent monitoring systems along with specifications and/or manufacturer's cut sheets

Implementation

- Sample surveys are available at
  The University of California, Berkeley's Center for the Built Environment website:
  www.cbesurvey.org
  The Usable Buildings Trust website:
  www.usablebuildings.co.uk

- Residential projects are not eligible for this credit.

- Teams attempting EQ C7.2 must also attempt/achieve EQ C7.1.

- Note that LEED v3 has an additional requirement of providing a permanent thermal comfort monitoring system for NC projects. This was not the case in LEED v2.2. Using programmable thermostats or other devices capable of measuring more than one of the ASHRAE 55-2004 thermal comfort parameters such as temperature, air speed, humidity, etc. are some recommended solutions.

Registration

Stage I

Architect　Pre-SD

- Discuss with owner, architects, engineers, and lighting consultants the overall lighting design strategy and the feasibility of incorporating daylighting strategies in design.
- Incorporate daylighting goals in the owner's project requirements (OPR).

Design Application

Stage II

SD LEED Phase

Architect, Lighting　SD

- Design to incorporate daylighting strategies; utilize daylighting simulation models (Opt 1) or prescriptive measures (Opt 2) to inform design decisions.
- Carefully select windows, curtain walls, glazing and other components of daylight design.
- Plan interior spaces in such a way that maximum number of occupants derive benefits of daylighting (open spaces at perimeter and closed spaces in the core).
- Incorporate glare control strategies for openings.

DD LEED Phase

Architect, Lighting　DD

- Coordinate with energy simulation expert to ensure that the glazing specification balances both daylighting as well as heat gain aspects.
- Consider incorporating daylighting sensors that interact with artificial lighting systems to ensure proper light levels are maintained.
- Perform calculations to determine percentage of regularly occupied areas that meet the daylighting credit requirement.

CD LEED Phase

Architect, Lighting　CD

- Finalize drawings and specifications, showing window schedules, types, glazing specifications, etc.
- Update daylighting simulation model (Opt 1) or prescriptive calculations (Opt 2) based on construction drawings.
- Upload documentation to LEED Online.
- LEED PM to review documentation and submit to GBCI for Design Review.

Construction Application

Stage III

Constn. LEED Review

Construction

- For teams pursuing Opt 3, after construction is complete, take indoor light measurements and demonstrate that a daylight illumination level of 25 foot-candles (fc) has been achieved in at least 75% or 90% of all regularly occupied areas.

O&M

Responsibilities

LEAD

- Lighting Consultant (Lighting)

SUPPORT

- LEED PM
- Architect
- Owner

Documentation Checklist

DESIGN SUBMITTAL

- Online credit form
- Floor plans showing all regularly occupied areas and areas that are daylit
- Interior elevations; window schedules and specifications
- Opt 1: Summary output from the computer simulation for daylight
- Opt 2: Elevation drawings and calculations for both sidelighting and toplighting
- Opt 3: Drawings showing measurements taken on a 10 foot grid

Implementation

STRATEGIES

TIPS & NOTES

- Daylighting simulation is an effective approach not only to demonstrate compliance with the credit but also to make informed decisions regarding glazing type, window sizes etc. during early design phases. It is an especially useful tool for buildings with complex geometries.

Registration

Stage

Pre-SD

Design Application

SD LEED Phase

- Orient building on the site to incorporate desirable views.
- Plan interior spaces in such a way that maximum number of occupants have access to views, by means of vision glazing between 30 inches and 90 inches above the finish floor.
- Ensure there is access to this glazing at 42 inches height above the finish floor.

Architect

SD

Stage II

DD LEED Phase

- Perform preliminary *view area* calculations (LEED PM may assist with this):
  - Using floor plans draw lines of sight to perimeter glazing
  - Using building sections, draw a line at 42 inches across the section to establish eye height and any obstruction to perimeter glazing
  - Identify regularly occupied areas that meet both the plan and section criteria
  - Perform calculations to determine the percentage of regularly occupied spaces that have access to views

Architect, LEED PM

DD

CD LEED Phase

- Finalize drawings and specifications for construction.
- Update calculations of compliant *view area* based on updated construction drawings.

- LEED PM to review documentation and submit to GBCI for Design Review.

Architect, LEED PM

CD

Construction Application

Stage III

Constn. LEED Review

Construction

O&M

Responsibilities

- Architect or LEED PM

- LEED PM
- Architect
- Owner

Documentation Checklist

- Online credit form
- Floor plans showing all regularly occupied areas and areas that have access to views (with lines of sight drawn in plan)
- Sections showing unobstructed horizontal view (at 42" height from finish floor)
- Spreadsheet showing *view area* calculations

Implementation

- The *LEED Reference Guide* defines regularly occupied spaces as "areas where people sit or stand as they work. In residential buildings these spaces include all living and family rooms and exclude bathrooms, closets, and storage or utility areas."

Sample Documents

The following pages present examples of some *Indoor Environmental Quality* credit documents that were submitted to USGBC by the teams of the case study projects (discussed in Chapter 8). Excerpts of documents from the following credits have been included.

EQ Credit 3.1, Construction IAQ Management Plan—During Construction

EQ Credit 3.2, Construction IAQ Management Plan—Before Occupancy

EQ Credit 7.2, Thermal Comfort—Verification

EQ Credit 8.1, Daylight and Views—Daylight

EQ Credit 8.2, Daylight and Views—Views

SHAWMUT
Design and Construction

Indoor Air Quality Plan

Project: Harvard Divinity School ñ Rockefeller Hall Renovations

Owner: Harvard Divinity School

Contractor: Shawmut Design and Construction

Date: August 8, 2007

This plan describes the measures to be taken to provide good indoor air quality (IAQ) during construction and after construction is complete and the occupants have moved into the building. This plan is based on the SMACNA standard ìIAQ Guidelines for Occupied Buildings under Constructionî and the requirements of the LEED New Construction version 2.2 Rating System.
It is not the intent of this document to replace or supersede OSHA regulations as to safe construction workplace practices. It remains the responsibility of Shawmut Design and Construction and the individual sub-contractors to maintain safe building and site operations.

The plan addresses construction IAQ procedures in five areas of concern, which in turn will allow the building to achieve one LEED program point:

.   •   Ventilation system protection
.   •   Contaminant source control
.   •   Pathway interruption
.   •   Housekeeping
.   •   Scheduling

The following describes the specific measures to be performed in each area of concern:

1.   Ventilation Protection

Demolition operations will not be ongoing during the time that return air systems are in use. Shawmut Design and Construction does not plan to use permanent HVAC systems while performing activities that produce high dust, such as drywall sanding, concrete cutting, masonry work, wood sawing and insulating.

During construction we will seal off the supply and return air system openings to prevent the accumulation of dust and debris in the permanent duct system, as well as, seal all ductwork during transit, storage, and installation.

Shawmut Design and Construction will keep mechanical rooms clean and neat and periodic duct inspections will be performed during construction. If the ducts become contaminated due to inadequate protection, Shawmut Design and Construction and its subcontractors will clean the ducts.

Shawmut Design and Construction will take photographs showing measures in place.

Sample Document: EQ Credit 3.1, Construction IAQ Management Plan—During Construction (page 1 of 7).
Credit: Harvard University, Office of Sustainability and Shawmut Construction

2.   Source Control

Low-emitting VOC products will be used as indicated by the specifications.

Volume of traffic will be restricted and idling of motor vehicles where emissions could be drawn into the building will be prohibited.

Shawmut Design and Construction will utilize electric or natural gas alternatives for gasoline and diesel equipment where possible and practical.

Equipment will be cycled off when not being used or needed.

Pollution will be exhausted to the outside with portable fan systems in order to prevent pollution from re-circulating back into the building.

Containers of wet products will be kept closed as much as possible. Containers will be covered and sealed to avoid release of odor or dust.

On-site materials will be stored to prevent exposure to weather and moisture; Materials will be wrapped with plastic and sealed tight to prevent moisture absorption.

Shawmut Design and Construction shall take photographs showing measures in place.

3.   Pathway Interruption

Pollutant sources will be located as far away as possible from supply ducts and areas occupied by workers when feasible.

Depending on weather, ventilation using 100% outside air will be used to exhaust contaminated air directly to the outside during installation of VOC emitting materials.

4.   Housekeeping

Regular cleaning will be provided with a concentration on ventilation equipment that removes contaminants from the building prior to occupancy.

All coils, air filters, fans and ductwork will remain clean during installation and, if required, will be cleaned prior to performing the testing, adjusting and balancing of the systems.

Dust will be suppressed and minimized with wetting agents or sweeping compounds. Utilize efficient and effective dust collecting methods such as a damp cloth, wet mop, and vacuum with particulate filters, or wet scrubber.

Accumulations of water will be removed from inside the building and porous materials will be protected from exposure to moisture (ex: insulation and gypsum wall board).

Shawmut Design and Construction will provide photographs of the above activities during construction to document compliance.

5.   Scheduling and Construction Activity Sequence

When possible, Shawmut Design and Construction will schedule high pollution activities that utilize high VOC level products (including paints, sealers, insulation, adhesives, caulking and cleaners) to take place prior to installing highly absorbent materials (such as ceiling tiles, gypsum wall board, fabric furnishings, carpet and insulation, for example).

6.   Planning & Inspection Checklists

Included in this document is an IAQ checklist that will be completed at least once every three months to confirm the IAQ management plan is being followed. At the time of inspection, photographs will be taken to support the checklist.

Sample Document: EQ Credit 3.1, Construction IAQ Management Plan—During Construction (page 2 of 7).

Credit: Harvard University, Office of Sustainability and Shawmut Construction

---

IAQ Management Plan Checklist

Project: Harvard Divinity School ñ Rockefeller Hall Renovations

Owner: Harvard Divinity School

Contractor: Shawmut Design and Construction

Date:

To be completed by Shawmut Design and Construction Superintendent every three months.

1. Ventilation Protection

☐   Seal return openings air during demolition

☐   Seal supply and return air openings during construction

☐   Mechanical rooms clean and neat

☐   Periodic duct inspections during construction.

☐   General Contractor to document with photographs

2. Source Control

☐   VOC products as indicated by specifications

☐   Restrict vehicle traffic volume and prohibit idling

☐   Utilize electric or natural gas alternatives for gasoline and diesel

☐   Cycle equipment off when not being used or needed

☐   Exhaust pollution sources to the outside

☐   Keep containers of wet products closed

☐   Cover or seal containers of waste materials

☐   Protect absorptive building materials from weather and moisture

☐   General Contractor to document with photographs

3. Pathway Interruption

☐   Locate pollutant sources as far away as possible from supply ducts and areas occupied by

     workers

☐   General Contractor to document with photographs

☐   When using VOC emitting materials ventilate using 100% outside air

☐   General Contractor to document with photographs

Sample Document: EQ Credit 3.1, Construction IAQ Management Plan—During Construction (page 3 of 7).

Credit: Harvard University, Office of Sustainability and Shawmut Construction

4. Housekeeping

☐   Provide regular cleaning, including ventilation equipment

☐   If necessary clean ventilation equipment prior to testing, adjusting and balancing the systems

☐   Suppress and minimize dust with wetting agents or sweeping compounds

☐   Remove accumulations of water inside the building

☐   Protect porous materials

☐   General Contractor to document with photographs

5. Scheduling and Construction Activity Sequence

☐   Schedule high pollution activities prior to installing absorbent materials

☐   General Contractor to document with photographs

I confirm the checked activities to be proceeding according to the Construction Indoor Air Quality Plan. Items that are not checked will be addressed, initialed and dated once corrective actions have been taken. Items that are not applicable are labeled as such.

Signed: _____   Date:_____

Sample Document: EQ Credit 3.1, Construction IAQ Management Plan—During Construction (page 4 of 7).
Credit: Harvard University, Office of Sustainability and Shawmut Construction.

HARVARD UNIVERSITY
Harvard Green Campus Initiative

46 Blackstone Street South
Fourth Floor
Cambridge, MA 02139

Phone   617.495.0551
Fax   617.495.9409
www.greencampus.harvard.edu

Harvard Divinity School
Rockefeller Hall
Indoor Environmental Quality credit 3.1 – IAQ Management Plan During Construction
Pictures from walkthrough 9/28/07

Sealed Ducts – before installation

Sealed Duct – after installation

Sample Document: EQ Credit 3.1, Construction IAQ Management Plan—During Construction (page 5 of 7).
Credit: Harvard University, Office of Sustainability and Shawmut Construction.

Insulation ready to be installed, kept off the ground

Insulation material kept sealed in plastic prior to installation

Sample Document: EQ Credit 3.1, Construction IAQ Management Plan—During Construction (page 6 of 7).

Credit: Harvard University, Office of Sustainability and Shawmut Construction.

Sweeping compound used to control dust

Temporary ventilation fan used during construction

Sample Document: EQ Credit 3.1, Construction IAQ Management Plan—During Construction (page 7 of 7).
Credit: Harvard University, Office of Sustainability and Shawmut Construction.

Harvard Divinity School, Rockefeller Hall
Indoor Environmental Quality credit 3.2

Narrative

The flushout of Rockefeller Hall began on 6/8/08. All construction activities were complete, furniture was installed, and new filtration media was installed in all air handling units. Building ventilation systems were run at 100% outside air. The Siemens BAS was set to track temperature, humidity, and volume of air flushed through the building. Any periods when humidity exceeded 60%, or temperature was below 60°F, were subtracted from the total amount of air flushed through the space. The flushout requirement was 374 million CF (26,715 SF * 14,000 CF/SF). Daily flushout air volumes were totaled cumulatively (see chart below) until the total flushout exceeded the requirement. The flushout continued until 7/31/2008, when filtration media were again replaced and the building was made available for occupancy.

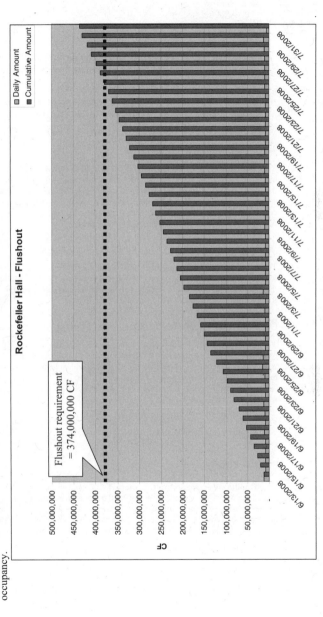

Sample Document: EQ Credit 3.2, Construction IAQ Management Plan—Before Occupancy.

Credit: Harvard University, Office of Sustainability.

HARVARD UNIVERSITY OFFICE FOR
SUSTAINABILITY

46 Blackstone Street South
Fourth Floor
Cambridge, MA 02139

Phone   617.495.0551
Fax   617.495.9409
www.greencampus.harvard.edu

Rockefeller Hall – Harvard Divinity School
Thermal Comfort Survey Plan

In complying with LEED New Construction v2.2 credit 7.2 *Thermal Comfort, Verification,* the Harvard Divinity School (HDS) facilities team in Rockefeller Hall plans to administer occupant surveys and adjust HVAC parameters to respond to comfort issues. This Thermal Comfort Survey Plan (TCP) addresses occupant surveys, the scheduling of surveys, and how HDS will respond to survey results.

Occupant Thermal Comfort Survey

The HDS survey is based on requirements of the LEED New Construction version 2.2 Reference Guide credit 7.2 *Thermal Comfort, Verification* as well as section 7.6 of the 2004 version of ANSI/ASHRAE Standard 55-2007, *Thermal Environmental Conditions for Human Occupancy* and modeled after the ASHRAE 55-2007 Informative Appendix E, *Thermal Environment Survey*. The survey addresses the comfort experience in the offices and meeting rooms, the layers of clothing that the occupants are wearing, and asks causes of discomfort , if applicable.

Administration of Survey

These surveys will be administered online via an icommons poll tracker. This service allows poll takers to anonymously submit their answers to the poll questions. An HDS staff person will email a link to the poll to the occupants of the building along with a cover letter explaining the intent of the survey. Results will be compiled an analyzed by HDS staff.

Responses to Survey Results

The operations team will view survey results for each residence floor. If greater than 20% of the survey respondents are dissatisfied, then the team will identify problem areas within the building. The team will identify the cause of discomfort in each situation and choose from a variety of options to correct the discomfort. For the purposes of this survey, "dissatisfied" will be defined as any responses of "somewhat dissatisfied," "dissatisfied," or "very dissatisfied" to question number 7, "General thermal comfort."

Rockefeller Hall can be monitored via the building management system (a Siemens direct digital control system) which is tied into temperature sensors in each floor. VAV boxes can be adjusted to affect supply air velocity and cooling. Heating of the offices is provided by fancoil units.

Schedule of Surveys

Because different seasons activate different HVAC systems in the building, the survey will be administered once per season. HDS staff will target especially cold winter days, mild spring days, and hot summer/fall days at the beginning of the fall semester when choosing survey dates. Should HDS determine corrective action is required to improve occupant comfort, the survey may be issued an additional time within the same season to ensure the effectiveness of the corrective action.

Sample Document: EQ Credit 7.2, Thermal Comfort—Survey (page 1 of 2).
Credit: Harvard University, Office of Sustainability.

Thermal Environmental Survey

To Be Filled By Occupant

1. **Occupant's Room Number (Optional):**

2. **Date:**

3. **Time:**

4. **Occupants Clothing**

Please refer to Table 1. Place a check mark next to the articles of clothing that you are currently wearing as you fill out this sheet. If you are wearing articles of clothing not listed in the table, please enter them into the space provided below

Article:

Article:

5. **Occupant Activity Level** (Check the one that is most appropriate)
   1. ☐ Reclining
   2. ☐ Reading Seated, Keyboarding or other light physical activity
   3. ☐ Standing, Relaxed
   4. ☐ Light Activity, Standing
   5. ☐ Medium Activity, Standing
   6. ☐ High Activity

6. **Equipment** (Equipment adding or taking away from the heat load.)

Item (copiers, fans, etc.)	Quantity

7. General Thermal Comfort (Check the one that is most appropriate)
   +3 ☐ Very satisfied
   +2 ☐ Satisfied
   +1 ☐ Somewhat satisfied
    0 ☐ Neutral
   -1 ☐ Somewhat dissatisfied
   -2 ☐ Dissatisfied
   -3 ☐ Very dissatisfied

General Environment Comments:

TABLE 1
Clothing Ensembles

Description	
Trousers, short-sleeve shirt/blouse	
Trousers, long-sleeve shirt/blouse	
Trousers, long-sleeve shirt/blouse plus suit jacket	
Trousers, long-sleeve shirt/blouse plus suit jacket, vest, T-shirt	
Trousers, long-sleeve shirt/blouse plus long sleeve sweater or other 2nd top layer, T-shirt	
Trousers, long-sleeve shirt/blouse plus long sleeve sweater or other 2nd top layer, T-shirt pus suit jacket, long underwear bottoms	
Knee-length skirt/shorts, short-sleeve shirt/blouse (sandals)	
Knee-length skirt/shorts, long-sleeve shirt	
Knee-length skirt/shorts, long-sleeve shirt, long-sleeve sweater or other 2nd top layer	
Ankle-length skirt/shorts, long sleeve shirt, suit jacket	
Athletic sweat pants, long-sleeve sweatshirt	
Other:	

Sample Document: EQ Credit 7.2, Thermal Comfort—Survey (page 2 of 2).

Credit: Harvard University, Office of Sustainability.

PERKINS + WILL

GREAT RIVER ENERGY HEADQUARTERS
LEED-NC 2.2 EQ 8.1

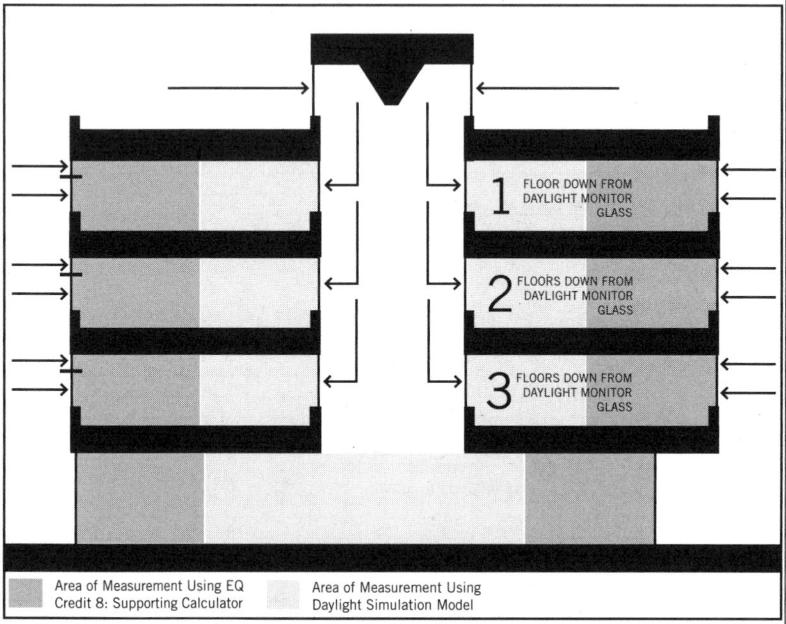

Fig. 01-A: Building Section at Daylight Atrium

EQ 8.2: ANALYSIS METHODOLOGY

DL-001

Sample Document: EQ Credit 8.1, Daylight and Views—Daylight (page 1 of 3).
Credit: Perkins + Will.

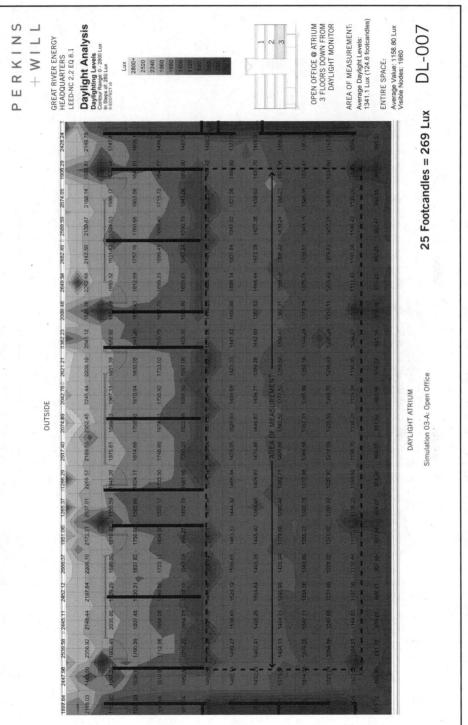

Sample Document: EQ Credit 8.1, Daylight and Views—Daylight (page 2 of 3).
Credit: Perkins + Will.

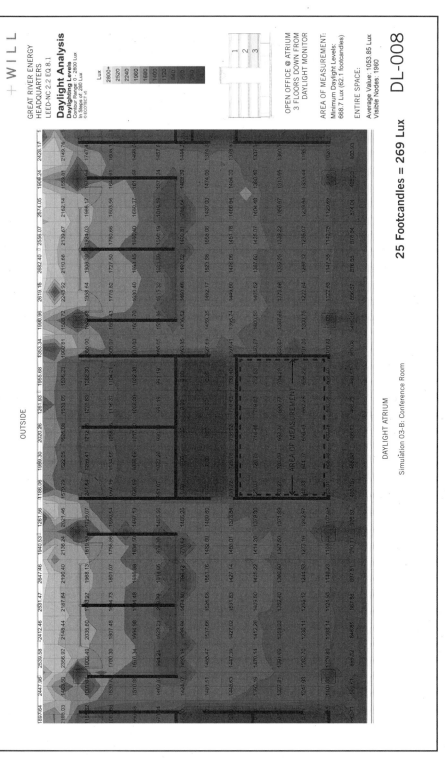

Sample Document: EQ Credit 8.1, Daylight and Views—Daylight (page 3 of 3).
Credit: Perkins + Will.

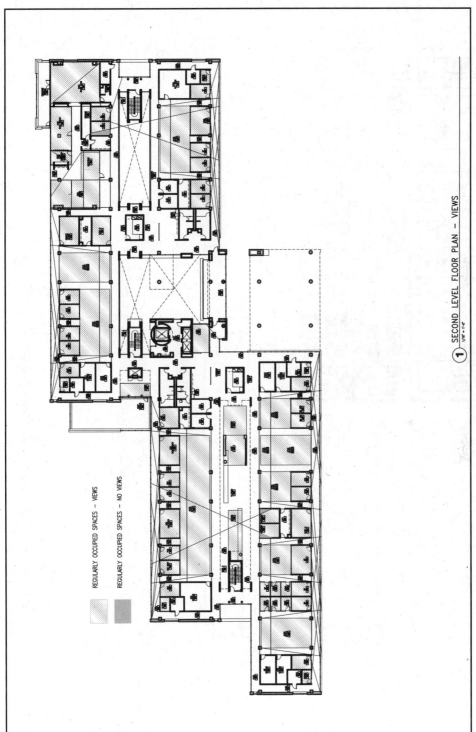

REGULARLY OCCUPIED SPACES – VIEWS

REGULARLY OCCUPIED SPACES – NO VIEWS

1   SECOND LEVEL FLOOR PLAN – VIEWS

Sample Document: EQ Credit 8.2, Daylight and Views—Views (page 1 of 2).

Credit: Perkins + Will.

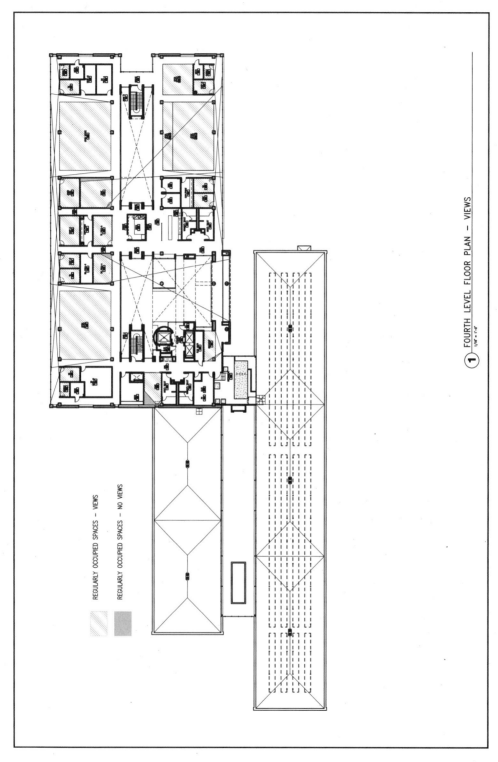

Sample Document: EQ Credit 8.2, Daylight and Views—Views (page 2 of 2).

Credit: Perkins + Will.

CREDIT IMPLEMENTATION— INNOVATION IN DESIGN

Introduction

The *Innovation in Design* (ID) credits offer projects a chance to gain extra points (up to 6) in addition to the credit points offered in the previous categories. There are two ways by which these credits can be achieved: one is by demonstrating the use of innovative strategies to improve the building's performance beyond what is required by the LEED credit (these are referred to as *exemplary performance* points) and the second way is by incorporating innovative ideas that are not specifically addressed by any credit in the rating system. There is also one point available for having a LEED Accredited Professional (LEED AP) on the project team.

Team members should explore various innovation opportunities that may be implemented during the design, construction, and operations of the building in order to achieve points in this category. Because the nature of ID credits will vary from project to project, no specific implementation sheet has been provided. However, some sample ID credit submittals have been included in this section. The intent is to provide ideas that project teams could potentially use in their projects.

Sample Documents

The following pages present excerpts from some ID credit submittals made to USGBC by teams of the case study projects (discussed in Chapter 8). Excerpts from the following credit submittal have been included.

ID Credit 1.1, Innovation in Design—Commuter Choice

HARVARD UNIVERSITY
Harvard Green Campus Initiative

46 Blackstone Street
First Floor
Cambridge, MA 02139

Phone   617.496.1278
Fax   617.495.9409
www.greencampus.harvard.edu

Rockefeller Hall
Harvard Divinity School
Innovation in Design Credit 1.1

Sample Document: ID Credit 1.1, Innovation in Design—Commuter Choice (page 1 of 17). *Credit: Harvard University, Office of Sustainability.*

HARVARD UNIVERSITY
Harvard Green Campus Initiative

46 Blackstone Street
First Floor
Cambridge, MA 02139

Phone   617.496.1278
Fax   617.495.9409
www.greencampus.harvard.edu

Credit 1.1 Innovation in Design Process

Innovation in Design – Commuter Choice Program Credit Description

Intent
Harvard University currently funds a commuter choice program that actively promotes the use of public, shared, and non-motor transportation for Harvard affiliate commutes. This program is responsible for achieving a drive alone commuter trip (DACT) rate of 21.26%; 10% above what is required of Harvard by the MA EPA 310 CMR 7.16 regulation.

The intent of this credit is to achieve a DACT of 21.26% at the Harvard Divinity School Rockefeller Hall by instituting the commuter choice program into its administration.

Requirements
Implement a program to achieve a DACT of 21.26% at Rockefeller Hall through the following actions:

- Show calculations of a DACT of 21.26% at the Institutional level

- Install a Commuter Choice Kiosk in the Lobby of Rockefeller Hall

- Subsidize all Massachusetts Bay Transportation Authority (MBTA) bus, subway and boat passes by 40%, up to the current limit established by the IRS

- Allow purchase of public transportation passes on a pre-tax basis, further reducing the cost to employees by an average of 28%

- Establish a website devoted to commuter choice options at the facility and provide assistance to employees seeking alternatives to single passenger vehicles

- Arrange discounts on Zip Cars as an option for employees who may need access to a vehicle during the day, but can otherwise make use of alternative means of transportation

- Provide an Emergency Rides Home program so that employees are not penalized for not having a vehicle if they, a member of their family, or a member of their car- or vanpool need to leave work for an emergency

- Arrange discounts on bicycle accessories and maintenance at local bike shops

- Provide maps detailing preferred bike routes

1

Sample Document: ID Credit 1.1, Innovation in Design—Commuter Choice (page 2 of 17).   *Credit: Harvard University, Office of Sustainability.*

- Designate a Transportation Representative who is responsible for promoting these commuting options. The Transportation Representative is also responsible for distributing monthly "commuter choice tips" from the Commuter Choice Office

Submittals
- Provide LEED Letter Template signed by the owner or other responsible party stating that all of the requirements are being implemented

- Show that Harvard has reduced the number of single passenger vehicles by 35% since 1998, an additional 10% beyond what is required by applicable state regulations.

- Show a drawing outlining the location of the Commuter Choice Kiosk in the lobby

- Provide literature that is distributed to staff at Rockefeller Hall outlining the T-pass program. Include information on the subsidy and pre-tax benefit, as well as a list of frequently asked questions

- Provide screen shots of the webpage devoted to the commuter choice program and sub programs (i.e.: Emergency Ride Home, Bicycling, Metro Pass Discount Tickets)

- Provide information on Zip card discounts to Harvard students, faculty and staff

- Provide information on Harvard's Emergency Ride Program

- Provide information on Harvard's Bike Program

- Provide maps detailing preferred bike routes

- Provide the name of the Transportation Coordinator

2

Sample Document: ID Credit 1.1, Innovation in Design—Commuter Choice (page 3 of 17). *Credit: Harvard University, Office of Sustainability.*

HARVARD UNIVERSITY
Harvard Green Campus Initiative

46 Blackstone Street
First Floor
Cambridge, MA 02139

Phone   617.496.1278
Fax   617.495.9409
www.greencampus.harvard.edu

Exhibit A: LEED Letter Template

ID Credit 1: Innovation in Design

I, Jessica Parks, submit the following innovation credit proposal for exceptional performance above the existing LEED requirements and/or innovative green building performance of a type that is not currently addressed within LEED. The following documentation has been provided:

    ID Credit 1.1 (1 point)   Title: <u>Harvard Commuter Choice Program</u>

For each innovation credit proposal, I have attached a sheet that succinctly identifies the intent of the proposed innovation credit, the proposed requirements for compliance, the proposed submittals to demonstrate compliance, and the design approach (technologies and strategies) that might be used to meet the requirements. I have also attached the relevant submittals as evidence of compliance.

ID Cr 1 (4 points possible): Innovation in Design

    Name: Jessica Parks
    Organization: Harvard Green Campus Initiative
    Role in Project: Consultant

    Signature:

    Date: 10/10/2008

3

Sample Document: ID Credit 1.1, Innovation in Design—Commuter Choice (page 4 of 17). *Credit: Harvard University, Office of Sustainability.*

HARVARD UNIVERSITY
Harvard Green Campus Initiative

46 Blackstone Street
First Floor
Cambridge, MA 02139

Phone   617.496.1278
Fax   617.495.9409
www.greencampus.harvard.edu

Exhibit B: Drive Alone Commute Trip (DACT) Reduction at Harvard

MA DEP 310 CMR 7.16 requires Harvard to reduce the number of drive-alone commuters by 25% from the baseline information collected in 1998. The 2002 Harvard University Ridesharing Update Report shows that in 1998, 33.75% of the eligible employees and students used single occupant vehicles to get to Harvard (28,559 drive alone trips out of 84,611 total trips). The goal set by the EPA was to reduce that to 25.31%. In 2002, there were 20,202 drive alone trips out of 95,025, or 21.26%. This is a 37% reduction in the percentage of single passenger vehicles in the 4 years from when the baseline data was collected, and a 12% better improvement than the EPA goal.

Harvard Yearly DACT Reductions

	1998	**2002**	**EPA Goal**
DACT	28,559	20,202	21,419
Total Trips	84,611	95,025	N/A
DACT % of All Trips	33.75%	21.26%	25.31%

EPA Target Improvement	**Percent Improvement**	**DACT Reduction Below Target**
25%	37.01%	1,217

4

Sample Document: ID Credit 1.1, Innovation in Design—Commuter Choice (page 5 of 17).   *Credit: Harvard University, Office of Sustainability.*

HARVARD UNIVERSITY
Harvard Green Campus Initiative

46 Blackstone Street
First Floor
Cambridge, MA 02139

Phone   617.496.1278
Fax   617.495.9409
www.greencampus.harvard.edu

Exhibit C: Location of Commuter Choice Kiosk

The Rockefeller Hall Commuter Choice Kiosk (in red) is located in the front entrance vestibule.

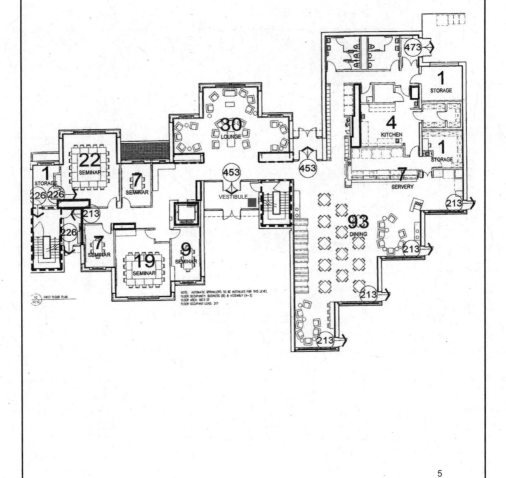

5

HARVARD UNIVERSITY
Harvard Green Campus Initiative

46 Blackstone Street
First Floor
Cambridge, MA 02139

Phone   617.496.1278
Fax   617.495.9409
www.greencampus.harvard.edu

Exhibit D: Commuter Choice Web Page
www.commuterchoice.harvard.edu

HARVARD UNIVERSITY • UNIVERSITY OPERATIONS SERVICES

Transportation Services

Search

Emergencies
Medical   Dial 911
Harvard Police   617.495.1212
Operations Center   617.495.5560
Emergency Guide (pdf)

COMMUTERCHOICE PROGRAM

617 495 5560 **24 HOURS**

MBTA Pass   Students   ADA Ride   Bicycling   Ridesharing   Transit   Walking   ZipCar   Contact

Services

Reference & Resources

Bike Guidelines
Mail Services FAQ
MBTA Pass Program FAQ
Online Forms
Parking Maps
Parking Services FAQ
Shuttle Schedule
Transportation Strategic Plan
Zip Code Address Verification

Online Tools

UOS Departments

Engineering & Utilities
Environmental Health & Safety
Facilities Maintenance Operations
Harvard Green Campus
Operations Center
Transportation Services
    CommuterChoice
    Fleet Management Services
    Parking Services
    Passenger Transport Services
    Mail Services
Administration & Finance
Applied Technologies

CommuterChoice

COMMUTERCHOICE PROGRAM

The *CommuterChoice* Program is committed to providing the best information and planning services for Harvard employees who work in the Cambridge Allston/Brighton campuses, regarding their commute to work. **(for Longwood Medical Area commuting information click here)**

Discounted MBTA Pass Program

Harvard University offers to its benefits eligible employees a 50% pre-tax subsidy for the MBTA (Massachusetts Bay Transit Authority) public transit system passes. This includes MBTA buses, subways, commuter rail lines and boats.

Buy Online

Employees
Buy your MBTA Pass On-line or **Manage Your Account**

If you are an undergraduate or graduate Student **click here**

For more information please visit our MBTA Pass Program FAQs or call 1 800 462-2235

Paper Purchasing Forms

- **English**
- **Haitian Creole**
- **Portuguese**
- **Spanish**

Non-received or defective CharlieCard?
Please call Crosby Benefits at 1-800-462-2235 #6 (Monday – Thursday, 8am – 6 pm and Friday, 8am – 5pm) or email:
servicecenter@crosbybenefits.com

Non-received/Incorrect Commuter Rail or Boat Pass Form

Whether you are a new employee, changing jobs within the University or moving to a new home, CommuterChoice can assist you in planning your commute.

News

ADA Ride Assistance
Information about the program and how to schedule service

Bicycling
Routes, Maps, and Programs for the Biking Community

Ridesharing
Carpools, Vanpools, GoLoco

Transit
Public, Private, Shuttles - Harvard and others

Walking
Events, Groups and Information about walking to work.

Zipcar
Use a car without the hassles of owning or renting one.

Commute Cost Calculator
Events, Groups and Information about walking to work.

Request Information
Events, Groups and Information about walking to work.

Attention HU-HUCTW Members:
Apply for the transportation fund >>

COMMUTERCHOICE PROGRAM
617-384-RIDE
commuterchoice@harvard.edu
Kris Locke - *Manager*
Carolyn Pisarczyk - *Administrative Coordinator*

6

Sample Document: ID Credit 1.1, Innovation in Design—Commuter Choice (page 7 of 17).   *Credit: Harvard University, Office of Sustainability.*

HARVARD UNIVERSITY
Harvard Green Campus Initiative

46 Blackstone Street
First Floor
Cambridge, MA 02139

Phone   617.496.1278
Fax   617.495.9409
www.greencampus.harvard.edu

Exhibit E: T-Pass Narrative

http://www.commuterchoice.harvard.edu/transit/mbta/buy_pass.shtml

Harvard offers its employees a 50 percent discount on monthly bus, subway and commuter rail passes. Employees can purchase a pass online, and only need to sign up once for automatic monthly renewal. Payment is made through pre-tax payroll deduction, so buyers save even more on top of Harvard's subsidy.

CommuterChoice | Transit | MBTA

BUY YOUR MBTA PASS ONLINE

Please have your Harvard ID and PIN numbers available, so you may access the on-line ordering system for the Discounted MBTA Pass Program.

 Buy your **MBTA Pass** Online or **Manage Your Account**.
View the pricing of the Discounted MBTA Pass program

New Employees

New employees will encounter a waiting period prior to being able to access the system, anywhere from 5-9 weeks after date of hire. Additionally, new employees may need to wait until receipt of at least one paycheck before being able to place an order, as this is a payroll debited system.

Discounted MBTA Passes and Harvard Parking Permits

Harvard does not allow employees to purchase both a subsidized MBTA Pass and an annual parking permit. A subsidized MBTA Pass may be purchased in conjunction with the following Harvard permits: an Evening Commuter permit, a Motorcycle permit or a Daily permit.

Log On Error Messages

If you should encounter an error message when attempting to enroll in this program, please contact Crosby Benefits at 800 462-2235, so they may help you determine why this is occurring.

MBTA Pass Mailing Dates

MBTA Passes are typically mailed between the 18th and the 22nd of each month in a non-descript envelope with the Crosby Benefits address listed in the return address window. If you do not receive your MBTA Pass by the 25th of the month, please contact Crosby Benefits at 800 462-2235 to check on your mailing.

Types of Passes

If you are ordering a SENIOR/TAP, Bus or LinkPass you will receive a durable plastic CharlieCard that is good for up to five years, so please do not throw this card away or punch a hole in it. If you are ordering a Commuter Rail Line Pass, you will receive a monthly mailing of a CharlieTicket, which also provides unlimited travel on the subway and buses.

7

Sample Document: ID Credit 1.1, Innovation in Design—Commuter Choice (page 8 of 17). *Credit: Harvard University, Office of Sustainability.*

HARVARD UNIVERSITY
Harvard Green Campus Initiative

46 Blackstone Street
First Floor
Cambridge, MA 02139

Phone   617.496.1278
Fax   617.495.9409
www.greencampus.harvard.edu

Exhibit F: Discounted MBTA Pass – Harvard Rate Table

http://www.commuterchoice.harvard.edu/transit/mbta/buy_pass.shtml

TYPE OF PASS	MBTA PRICE	EMPLOYEE COST
SENIOR/TAP	$20.00	$ 10.00
BUS	$40.00	$ 20.00
LINKPASS	$59.00	$ 29.50
ZONE 1	$135.00	$ 67.50
Zone 2	$151.00	$ 75.50
ZONE 3	$163.00	$ 81.50
ZONE 4	$186.00	$ 93.00
ZONE 5	$210.00	$105.00
ZONE 6	$223.00	$111.50
ZONE 7	$235.00	$117.50
ZONE 8	$250.00	$125.00
BOAT	$198.00	$ 99.00
INNER EXPRESS BUS	$89.00	$ 44.50
OUTER EXPRESS BUS	$129.00	$ 64.50

8

Sample Document: ID Credit 1.1, Innovation in Design—Commuter Choice (page 9 of 17).   *Credit: Harvard University, Office of Sustainability.*

HARVARD UNIVERSITY
Harvard Green Campus Initiative

46 Blackstone Street
First Floor
Cambridge, MA 02139

Phone   617.496.1278
Fax   617.495.9409
www.greencampus.harvard.edu

Exhibit G: Emergency Ride Home
www.commuterchoice.harvard.edu

Commuter*Choice* at Harvard University

EMERGENCY RIDE HOME GUIDELINES

Harvard University employees participating in some form of ridesharing program (carpool or vanpool) five days a week are eligible for the Emergency Ride Home (ERH) Program. All employees must register in advance with the *CommuterChoice* Office to utilize this benefit.

Procedure
When the need for emergency transportation arises, the registered employee must notify his/her supervisor or authorized staff member. The supervisor or authorized staff member must sign the *Emergency Ride Home* confirmation form to verify the emergency.

The rider must contact their designated Emergency Ride Home transport. If using the Ambassador Brattle/Yellow Cab Company call **617 492-1110** for a pick up. Please give the cab driver the taxi voucher for your Emergency Ride Home, and take a receipt. For the participants in this program that live outside the 17-mile radius from work, please call Lifestyle Limo at **1 800 228-6894** and give them your account # **2071**.

The rider or supervisor **must** notify the *CommuterChoice* Office that the service is being used; a message can be left at *CommuterChoice*, 617-384-RIDE. After using the ERH service, please complete and return the Confirmation form with the receipt to the *CommuterChoice* Program, 46 Blackstone Street, 1st floor, Cambridge, MA. 02139.

Emergency Ride Home service will be supplied during the following situations, when regular transportation is not available:

- Illness or crisis of the participant or that of a family member.
  (note: this does not include injuries sustained at work that would fall under a Worker's Compensation Claim)

- Unexpected request of a supervisor to work past regular schedule without advance notice. Unexpected is defined as not knowing before the morning of the request.

- If you become stranded at work because the driver of your carpool or vanpool had to leave because of an emergency. If the driver of your carpool is unable to drive home, you may obtain an emergency ride home by following the guidelines listed above. If the driver of your vanpool is unable to drive home, the driver will receive an emergency ride home and a designated alternate driver will drive remaining van riders home.

Emergency Ride Home service will not be supplied during the following situations:

- Personal errands, pre-planned medical appointments, work-related travel, working late without manager's request, early departure or delays due to inclement weather or utility failure.

- *CommuterChoice* reserves the right to deny the use of ERH in certain other situations. Inappropriate use will result in reimbursement of expenses by rider.

Acceptable Destinations

- Home or other destination related to emergency.

- A public transportation connection point.

- A hospital or doctor's office in the event of sudden illness.

9

Sample Document: ID Credit 1.1, Innovation in Design—Commuter Choice (page 10 of 17).   *Credit: Harvard University, Office of Sustainability.*

HARVARD UNIVERSITY
Harvard Green Campus Initiative

46 Blackstone Street
First Floor
Cambridge, MA 02139

Phone   617.496.1278
Fax   617.495.9409
www.greencampus.harvard.edu

Exhibit H: Commuter Choice Bicycling Program
http://www.commuterchoice.harvard.edu/bicycling/

MBTA Pass   Students   ADA Ride   Bicycling   Ridesharing   Transit   Walking   ZipCar   Contact

CommuterChoice | Bicycling

BICYCLING

Register your bike

Departmental Bike Program

Consider purchasing one or more of your bicycles for members of your department to use around campus as an environmentally sustainable and health means of transport during the work week. **Learn More** »

Broadway Garage Bike Parking

1. Register your bike with the **Harvard University Police Department (HUPD)**
2. Complete the Parking Services Broadway Garage Bicycle Parking Form
3. Bring the completed form and your HUPD sticker number to the office of Parking Services, 46 Blackstone Street, Cambridge, and complete the registration process for access to the Broadway Garage bike parking area

Registration for Bike Groups

Bicycle Users Groups (BUG)
Harvard Cycling Group
MassBike

Register your bike with the Harvard University Police Department (HUPD)
Bike Routes and Amenities
Need a new or used bicycle?

Check out the **local bike directory** or visit **Quad Bikes**
Need Gear?

Check out Harvard discounts at local bike shops
Bicycle Repair & Maintenance

Quad Bikes, the non-profit bicycle shop for the Harvard community, offers $30 basic tune-ups
Bike Storage

Bike Racks at Harvard are intended for active use. If you need to store your bike, Quad Bikes has indoor secure bike storage for $8 per month.
Bike Racks on Harvard Shuttles

Put your bike on a Harvard Shuttle if you get caught in the dark, in the rain, or with a flat tire. **More** »
Map of bike racks on campus

10

Sample Document: ID Credit 1.1, Innovation in Design—Commuter Choice (page 11of 17).   *Credit: Harvard University, Office of Sustainability.*

HARVARD UNIVERSITY
Harvard Green Campus Initiative

46 Blackstone Street
First Floor
Cambridge, MA 02139

Phone   617.496.1278
Fax   617.495.9409
www.greencampus.harvard.edu

Exhibit I: Map of Preferred Bike Routes
http://www.commuterchoice.harvard.edu/bicycling/bike_routes.shtml

Below is a close up of the area around the Harvard Divinity School and Rockefeller Hall. The
following page shows the complete Cambridge & Allston map.

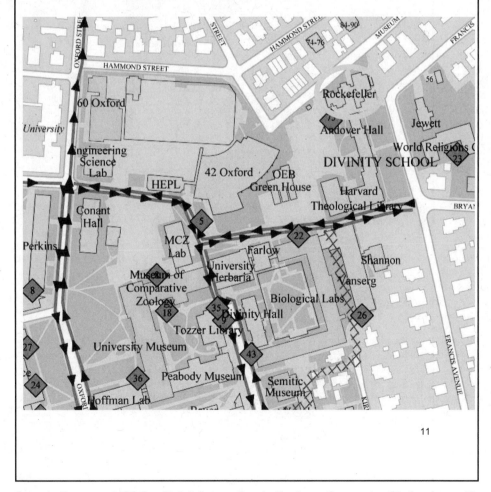

11

**Sample Document: ID Credit 1.1, Innovation in Design—Commuter Choice (page 12
of 17).**   *Credit: Harvard University, Office of Sustainability.*

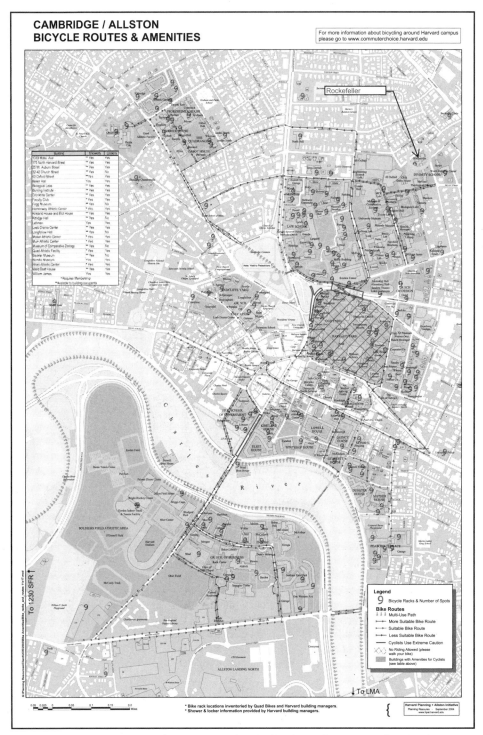

Sample Document: ID Credit 1.1, Innovation in Design—Commuter Choice (page 13 of 17).   *Credit: Harvard University, Office of Sustainability.*

HARVARD UNIVERSITY
Harvard Green Campus Initiative

46 Blackstone Street
First Floor
Cambridge, MA 02139

Phone   617.496.1278
Fax   617.495.9409
www.greencampus.harvard.edu

Exhibit J: Zip Car Program
http://www.commuterchoice.harvard.edu/zipcar.shtml

COMMUTERCHOICE
PROGRAM

MBTA Pass   Students   ADA Ride   Bicycling   Ridesharing   Transit   Walking   ZipCar   Contact

CommuterChoice

ZIPCAR AT HARVARD

The Boston area car share provider, Zipcar, offers cars throughout the greater Boston and Cambridge areas.

Zipcar offers a substantial membership discount to eligible Harvard affiliates. **Zipcar eligibility requirements.**

Car sharing allows you to use a car when you need it with incurring the fixed costs of ownership. Car share makes cars available on a per-use basis. Think of it as a neighborhood-based car rental that allows you to use vehicles when needed (for as little as one hour), and pay based on how much you drive.

Individual Membership: Harvard faculty, staff and students 21 years and older can join Zipcar for the annual membership fee of $25.

Departmental Membership Accounts: Need a car for work errands? Your department could set up a departmental account with Zipcar and save monthly, especially relative to taxi costs.

For more information, visit **zipcar.com/crimson**

12

Sample Document: ID Credit 1.1, Innovation in Design—Commuter Choice (page 14 of 17). *Credit: Harvard University, Office of Sustainability.*

HARVARD UNIVERSITY
Harvard Green Campus Initiative

46 Blackstone Street
First Floor
Cambridge, MA 02139

Phone   617.496.1278
Fax   617.495.9409
www.greencampus.harvard.edu

Exhibit K: Transportation Representative

Ms. Kristine Locke
Manager, Commuter Choice
Harvard University Operations Services

46 Blackstone Street
Cambridge, MA 02139
(t) 617.496.5354
(f) 617.495.9119
kris_locke@harvard.edu

13

Sample Document: ID Credit 1.1, Innovation in Design—Commuter Choice (page 15 of 17).   *Credit: Harvard University, Office of Sustainability.*

HARVARD UNIVERSITY
Harvard Green Campus Initiative

46 Blackstone Street
First Floor
Cambridge, MA 02139

Phone   617.496.1278
Fax   617.495.9409
www.greencampus.harvard.edu

Exhibit L: Credit Interpretation Request
Note: This CIR was originally submitted for the Harvard University 60 Oxford Street Project, but is applicable to all Harvard University projects in compliance with requirements for the Innovation in Design – Commuter Choice Program Credit.

11/17/2004 - Credit Interpretation Request
Commuter Choice Program

We are a large academic institution. We were required under Massachusetts EPA 310 CMR 7.16 regulation to reduce our Drive Alone Reduction Trip (DACT) rate from 33.75% in 1998 to 25.31% in 2002. We responded by creating a unique Commuter Choice program to comply with this regulation. By instituting several initiatives under this program we have reduced our DACT to 21.26%; 12% below what is required.

Our ID credit states that we will match the institutional DACT of 21.26% in our new building by ensuring that the commuter choice program is instituted in the administration of our building. Our requirements for this credit are as follows:

- Show calculations of a DACT of 21.26% at the Institutional level.

- Install a Commuter Choice Kiosk in the Lobby

- Subsidize all Massachusetts Bay Transportation Authority (MBTA) bus, subway and boat passes by 40%, up to the current limit established by the IRS.

- Allow purchase of public transportation passes on a pre-tax basis, further reducing the cost to employees by an average of 28%.

- Establish a website devoted to commuter choice options at the facility and provide assistance to employees seeking alternatives to single passenger vehicles.

- Arrange discounts on Zip Cars as an option for employees who may need access to a vehicle during the day, but can otherwise make use of alternative means of transportation.

- Provide an Emergency Rides Home program so that employees are not penalized for not having a vehicle if they, a member of their family, or a member of their car- or vanpool need to leave work for an emergency.

- Arrange discounts on bicycle accessories and maintenance at local bike shops.

- Provide maps detailing the locations of bicycle racks, lockers and showers and detailing preferred bike routes as well as areas where bicyclists should use caution.

14

Sample Document: ID Credit 1.1, Innovation in Design—Commuter Choice (page 16 of 17). *Credit: Harvard University, Office of Sustainability.*

- Provide maps detailing the locations of bicycle racks, lockers and showers and detailing preferred bike routes as well as areas where bicyclists should use caution.
- Designate Transportation Representative at 60 Oxford Street who is responsible for promoting these commuting options. The Transportation Representative is also responsible for distributing monthly commuter choice tips from the Commuter Choice Office.

We are aware that the intent of Alternative Transportation credits under Sustainable Sites is to reduce the amount of vehicles traveling to the building. We feel, however, that the commuter choice program goes above and beyond the requirements listed in LEED (we are expecting to achieve credits SS 4.1, 4.2, 4.4) and the EPA by instituting unique, generous incentives. Further, we have set a measurable goal to achieve a DACT of 21.6%.

Would this ID credit be likely to pass?

12/16/2004 - Ruling
This CIR is being submitted based on the merits of a Commuter Choice program that will be instituted in the administration of the building. The program is part of the parent institution's efforts to significantly reduce the Drive Alone Reduction Trip (DACT) rate.

This inquiry is similar in nature to a previous IDc1.1 CIR ruling dated 5/9/2003. An applicant will be awarded an innovation point for exemplary performance in alternative transportation by instituting a comprehensive transportation management plan (or program as described above), provided the project achieves three out of the four SS Credit 4 subcredits, AND is able to demonstrate that the requirements are met and all commitments are adequately and officially documented.

15

Sample Document: ID Credit 1.1, Innovation in Design—Commuter Choice (page 17 of 17). *Credit: Harvard University, Office of Sustainability.*

ACRONYMS

AIA COTE	American Institute of Architects' Committee on the Environment
ANSI	American National Standards Institute
ASHRAE	American Society of Heating Refrigerating and Air-Conditioning Engineers
BIM	Building information modeling
BMP	Best management practices
BOD	Basis of design
BREEAM	Building Research Establishment Environmental Assessment Method
CASBEE	Comprehensive Assessment System for Building Environmental Efficiency
CBE	Center for the Built Environment
CBECS	Commercial Building Energy Consumption Survey
CD	Construction documents
CDCR	Combined Design and Construction Review
CIBSE	Chartered Institution of Building Services Engineers
CIR	Credit Interpretation Ruling
CRI	Carpet and Rug Institute
CSI	Construction Specification Institute
CxA	Commissioning Authority
DD	Design development
DOE	Department of Energy
DSIRE	Database of State Incentives for Renewables and Efficiency
EA	Energy and Atmosphere

EIA	Energy Information Administration
EPA	Environmental Protection Agency
EPAct	Energy Policy Act
EQ	Indoor Environmental Quality
ESA	Environmental site assessment
EUI	Energy use intensity
FEMA	Federal Emergency Management Agency
FSC	Forest Stewardship Council
FTE	Full-time equivalent
GBCI	Green Building Certification Institute
GC	General contractor
GHG	Greenhouse gas
HAP	Hourly Analysis Program
HVAC	Heating, ventilation and air-conditioning
IAQ	Indoor air quality
ID	Innovation in Design
IDP	Integrated design process
IECC	International Energy Conservation Code
IESNA	Illuminating Engineering Society of North America
IGCC	International Green Construction Code
ISO	International Standards Organization
KW	Kilowatts
LBNL	Lawrence Berkeley National Laboratory
LEED	Leadership in Energy and Environmental Design
LEED AP	LEED Accredited Professional
LEED-EBOM	LEED for Existing Buildings: Operations and Maintenance
LEED v3	LEED–version 3 also known as LEED 2009
LEED-CS	LEED for Core & Shell
LEED-NC	LEED for New Construction
LEED-ND	LEED for Neighborhood Development
LPD	Lighting power density
M&V	Measurement and Verification
MDF	Medium density fiberboard

MEP	Mechanical, electrical and plumbing
MERV	Minimum efficiency reporting value
MR	Materials and Resources
NBI	New Building Institute
NPV	Net present value
OPR	Owner's project requirements
PM	Project manager
PV	Photovoltaics
RECs	Renewable energy credits
RFP	Request for proposal
RP	Regional priority
SBS	Sick building syndrome
SD	Schematic design
SDCR	Split Design and Construction Review
SHGC	Solar Heat Gain Coefficient
SIPs	Structurally insulated panels
SMACNA	Sheet Metal and Air-Conditioning Contractors' National Association
SRI	Solar Reflectance Index
SS	Sustainable Sites
UFAD	Underfloor air distribution system
USGBC	United States Green Building Council
VFDs	Variable frequency drives
VOCs	Volatile organic compounds
WE	Water Efficiency

GLOSSARY

Baseline building performance   Baseline building performance is defined as the annual energy cost for a building intended for use as a baseline for rating the proposed design. (ASHRAE 90.1-2007)

Basis of design (BOD)   The BOD is a written document that identifies design information for the systems to be commissioned according to the owner's project requirements. It includes primary design assumptions, applicable codes, standards, and narrative descriptions of performance criteria for systems to be commissioned.

Bioswales   Bioswales are landscape elements designed to remove silt and pollution from surface runoff water. They consist of a swaled drainage course with gently sloped sides (less than six percent) and filled with vegetation, compost, and/or riprap. The water's flow path, along with the wide and shallow ditch, is designed to maximize the time water spends in the swale, which aids the trapping of pollutants and silt.

British thermal unit (Btu)   A Btu is the quantity of heat required to raise the temperature of one pound of water by one degree Fahrenheit at a specified temperature.

Brownfield   Brownfields are real property, the expansion, redevelopment, or reuse of which may be complicated by the presence or potential presence of a hazardous substance, pollutant, or contaminant. Cleaning up and reinvesting in these properties protects the environment, reduces blight, and takes development pressures off greenspaces and working lands. (U.S. Environmental Protection Agency)

Building conditioned area   Total interior floor area of a building's conditioned spaces, measured from the inside surface of the exterior walls or from the interior surface of walls of adjoining buildings. The areas of interior walls, columns, and pillars are included in this measurement. This metric is measured on a floor-by-floor basis and consists of all conditioned spaces, including the area of interior walls, basements, mezzanines, penthouses, equipment rooms, and vertical penetrations on each floor (such as elevator shafts, and stairwells). It does not include interior parking garages, open covered walkways, courtyards with no roof, balconies, and canopies. (National Renewable Energy Laboratory)

Building footprint area   A building footprint is the outline of the total area of a lot or site that is surrounded by the exterior walls of a building or portion of a building, exclusive of courtyards. In the absence of surrounding exterior walls, the building footprint shall be the area under the horizontal projection of the roof.

Building gross area   Total floor area of a building's enclosed spaces, measured from the outside face of exterior walls or from the centerline of walls of adjoining buildings. This metric is measured on a floor-by-floor basis and includes the area of interior walls, basements, mezzanines, penthouses, equipment rooms, and interior parking. Vertical penetrations on each floor (such as elevator shafts and stairwells) should be included based on the floor area they penetrate. It does not include open covered walkways, courtyards with no roof, balconies, and canopies. Structures that extend beyond the plane of the outside face of the exterior wall, such as cornices, pilasters, buttresses, and overhangs, are not included. (National Renewable Energy Laboratory)

Building information modeling (BIM)   BIM refers to the digital representation of physical and functional characteristics of a facility that portrays the building as a full, 3-D computer model stored as a database with all information about the building. It allows architects, engineers, and building owners to evaluate various design options and their effects on energy consumption, daylighting, thermal comfort, and more.

Commissioning (Cx)   Commissioning is defined as the systematic process of ensuring that a building's complex array of systems is designed, installed, and tested to perform according to the basis of design and the owner's project requirements. The commissioning process starts from the beginning of design until the building is occupied. The process involves establishing expectations for the performance of the building systems and also procedures to determine whether those expectations have been met. (Energy Design Resources)

Commissioning Authority (CxA)   The CxA is the individual designated to lead the commissioning process activities.

Composite wood and agrifiber products   These are products that are manufactured by bonding wood or plant particles or agricultural waste fibers with synthetic resins or binders. Common examples include: particleboard, medium density fiberboard (MDF), plywood, wheatboard, strawboard, panel substrates, and door cores. (*LEED Reference Guide*)

Conditioned space   An enclosed space whose thermal conditions are intentionally controlled using natural, electrical, or mechanical means. Spaces that do not have heating or cooling systems but rely on natural or mechanical flow of thermal energy from adjacent spaces to maintain thermal conditions are considered conditioned spaces. Examples include restrooms that use exhaust fans to draw in conditioned air to maintain thermal conditions and atria that rely on natural convection flow to maintain thermal conditions. (National Renewable Energy Laboratory)

Credit Interpretation Rulings (CIR) A credit interpretation ruling process has been established by GBCI to address specific issues or challenges that a project team may encounter when interpreting the requirements of a prerequisite or credit for their project. Project teams have the option to submit throughout the application process what are known as Credit Interpretation Requests. The feedback received from GBCI are known as Credit Interpretation Rulings (CIRs).

Daylighting Daylighting is the practice of using natural light to illuminate building spaces. Rather than relying solely on electric lighting during the day, daylighting brings indirect natural light into the building. Daylighting reduces the need for electric lighting and connects people to the outdoors. (Daylighting Collaborative)

Development footprint The development footprint is the area affected by development or by project site activity. Hardscape, access roads, parking lots, nonbuilding facilities, and the building itself are all included in the development footprint. (*LEED Reference Guide*)

Ecosystem An ecosystem is a basic unit of nature that includes a community of organisms and their nonliving environment linked by biological, chemical, and physical processes. (*LEED Reference Guide*)

Endangered species An endangered species is one that is in danger of extinction throughout all or a significant portion of its range. A threatened species is one that is likely to become endangered in the foreseeable future. (U.S. Fish and Wildlife Service)

Energy efficiency Energy efficiency improvements refer to a reduction in the energy used for a given service (heating, lighting, etc.) or level of activity. (World Energy Council)

Energy simulation Energy simulation is the process of simulating a proposed building using computer programs to predict the annual energy costs when the building will be in operation.

Energy use intensity (EUI) EUI is a measure of the total energy use of a building normalized for floor area. This is used to compare the energy use of different buildings. Whole-building energy use is measured in KBtu (1000 British Thermal Units) per square foot, per year, to standardize units between fuels. (Energy Information Administration)

Environmental tobacco smoke (ETS) Smoke that is emitted from the burning end of cigarettes, pipes, and cigars, and is exhaled by smokers. (*LEED Reference Guide*)

Foot-candle (fc)   A foot-candle (sometimes foot candle; abbreviated fc, lm/ft², or sometimes ft-c) is a non-SI unit of illuminance or light intensity. The unit is defined as the amount of illumination the inside surface of a 1-foot radius sphere would be receiving if there were a uniform point source of one candela in the exact center of the sphere. Alternatively, it can be defined as the illuminance on a 1-square-foot surface of which there is a uniformly distributed flux of one lumen. This can be thought of as the amount of light that actually falls on a given surface. The foot-candle is equal to one lumen per square foot.

Formaldehyde   Formaldehyde is a naturally occurring VOC found in small amounts in animals and plants. It is carcinogenic and an irritant to most people when present in high concentrations, causing headaches, dizziness, mental impairment, and other symptoms. When present in the air at levels above 0.1 ppm, it can cause watery eyes; burning sensations in the eyes, nose, and throat; nausea; coughing; chest tightness; wheezing; skin rashes; and asthmatic and allergic reactions. (*LEED Reference Guide*)

Full-time equivalent (FTE)   FTE represents a building occupant who spends 40 hours per week (8 hour days) in a building. Part-time occupants have an FTE value depending on the number of hours they work divided by 40. (*LEED Reference Guide*)

Greenhouse gases (GHG)   Gases that trap heat in the atmosphere are often called greenhouse gases. Some greenhouse gases such as carbon dioxide occur naturally and are emitted to the atmosphere through natural processes and human activities. Other greenhouse gases (e.g., fluorinated gases) are created and emitted solely through human activities. The principal greenhouse gases that enter the atmosphere because of human activities are: carbon dioxide (CO_2), methane (CH_4), nitrous oxide (N_2O) and fluorinated gases, such as hydrofluorocarbons, perfluorocarbons, and sulfur hexafluoride that are emitted from a variety of industrial processes. (U.S. Environmental Protection Agency)

Greenfield   A term used to describe a piece of previously undeveloped land, in a city or rural area.

Greywater (also graywater)   According to the International Plumbing Code (IPC), greywater is waste discharged from lavatories, bathtubs, showers, clothes washers, and laundry trays. According to the Uniform Plumbing Code (UPC), greywater is untreated household waste water that has not come into contact with toilet waste. Greywater includes used water from bathtubs, showers, and bathroom wash basins, as well as water from clotheswashers and laundry tubs. It does not include wastewater from kitchen sinks or dishwashers.

Heat island   The term *heat island* is used to describe built-up areas that are hotter than nearby rural areas. The annual mean air temperature of a city with 1 million people or more can be 1.8–5.4°F (1–3°C) warmer than its surroundings. In the evening, the difference can be as high as 22°F (12°C). Heat islands can affect communities by increasing summertime peak energy demand, air-conditioning costs, air pollution and greenhouse gas emissions, heat-related illness, and mortality. (U.S. Environmental Protection Agency)

Integrated design process (IDP)   IDP is an approach to building design that seeks to achieve high performance on a wide variety of well-defined environmental and social goals while staying within budgetary and scheduling constraints. It relies upon a multidisciplinary and collaborative team whose members make decisions together based on a shared vision and a holistic understanding of the project.

LEED Online   LEED Online is a web-based resource for managing the LEED documentation process. Through LEED Online, project teams can manage project details, complete documentation requirements for LEED credits and prerequisites, upload supporting files, submit applications for review, receive reviewer feedback, and ultimately earn LEED certification.

LEED open space area   Open space area, for the purpose of LEED calculations, is defined as the area of the site that is vegetated and pervious and is calculated by subtracting the development footprint from the site area.

LEED project boundary   The LEED project boundary is the portion of the project site that will be submitted for LEED certification. Typically, for most single-building developments, the entire site is submitted for LEED certification, requiring no separate definition of a LEED project boundary. In such cases, the *LEED site area* is the same as the *site area*. However, in some scenarios, such as multiple building developments, the design team is allowed to determine the limits of the project submitted for LEED certification differently from the overall site boundaries.

LEED site area   The LEED site area is defined as the area within the LEED project boundary of the site.

Life-cycle cost analysis (LCCA)   Life-cycle cost analysis (LCCA) is a method for assessing the total cost of facility ownership. It takes into account all costs of acquiring, owning, and disposing of a building or building system. LCCA is especially useful when project alternatives that fulfill the same performance requirements, but differ with respect to initial costs and operating costs, have to be compared in order to select the one that maximizes net savings. (Whole Building Design Guide)

Light pollution   Light pollution is any adverse effect of artificial light, including sky glow, glare, light trespass, light clutter, decreased visibility at night, and energy waste. Light pollution wastes energy, affects astronomers and scientists, disrupts global wildlife and ecological balance, and has been linked to negative consequences in human health. (International Dark-Sky Association)

Lighting power density (LPD)   LPD is the lighting power installed in the building per unit area. It is typically represented as watts per square foot (W/SF).

Low-emitting materials   Low-emitting materials are products that do not release significant pollutants into the indoor environment. These products contain zero- and low-volatile organic compounds (VOCs).

Measurement and Verification (M&V)   M&V refers to the process of measuring and verifying whether a building's performance, after it is in operation, is in accordance to the design intent.

Mechanical ventilation   Ventilation provided by mechanically powered equipment, such as motor-driven fans and blowers, but not by devices, such as wind-driven turbine ventilators and mechanically operated windows. (*LEED Reference Guide*)

Natural ventilation   Ventilation provided by thermal, wind, or diffusion effects through doors, windows, or other intentional openings in the building. (*LEED Reference Guide*)

Net present value (NPV)   NPV represents the present value of an investment's future financial benefits minus any initial capital costs. The investment should be made only for positive NPV projects. The formula for NPV is:

$$\text{NPV} = \sum_{j=1}^{n} \frac{values_j}{(1 + rate)^j}$$

where *values* represent payments and income and *rate* is the rate of discount over the length of one period.

Owner's project requirements (OPR)   OPR is a written document that identifies the various criteria and requirements that are determined by the owner to be important for the success of the project.

Photovoltaics (PV)   Solar cells, also called photovoltaic (PV) cells by scientists, convert sunlight directly into electricity. PV gets its name from the process of converting light (photons) to electricity (voltage), which is called the PV effect. (National Renewable Energy Laboratory)

Post-occupancy evaluation   Post-occupancy evaluation involves systematic evaluation of opinion about buildings in use, from the perspective of the people who use them. It assesses how well buildings match users' needs and identifies ways to improve building design, performance, and fitness for purpose.

Postconsumer recycled content   Percentage of material in a product that was derived from consumer waste—household, commercial, industrial, or institutional end-user's— and that can no longer be used for its original purpose. Some examples include construction and demolition debris, materials collected through recycling programs, discarded products like furniture and cabinetry, as well as landscaping waste like leaves, grass clippings, etc. (*LEED Reference Guide*)

Potable water   Water that is fit for consumption by humans and other animals. It is also called drinking water, in a reference to its intended use.

Pre-consumer recycled content   Percentage of material in a product that is recycled from manufacturing waste. Examples include planer shavings, sawdust, trimmed materials, and obsolete inventories. (*LEED Reference Guide*)

Preferred parking   Preferred parking refers to designated spaces close to the building (aside from designated handicapped spots), designated covered spaces, discounted parking passes, and guaranteed passes in a lottery system. (*LEED Reference Guide*)

Previously developed sites   These are sites that have been disturbed by human activity either by building, clearing, or grading on-site.

Prime farmland   Prime farmland is land that has the best combination of physical and chemical characteristics for producing food, feed, forage, fiber, and oilseed crops and is also available for these uses (the land could be cropland, pastureland, rangeland, forestland, or other land, but not urban built-up land or water). (*LEED Reference Guide*)

Proposed building performance   Proposed building performance is defined as the annual energy cost calculated for a proposed design. (ASHRAE 90.1-2007)

Rainwater harvesting   Rainwater harvesting is the gathering, or accumulating and storing, of rainwater to provide drinking water, water for livestock, water for irrigation, or to refill aquifers in a process called groundwater recharge.

Rapidly renewable materials   Rapidly renewable materials are agricultural products, both fiber and animal, that take 10 years or less to grow or raise and can be harvested in a sustainable fashion. (*LEED Reference Guide*)

Regularly occupied spaces   Regularly occupied spaces are areas where workers are seated or standing as they work inside a building. In residential applications, these areas include all spaces except bathrooms, utility areas, and closets or other storage rooms. In schools, they include areas where students, teachers, or administrators are seated or standing as they work or study inside a building. (*LEED Reference Guide*)

Renewable energy credits (RECs)   RECs, also known as green tags, renewable electricity certificates, or tradable renewable certificates (TRCs), are tradable, non-tangible energy commodities in the United States that represent proof that electricity was generated from an eligible renewable energy resource. These certificates can be sold and traded or bartered, and the owner of the REC can claim to have purchased renewable energy.

Renewable energy   Renewable energy is energy that comes from natural resources, such as sunlight, wind, rain, tides, and geothermal heat, which are renewable (naturally replenished).

Salvaged, refurbished, or reused materials   Salvaged materials or reused materials are construction materials recovered from existing buildings or construction sites and reused. Common salvaged materials include structural beams and posts, flooring, doors, cabinetry, brick, and decorative items. (*LEED Reference Guide*)

Sick building syndrome (SBS)   SBS is used to describe situations in which building occupants experience acute health and comfort effects that appear to be linked to time spent in a building, but no specific illness or cause can be identified. The complaints may be localized in a particular room or zone, or may be widespread throughout the building. Typical indicators of SBS inlcude: occupants complain of symptoms associated with acute discomfort, e.g., headache; eye, nose, or throat irritation; dry cough; dry or itchy skin; dizziness and nausea; difficulty in concentrating; fatigue; and sensitivity to odors. The following have been cited as causes of or contributing factors to sick building syndrome: inadequate ventilation, chemical contaminants from indoor sources and outdoor sources, and biological contaminants. (U.S. Environmental Protection Agency)

Site area   Site area is defined as the area in square feet within the legal property boundary of the site.

Site assessment   A site assessment is an evaluation of a site's aboveground and subsurface characteristics, including its geology and hydrology.

Solar Heat Gain Coefficient   The Solar Heat Gain Coefficient (SHGC) measures how well a window blocks heat from sunlight. The SHGC is the fraction of the heat from the sun that enters through a window. SHGC is expressed as a number between 0 and 1. The lower a window's SHGC, the less solar heat it transmits.(Building Energy Codes Research Center)

Solar Reflectance Index (SRI)   The SRI is a measure of a material's ability to reject solar heat, as shown by a small temperature rise. It is defined so that a standard black (reflectance 0.05, emittance 0.90) is 0 and a standard white (reflectance 0.80, emittance 0.90) is 100. For example, the standard black has a temperature rise of 90°F (50°C) in full sun, and the standard white has a temperature rise of 14.6°F (8.1°C). Once the maximum temperature rise of a given material has been computed, the SRI can be computed by interpolating between the values for white and black. (Lawrence Berkely National Laboratory Cool Roofing Materials Database)

Transient users   Occupants who do not use a facility on a consistent, regular, daily basis. Examples include students in higher education settings, customers in retail settings, and visitors in institutional settings. (*LEED Reference Guide*)

Uunderfloor air distribution (UFAD)   UFAD is a method of delivering space conditioning in offices and other commercial buildings in which the open space (underfloor plenum) between the structural concrete slab and the underside of a raised access floor system is used to deliver conditioned air directly into the occupied zone of the building. Air can be delivered through a variety of supply outlets located at floor level (most common), or as part of the furniture and partitions. UFAD systems have several potential advantages over traditional overhead systems, including improved thermal comfort, improved indoor air quality, and reduced energy use. By combining a building's heating, ventilating, and air-conditioning (HVAC) system with all major power, voice, and data cabling into one easily accessible service plenum under the raised floor, significant improvements can be realized in terms of increased flexibility and reduced costs associated with reconfiguring building services. (Center for the Built Environment, University of California, Berkeley)

Urea-formaldehyde   A combination of urea and formaldehyde that is used in some glues and may emit formaldehyde at room temperature. (*LEED Reference Guide*)

Variable-frequency drives (VFDs)   A VFD is a system for controlling the rotational speed of an alternating current (AC) electric motor by controlling the frequency of the electrical power supplied to the motor. Variable-frequency drives are also known as adjustable-frequency drives (AFD), variable-speed drives (VSD), AC drives, micro-drives or inverter drives. In ventilation systems for large buildings, variable-frequency motors on fans save energy by allowing the volume of air moved to match the system demand.

Volatile organic compounds (VOC)   VOCs are emitted as gases from certain solids or liquids. VOCs include a variety of chemicals, some of which may have short- and long-term adverse health effects. Concentrations of many VOCs are consistently higher indoors (up to ten times higher) than outdoors. VOCs are emitted by a wide array of products numbering in the thousands. Examples include: paints and lacquers, paint strippers, cleaning supplies, pesticides, building materials and furnishings, office equipment such as copiers and printers, correction fluids and carbonless copy paper, graphics and craft materials including glues and adhesives, permanent markers, as well as photographic solutions. (U.S. Environmental Protection Agency)

Wetlands   Wetlands consist of areas that are inundated or saturated by surface or groundwater at a frequency and duration sufficient to support, and that under normal circumstances do support, a prevalence of vegetation typically adapted for life in saturated soil conditions. (*LEED Reference Guide*)

BIBLIOGRAPHY

American Institute of Architects (AIA). "Local Leaders in Sustainability: Green Incentives." aia.org. http://www.aia.org/aiaucmp/groups/aia/documents/pdf/aias076934. pdf (accessed April 1, 2010).

Architecture 2030. "Climate Change, Global Warming, and the Built Environment." Architecture2030.org. http://www.architecture2030.org/current_situation/science.html (accessed March 15, 2010).

Burr, Andrew C. "CoStar Study Finds Energy Star, LEED Bldgs. Outperform Peers." costar.com, March 26, 2008. http://www.costar.com/News/Article.aspx?id=D968F1E0DC F73712B03A099E0E99C679 (accessed March 15, 2010).

Busby Perkins + Will. "Roadmap for Integrated Design Process." Report prepared for BC Green Building Roundtable, 2007.

Callan, David P. "Studies Relate IAQ and Productivity." facilitiesnet.com, November 2006. http://www.facilitiesnet.com/green/article/Studies-Relate-IAQ-and-Productivity—5581 (accessed February 10, 2010).

Center for the Built Environment. *Underfloor Air Technology: Overview.* cbe.berkley. edu. http://www.cbe.berkeley.edu/underfloorair/default.htm (accessed February 10, 2010).

Energy Design Resources. *Design Brief: Building Commissioning.* energydesignresources. com, August 02, 2002. http://www.energydesignresources.com/Design/Building Commissioning/tabid/90/articleType/ArticleView/articleId/109/Design-Briefs-Building-Commissioning.aspx (accessed April 1, 2010).

—. *Design Brief: Energy Simulation.* energydesignresources.com, January 1, 2002. http://www.energydesignresources.com/Resources/Publications/DesignBriefs/tabid/74 /articleType/ArticleView/articleId/3/Design-Briefs-Building-Simulation.aspx (accessed February 10, 2010).

—. *Design Brief: Indoor Air Quality*. energydesignresources.com, June 02, 2003. http://www.energydesignresources.com/Resources/Publications/DesignBriefs/tabid/74 articleType/ArticleView/articleId/123/Design-Briefs-Indoor-Air-Quality.aspx (accessed April 1, 2010).

Fisk, William J. "Health and Productivity Gains from Better Indoor Environments." In *The Role of Emerging Energy-Efficient Technology in Promoting Workplace Productivity and Health*. Lawrence Berkeley National Laboratory, 2002.

Gail, Lindsey, Joe Ann Todd, Sheila J. Hater, and Peter G. Ellis. *A Handbook for Planning and Conducting Charrettes for High-Performance Projects,* 2nd Edition. National Renewable Energy Laboratory (NREL).

GBCI. "GBCI: Certification Bodies." gbci.org. http://www.gbci.org/org-nav/about-gbci/about-gbci.aspx (accessed July 1, 2010).

—. *LEED 2009 Minimum Program Requirements*. gbci.org. http://www.usgbc.org/ShowFile.aspx?DocumentID=6715 (accessed July 1, 2010).

Goffman, Ethan. "Green Buildings: Conserving the Human Habitat." csa.com, October 2006. http://www.csa.com/discoveryguides/green/review2.php (accessed February 8, 2010).

Great River Energy. "A White Paper on Building for Platinum LEED Certification: The planning, design and construction of Great River Energy's headquarters building in Maple Grove, Minnesota." greatriverenergy.com, 2009. http://www.greatriverenergy. com/aboutus/ourleedbuildings/leedwhitepaper.pdf (accessed July 2010)

Greg, Kats. "The Costs and Financial Benefits of Green Buildings." A Report to California's Sustainable Building Task Force, 2003.

Harvard University Office of Sustainability. "Rockefeller Hall, HDS Case Study." green.harvard.edu, 2009. http://green.harvard.edu/theresource/case-studies/index.php?nid=630 (accessed March 1, 2010).

Heschong, Lisa. "Daylighting and Human Performance." In *The Role of Emerging Energy-Efficient Technology in Promoting Workplace Production and Health*. Lawrence Berkeley National Laboratory, 2002.

Matthiessen, Lisa Fay, and Peter Morris. "The Cost of Green Revisited: Reexamining the Feasibility and Cost Impact of Sustainable Design in the Light of Increased Market Adoption." Davis Langdon, 2007.

McDonough, William, and David Gissen (eds.). *Big and Green: Toward Sustainable Architecture in the 21st Century*. New York: Princeton Architectural Press, 2002.

McKinsey & Company. *Unlocking Energy Efficiency in the U.S. Economy*. Milton, VT: Vilanti & Sons, Printers, Inc., 2009.

Miller, Norm G., and Dave Pogue. *Do Green Buildings Make Dollars and Sense?* Burnham-Moores Center for Real Estate, University of San Diego; CB Richard Ellis, 2009.

National Institute of Building Sciences Task Group on Building Rating and Certification. "Building Rating and Certification in the U.S. Building Community." 2009.

Oppenheimer, Stephen, et al. *The Cost of LEED: A Report on Cost Expectations to Meet LEED 2009 for New Construction and Major Renovations (NC v2009).* Brattleboro, VT: BuildingGreen, LLC, 2010.

Public Technology Inc. and USGBC. *Sustainable Buildings Technical Manual.* Public Technology Inc., 1996.

Stephens, Jeff. "Modeling for Good Performance: Building Information Modeling Facilitates Greener Buildings." *EcoStructure*, 2008.

Syal, Dr. M. G. "Impact of LEED-NC Projects on Constructors." Michigan State University Research, 2007.

Syphers, Geof, Mara Baum, Darren Bouton, and Wesley Sullens. "Managing the Cost of Green Buildings." Report partnered with State of California's Sustainable Building Task Force, the California State and Consumer Services Agency, and the Alameda County Waste Management Authority, 2003.

Turner, Cathy, and Mark Frankel. *Energy Performance of LEED for New Construction Buildings.* NBI; USGBC, 2008.

U.S. General Services Administration. *Project Planning Guide.* 2005.

USGBC Chicago Chapter. "Regional Green Building Case Study Project: A post-occupancy study of LEED projects in Illinois." Year I Report, Chicago, 2009.

USGBC. *LEED for Core & Shell.* usgbc.org. http://www.usgbc.org/DisplayPage.aspx?CMSPageID=295 (accessed January 10, 2010).

—. *LEED Reference Guide for Green Building Design and Construction.* Washington, DC: USGBC, 2009.

—. *Portfolio Program.* usgbc.org. http://www.usgbc.org/DisplayPage.aspx?CMSPageID=1729 (accessed January 10, 2010).

—. *LEED Rating Systems.* usgbc.org. www.usgbc.org/DisplayPage.aspx?CMSPageID=222 (accessed February 5, 2010).

Walsh, Valerie. "How-To Guide to LEED Certification for New Mexico Buildings." emnrd.state.nm.us. February 2007. http://www.emnrd.state.nm.us/ECMD/cleanenergytaxincentives/documents/LEED Guidebook_001.pdf (accessed January 15, 2010).

Watson, Rob. *Green Building Impact Report.* Greener World Media, Inc., 2009.

Whole Building Design Guide (WBDG). "Project Planning, Management and Delivery." wbdg.org, February 24, 2010. http://www.wbdg.org/project/pm.php (accessed March 15, 2010).

Wilson, Alex. "Building Materials: What Makes a Product Green?" *Environmental Building News*, 2006.

INDEX